Late-Stage Functionalization and Diversification in Organic Synthesis

Late-Stage Functionalization and Diversification in Organic Synthesis

Methods and Applications

Edited by Tatiana Besset

WILEY-VCH

Editor

Dr. Tatiana Besset
INSA Rouen Normandie, Univ Rouen Normandie, Univ Caen Normandie ENSICAEN, CNRS, Institut CARMeN (UMR 6064), 76000 Rouen, France

Cover Design: Wiley
Cover Image: © Dr. Tatiana Besset, CNRS Normandy Rouen University, Mont-Saint-Aignan, France

Typesetting: Straive, Chennai, India
Printing and Binding: CPI Group (UK) Ltd, Croydon, CR0 4YY

Library of Congress Card No.: applied for

British Library Cataloguing-in-Publication Data A catalogue record for this book is available from the British Library.

Bibliographic information published by the Deutsche Nationalbibliothek
The Deutsche Nationalbibliothek lists this publication in the Deutsche Nationalbibliografie; detailed bibliographic data are available on the Internet at <http://dnb.d-nb.de>.

Print ISBN: 978-3-527-35418-4
ePDF ISBN: 978-3-527-84844-7
ePub ISBN: 978-3-527-84843-0
oBook ISBN: 978-3-527-84845-4

C9783527354184_060326

It is hard to fail, but it is worse never to have tried to succeed.

—Theodore Roosevelt

Contents

Foreword

Late-stage functionalization has emerged as one of the defining frameworks of modern synthetic chemistry. It asks a deceptively simple question: what chemistry remains available to us when a molecule is already complicated, valuable, and fragile? The answers to that question are reshaping how we think about reactivity, selectivity, and the very purpose of method development.

At its heart, late-stage functionalization is about being selective under constraint. Whether the periphery, a specific functional group, or the core of the molecule is altered, every successful transformation in this setting requires that one bond reacts while many others do not. There is one unifying challenge that runs through every chapter in late-stage functionalization, regardless of whether the tools involved are transition metals, light, electricity, flow reactors, or a combination thereof: chemoselectivity. What makes this moment particularly compelling is that these advances are not isolated. They form a coherent movement toward a chemistry that is more precise, more modular, and more responsive to real-world constraints. Catalysts, reagents, and activation modes are increasingly designed with complexity in mind. Mechanistic insight is being used not only to explain outcomes but also to anticipate selectivity and guide reaction design. The result is that late-stage functionalization is evolving from a collection of impressive case studies into a broadly enabling strategy. This book captures that expansion with clarity and shows how different activation strategies converge on the same goal and are mutually reinforcing. The chapters examine some of the most compelling examples of how diverse approaches have advanced the field. Aspects of reagents and catalysts that can deal with increasing complexity are discussed, as are impacts of energy delivery on selectivity. Photochemistry allows access to reactive intermediates under conditions that preserve sensitive functionality. Electrochemistry offers redox control without reagent-derived constraints. Flow chemistry provides temporal and spatial control over reactivity. Diversity no longer needs to be built step by step from simple starting materials. Increasingly, it can be accessed directly from molecules that are already complex, functional, and close to their ultimate application. Several areas benefit from late-stage functionalization, while others cannot do without it, such as the incorporation of unstable radioisotopes that simply do not sustain a lengthy stepwise approach.

The chapters that follow do not claim that all challenges have been solved. On the contrary, they make clear that chemoselectivity remains the central frontier. They also demonstrate that the field now possesses the conceptual tools, the experimental techniques, and the collective ambition to address that frontier directly. Late-stage functionalization is no longer defined by what cannot be done at a late stage. It is increasingly defined by what can be done. This book is therefore not only a record of progress but also an invitation. It invites the reader to think of synthesis less as the construction of molecules from scratch and more as the thoughtful transformation of molecules that already exist. It invites us to see complexity not as an obstacle to be avoided but as the arena in which meaningful chemistry now takes place. And it invites the community to continue refining the art of chemoselectivity because the ability to choose one bond among many is what gives late-stage functionalization its power, relevance, and enduring promise.

March 2026

Prof. Tobias Ritter
Max-Planck-Institut für
Kohlenforschung
Mülheim an der Ruhr
Germany

Preface

Late-stage functionalization (LSF) and diversity-oriented synthesis have provided unprecedented and straightforward tools to chemical synthesis, completely reshaping the field of organic chemistry. Modifying structurally and functionally complex molecules while ensuring a high control of regio-, chemo-, and site-selectivity represents a challenging task and has inspired researchers to provide innovative and creative solutions to address it. As a result, remarkable progress has been achieved in recent years, offering a comprehensive set of tools for the diversification in organic synthesis by leveraging the existing pool of molecules and by allowing for the straightforward synthesis of challenging and even previously impossible molecules via alternative synthetic paths.

This book provides an overview of the current state of the art in synthetic methodologies for LSF and diversity-oriented synthesis that continue to generate significant research momentum. This book is divided into three parts. Part I focuses on LSF strategies. It provides an overview of the recent advances made in selective C—H functionalization by transition metal catalysis (Chapter 1, *Wencel-Delord et al.*), light-driven transformations (Chapter 2, *Giustiniano et al.*), strategies using electrochemistry (Chapter 3, *Ackermann et al.*), and flow chemistry (Chapter 4, *Luisi et al.*) for LSF. Special attention was paid to a direct and important application of LSF with radioisotope insertion (Chapter 5, *Cailly et al.*). Part II is dedicated to major advances made in diversity-oriented synthesis, encompassing LSF strategies using C—O bond activation (Chapter 6, *Blieck*), C—N bond activation under redox-neutral conditions (Chapter 7, *Zhao*, *Szostak et al.*), and C—S bond activation (Chapter 8, *Huang et al.*). Finally, major achievements made in molecular editing and stereochemical editing (Chapter 9, *Besset et al.*) have been highlighted in Part III.

I believe this book will be of interest to researchers and professionals in both academia and industry, particularly those engaged in organic synthesis, natural products chemistry, medicinal chemistry, drug discovery, and molecular imaging. I hope you will enjoy reading this book as much as I have enjoyed preparing it.

Above all, I would like to warmly thank all the contributors (colleagues and friends) who kindly agreed to take part in this project and joined me on this journey. Their outstanding contributions have turned this book into a truly collective achievement. Moreover, I would like to sincerely thank Anne Brennführer, and

Priyadarshini Natarajan from Wiley-VCH for their trust in allowing me to edit this book, as well as their kind support, guidance, and invaluable assistance throughout this entire adventure.

September 2025

Dr. Tatiana Besset
Rouen, France

Part I

Late-Stage Functionalization Strategies

1

Late-Stage Functionalization by Transition Metal-Catalyzed C—H Functionalization

Erwan Brunard and Joanna Wencel-Delord

JMU Würzburg, Institute of Organic Chemistry, Faculty of Chemistry and Pharmacy, Am Hubland, 97074, Würzburg, Germany

1.1 Introduction

The 21st century is clearly marked by important environmental issues, global warming, expanding worldwide population, depletion of natural resources, and inflating energy needs. These issues clearly impact all domains of everyday life. However, chemistry is clearly the field that can provide real solutions to these major and urgent societal problems. Accordingly, the chemical industry has initiated the sustainable revolution, and extensive efforts are devoted to produce all types of chemicals in more sustainable manner, while decreasing the environmental footprint. However, the development of modern industrial process requires initial fundamental discoveries and the development of new synthetic tools, allowing the synthesis of desired organic products in more resource-, energy-, and step-efficient manner. Accordingly, the fundamental research in organic chemistry has been undergoing, since the last two decades, a real "sustainability revolution" [1, 2]. Various approaches have been proposed to allow more sustainable synthesis, among which visible-light-mediated photocatalysis, electrochemistry, continuous-flow reactions, and "benign solvent design" are selected examples. These strategies reached now the maturity to be considered valuable and general tools for organic synthesis. In parallel, C—H activation strategy profoundly changed the mindset of a synthetic chemist [3]. Indeed, a possibility of using non-prefunctionalized substrates in the transition-metal-catalyzed process presents a formidable advantage over the standard cross-coupling reactions. Over the past years, organic chemists have learned how to use latent and strong C—H bonds as functional motifs and directly convert them into desired functional groups. Over the years, the field of C—H activation expanded tremendously, allowing major breakthroughs and providing key improvements, such as efficient site-selectivity [4], use of abundant, cheap, and nontoxic metals [5], or the design of highly reactive catalytic systems able to promote the key C—H cleavage step under mild reaction conditions. Accordingly, C—H activation field, which was 15 years ago considered a scientific curiosity, now

Late-Stage Functionalization and Diversification in Organic Synthesis: Methods and Applications,
First Edition. Edited by Tatiana Besset.

reached a totally new level of industrial applications in both production [6] and R&D industrial sectors [3c]. The C—H activation field holds, in particular, an exceptional potential for speeding up research in pharmaceutical industry [7] by unlocking the door toward rapid diversification of lead drug candidates via C—H late-stage functionalization. Indeed, if a direct C—H functionalization of a complex molecule can be done in one step, the diversity of lead's congeners can be rapidly accessed by avoiding traditional *de novo*, multistep synthesis of each product.

While the first, general example of the LSF to directly install diversity of motifs on celecoxib was reported as early as in 2011 by the group of Yu [8], over the last year, late-stage functionalization via C—H activation has advanced tremendously [9]. Indeed, over the years, academic groups have devoted considerable effort to go beyond the development of new C—H activation reactions that are compatible with benchmark substrates and to showcase the synthetic usefulness of these reactions by demonstrating their compatibility with complex molecules, such as drugs and natural products.

In parallel, industrial chemists tend to implement LSF mindset, often combined with high-throughput experimentation (HTE), to expand a chemical space. In this chapter, key achievements in the LSF via C—H activation are present. First, the LSF of aromatic compounds will be discussed, followed by the aliphatic ones. Each section is organized following the nature of the C—H transformation, including direct alkylation, arylation, alkenylation, and C—X bond formation (C—O, C—N, C—B, halogenation). Attention will also be focused on the selectivity issues and the mechanistic reasons controlling the site-selectivity. It is worth noting that while remarkable progress has been achieved in LSF of peptides [10], this topic has been covered extensively and therefore is considered out of the scope for this chapter.

1.2 Late-Stage Functionalization of Aromatic Compounds

1.2.1 LSF via Direct Alkylation

LSF of drug candidates is particularly appealing when allowing the direct introduction of small groups, impacting significantly the biological activity of a drug molecule while not altering excessively the overall structure. Accordingly, LSF methylation is particularly attractive, allowing to rapidly explore the "magic methyl effect" on a given molecule. An early example of late stage methylation was published in 2013 by Yu [11]. The catalytic system combining $Pd(OAc)_2$ catalyst together with a monoprotected amino acid ligand and Li_2CO_3 was initially optimized to perform direct C—H methylation of aryl carboxylic acids using methyl tetrafluoroborate as the methylating agent (Figure 1.1a). The potential of this strategy could be further demonstrated by synthesizing a therapeutically relevant methylated derivative of BMS-98947-0550-1. The superiority of the carboxylate function as a directing group over the ester motif warrant fully site-selectivity of this reaction, furnishing the modified compound in synthetically useful

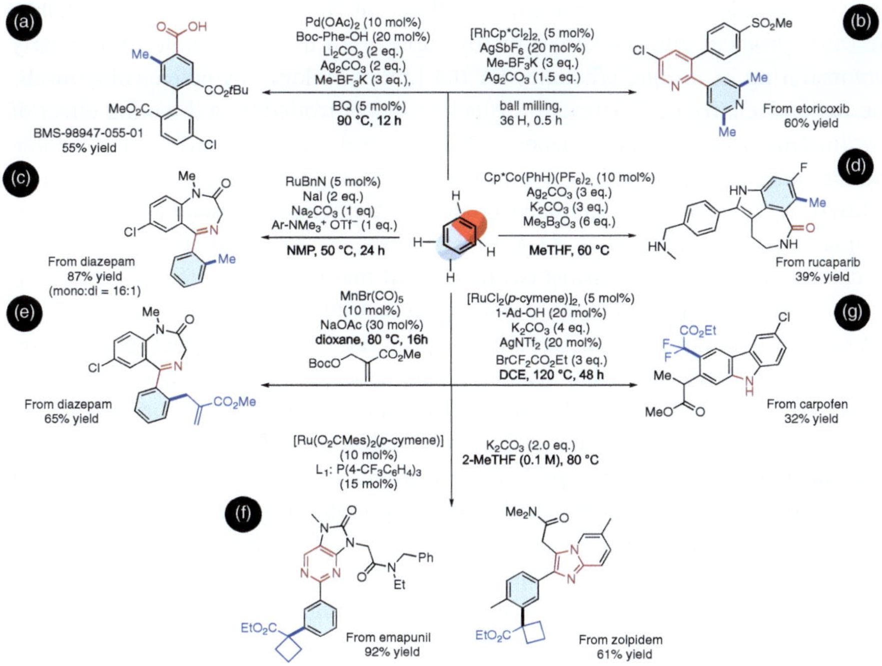

Figure 1.1 Late-stage functionalization via direct alkylation. a) Pd-catalyzed late-stage methylation of BMS-98974-0550-1 developed by Yu et al. [11] b) Rh-catalyzed methylation of etoricoxib developed by Pilarski et al. [12] c) Ru-catalyzed methylation of diazepam developed by Larrosa et al. [13] d) Co-catalyzed methylation of rucaparib developed by Johansson, Ackermann et al. [14] e) Mn-catalyzed methylation of diazepam developed by Ackermann et al. [15] f) Ru-catalyzed methylation of emapunil and zolpidem developed by Ackermann, Johansson et al. [16] g) Ru-catalyzed methylation of carprofen developed by Lan, Zhao et al. [17].

55% yield. Subsequently, Pilkarski [12] and Larrosa [13], designed, respectively, Rh and Ru-catalyzed direct methylation of phenyl pyridine and imine derivatives (Figure 1.1b, c). While RhCp* catalytic system was designed to promote direct methylation using methyl tetrafluoroborate reagent under mechanochemical, solvent-free conditions, the Ru-metallacyclic catalyst performed well in the presence of quaternary anilinium salt. Both protocols could be subsequently used for LSF of drugs, delivering, in high yields, methyl derivatives of etoricoxib and diazepam.

Regarding the great potential of LSF for R&D sector, several companies conduct their own research programs. Considering the "magic methyl effect," the researchers from AstraZeneca in collaboration with Ackermann focused on developing a general protocol for LS-methylation of complex molecules [14]. Use of HTE approach allowed them to rapidly screen hundreds of reaction conditions, aiming in a general protocol compatible with a large diversity of compounds incorporating distinctive functional groups that could play the role of DGs. This exceptional approach resulted in establishing a Co-based protocol, enhancing direct methylation of 26 structurally very different simple aromatics. Remarkably, the same protocol allowed direct methylation of 16 drug-like molecules, clearly demonstrating its unique efficiency for the LSF of biologically active compounds. The regioselectivity in this reaction is generally controlled by a directing effect of coordinating groups (amides, carbonyl, *N*-heterocycles), thus allowing a predictable and site-selective methylation (Figure 1.1d). Although in some cases rather low yields were reached, this approach enables a tremendous research acceleration compared to the traditional *de novo* synthesis, as exemplified on rucaparib. Co-catalyzed methylation furnished the desired methylated derivative in one step in 39% yield, whereas *de novo* synthesis would require a 12-step protocol.

While methylation reactions are highly appealing from drug-design perspective, LSFs via alternative alkylation reactions have also been described. The group of Ackermann explored Mn-catalyzed direct allylation reactions (Figure 1.1e) [15]. Although initially developed for C2-functionalization of indole scaffolds with remarkable applications for peptide modifications, this transformation performed equally well using diazepam as the starting material. The desired alkylated diazepam was thus isolated in 65% yield, and the excellent site-selectivity results from *ortho*-directing properties of the imine group.

A complementary, distal *meta*-selective direct alkylation of complex molecules was recently reported by Ackermann and Johansson (Figure 1.1f) [16]. The HTE approach allowed expedient evaluation of various reaction conditions, targeting a reaction protocol compatible with a diversity of aromatic substrates. After the extensive optimization, $[Ru(O_2CMes)_2(p\text{-cymene})]$ was determined as the optimal catalyst in the presence of a monophosphine ligand **L1**, allowing direct *meta*-coupling between aromatics bearing various DGs and bromoalkanes. Remarkably, while challenging this catalytic system with several aromatic drug derivatives, the desired alkylation proceeded smoothly, providing products such as emapunil or zolpidem derivatives in remarkably high 92% and 62% yields.

An additional example of unusual site-selectivity in LSF-difluoroalkylation was disclosed by Lan and Zhao. While focusing on Ru-catalyzed direct functionalization of aniline derivatives, they discovered a unique *para*-selective transformation (Figure 1.1g) [17]. After illustrating the compatibility of this strategy with diversity of aniline derivatives, including indoline and tetrahydroquinoline, LSF of carbazole-derived drug, carprofen, was accomplished competently.

1.2.2 LSF via Alkenylation

Direct alkenylation, i.e., oxidative Heck reaction, is one of the most developed direct functionalizations. Thus, non-surprisingly, several examples of such C—H functionalization performed on drug molecules have been reported. In 2019, Zhu disclosed direct *ortho*-selective olefination of free phenol derivatives [18]. Considering the ubiquitous presence of phenol rings in medicinal chemistry, the potential of the newly discovered protocol was rapidly validated in LSF (Figure 1.2a). Not only alkenylated estrone and estradiol but also ethinylestradiol could be obtained. The *ortho*-selectivity of this reaction is clearly controlled via coordinating

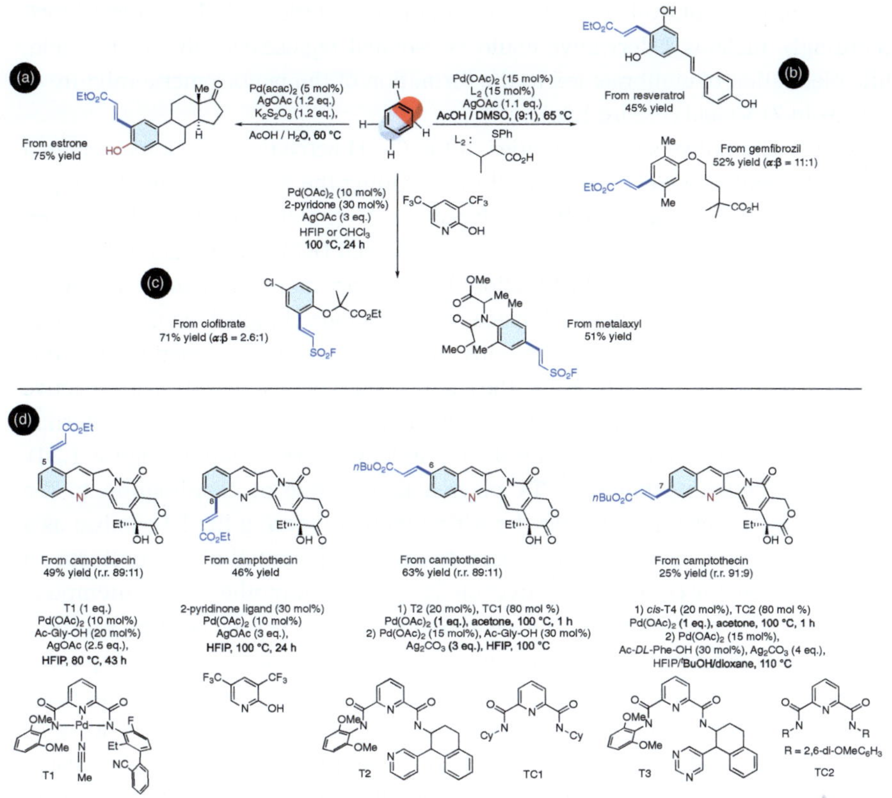

Figure 1.2 LS olefination of aromatic drug-type molecules (r.r. = regioisomeric ratio). a) Pd-catalyzed late-stage olefination of estrone developed by Zhu et al. [18] b) Pd-catalyzed late-stage olefination of resveratrol and gemfibrozil developed by Wu, Yao et al. [19] c) Pd-catalyzed late-stage olefination of metalaxyl and ciofibrate developed by Wang, Yu et al. [21] d) Pd-catalyzed late-stage olefination of camptothecin developed by Yu et al. [22, 23] and Yu, Houk et al. [24].

properties of the hydroxyl group, exercising the role of the DG and "preinstalling" the metal in proximity at the *meta*-positions.

An alternative example of LSF of phenol derivatives was reported in 2023 by Wu and Yao [19]. Capitalizing on the arylthiol acid ligands **L2** designed by Fernández-Ibáñez [20], they reported the undirected functionalization of several phenol- and ether-containing natural products and drug derivatives. The C—H activation occurs in general at the electronically favored positions, such as *ortho*- and *para*-positions, while steric effects cannot be obviated. For example, fully selective olefination of resveratrol was reached, while the mixture of regioisomeric products was isolated in the case of gemfibrozil (Figure 1.2b).

Non-directed olefination of phenol derivatives is not limited to the use of acrylates as coupling partners. Wang and Yu extended the applicability of this methodology to the use of vinyl sulfonyl fluorides [21]. The Pd-based catalytic system request addition of the 2-pyridinone ligand to warrant sufficient reactivity and the selectivity of the reaction if controlled by the combination of steric and electronic factors. Accordingly, metalaxyl derivative could be isolated regioselectively in 51% yield, while olefination of ciofibrate led to the formation of the regioisomeric mixture of products in 71% yield (Figure 1.2c).

Reaching alternative site-selectivities in the C—H activation reactions is often a formidable challenge. Several approaches, including the design of long, U-shaped Directing Groups, as well as Ru-catalyzed C—H functionalizations have been found in the literature. However, the use of U-shaped DG is tedious and difficult to predict, while meta-selective Ru-functionalizations are mainly limited to alkylation reactions. To expand the chemical space of compounds obtained via C—H activation, the group of Yu focused on alternative strategies promoting functionalizations at distant positions. In early 2017, they described a Pd-catalyzed *meta*-selective olefination [22]. The site-selectivity was dictated by the presence of an external, bifunctional nitrile template **T1**, used in a stoichiometric amount (Figure 1.2d). Such a bicoordinating template **T1** plays a double role of coordination to the DG, directing the metal to the distal position while rendering the Pd-catalyst as a sufficiently reactive catalyst. After demonstrating the potential of this protocol on various aza-aromatic compounds, direct olefination of camptothecin was attempted. Remarkably, selective functionalization at remote C5 benzocyclic position was thus achieved. Interestingly, a reversed site-selectivity was observed while exploring the potential of an alternative Pd-based catalytic system developed by the same group [23]. Interestingly, C8-selectivity, i.e., the functionalization at the most reactive, electronically driven position could be completed selectively while using "non-directed approach." In this case, the key difficulty lies in selecting a ligand providing the Pd-catalyst sufficient reactivity. Intensive optimization by the group of Yu established 2-pyridone **L2** as such privileged ligand for non-directed olefination of aromatics. Accordingly, the acrylate function could be introduced at the C8-position of camptothecin in excellent selectivity. Noteworthy, this non-directed functionalization could also be used to diversify other complex molecules, including fenofibrate, caffeine, estrone-OMe, and viloxazine-NTs.

Selective functionalization of C6 and C7 positions of a quinoline scaffold, such as in camptothecin, presents an additional tremendous challenge, resulting from the chemical inertness and electronic similarity of these two C—H bonds. Exceptional regiochemical precision toward functionalization of these distant yet non-activated positions were attained by Yu and Houk by designing two conceptually distinct directing templates [24]. The concept is based on the (1) initial coordination between the N-atom of the substrate and Pd-coordinated template, followed by the coordination of the second Pd-atom on the pyridine motif of the template, thus delivering the metal catalyst near the C—H bond to be cleaved. The discrimination between different positions is thus achieved via very precise spatial and chiral recognition. Accordingly, C7 position of camptothecin underwent oxidative Heck reaction in the presence of chiral **T2** and **TC1** (template chaperon used to turn over the directing template that now can be used in "catalytic" amount), while C7 functionalization required the use of **T3** template.

While the above present work perfectly illustrates the potential of LSF for site-selective functionalization of drug derivatives, the incredible challenge associated with the meticulous regioselectivity control is evident.

1.2.3 LSF via Arylation

Ortho-directed C—H arylation is certainly among the most developed C—H functionalization reactions. Therefore, it is not surprising that this approach has also been successfully used for the late-stage diversification of complex molecules. In 2018, Larrosa reported the cyclometallated Ru-complex-catalyzed direct arylation (Figure 1.3a) [25]. While the catalyst optimization and the mechanistic studies were conducted using largely explored phenyl pyridine substrate, the catalytic cycle turned out to be equally efficient for several complex molecules, featuring aza- or imine functionalities that can be used as directing groups. Accordingly, direct diversification of scaffolds such as zolimidine, flurazepam, or sulfaphenazole could

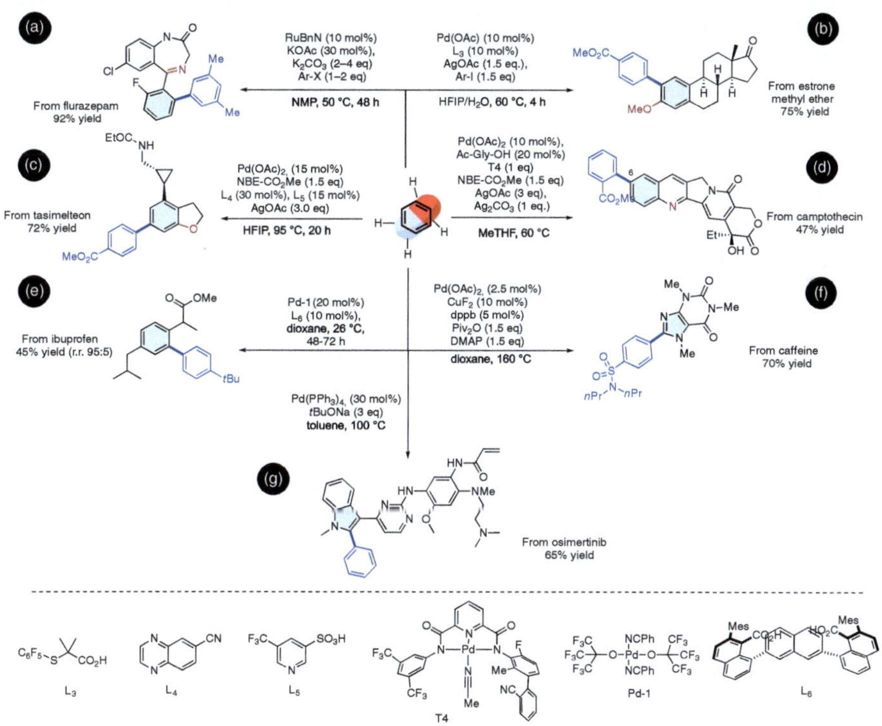

Figure 1.3 LSF of aromatic compounds via direct arylation. a) Ru-catalyzed late-stage arylation of flurazepam developed by Larrosa et al. [25] b) Pd-catalyzed late-stage arylation of estrone methyl ester developed by Fernández-Ibáñez et al. [26] c) Pd-catalyzed late-stage arylation of tasimelteon developed by Yu et al. [27] d) Pd-catalyzed late-stage arylation of camptothecin developed by Houk, Yu et al. [28] e) Pd-catalyzed late-stage arylation of ibuprofen developed by Čorić et al. [30] f) Pd/Cu-catalyzed late-stage arylation of caffeine developed by Hong, Szostak et al. [31] g) Pd-catalyzed late-stage arylation of osimertinib developed by Li, Zhang, Gao et al. [32].

be reached under very mild reaction conditions (at a temperature of 35 °C) and with excellent yields.

Ortho- and *para*-directed effect of OR group in electrophilic substitution is evident; however, accessing sufficiently reactive catalysts to promote direct metalation of phenol-derivatives is far from evident. Very recently, Fernández-Ibáñez reported an elegant example of non-directed C—H arylation of alkoxy aromatics. In this case, the use of S,O-ligand **L3** allowed to generate highly reactive Pd-species, prompt to promote electronically favored *ortho*-arylations (Figure 1.3b) [26]. The potential of this strategy toward LSF of complex molecules was illustrated while challenging the catalytic system with estrone methyl ether as a substrate.

While non-directed C—H activation generally occurs at *ortho-* and *para*-positions, alteration of this native reactivity of electron-rich alkoxy aromatics to reach *meta*-selectivity is extremely challenging. Elegant example of such non-directed *meta*-selective C—H arylation was reported in 2019 by Yu [27]. The key to reverse the conventional selectivity lies in the use of norbornene additive, able to displace the site-selectivity from the *ortho-* to the *meta*-position, as initially used in the Catellani reaction [28] (Figure 1.3c). Following this concept, a large diversity of biaryl phenol derivatives could be prepared using a complex catalytic system composed of $Pd(OAc)_2$, norbornene, and dual ligand system (quinoxaline derivative **L4** together with pyridine sulfonic acid derivative **L5**). Remarkably, direct *meta*-selective arylation of tasimelteon proceeded smoothly, delivering the desired product in 72% yield.

An additional example of a brilliant use of norbornene mediator to modify the selectivity of the C—H activation was reported by the same group in 2020 [29]. In this case, direct arylation of quinoline was investigated. While the use of the bifunctional template approach allows to selectively address C5 position, the functionalization could be displaced by one carbone, thus occurring at the C6 position, in the presence of a norbornene additive. This approach thus warrants access to the C6-arylated camptothecin in 47% yield (Figure 1.3d).

An additional interesting concept for controlling site-selectivity in non-directed C—H activation was proposed by Čorić in 2020 [30]. The author speculated that a fine design of carboxylic acid anionic ligand can allow spatial control over the regioselectivity, the metalation event. Accordingly, naphthyl-derived biscarboxylic acid **L6** was designed and explored in Pd-catalyzed direct arylation using hypervalent iodines as coupling partners (Figure 1.3e). While high reactivity was observed for various substrates under mild reaction conditions (room temperature), only moderate regioselectivity was generally observed. However, in the case of ibuprofen methyl ester as a substrate, featuring two sterically different substituents, selective diversification occurred, delivering the functionalized compound in 45% yield and high selectivity.

Selective non-directed C—H activation can frequently be achieved in case of heterocyclic compounds. Elegant example of LSF of caffeine was achieved by Hong and Szostak [31]. Bimetallic catalytic system composed of Cu and Pd was thus used to allow decarboxylative arylation of N-heterocyclic compounds. The key C—H activation step results from coordination of a heterocyclic substrate by Cu, followed by migration occurring via the C—H activation step, while the Pd-catalyst promotes the activation of the second aromatic partner via decarboxylation (Figure 1.3f). Under such finely designed catalytic conditions and using probenecid acid derivative as the coupling partner, direct arylation of caffeine was achieved in 70% yield.

The importance of LSF of heteroaromatic cores was also clearly demonstrated in the recent study by Li, Zhang, and Gao who explored LSF of osimertinib (Figure 1.3g) [32]. Fine optimization of the reaction conditions allowed determining a selective protocol for C2 functionalization of the indole core. A large library of the arylated osimertinib derivatives could thus be afforded, and their biological activity was directly evaluated.

1.2.4 LSF via C—O Bond Formation

LSF via introduction of O-atom may have a critical impact on the biological properties, including increased solubility and decreased *logP*. Besides, binding affinity of the oxygenated derivatives may also be improved, as new interactions may now be expected. Therefore, development of LS reactions allowing direct conversion of C—H bonds into C—O bonds seems extremely appealing. Following this analysis, the group of Yu developed carboxylate-directed hydroxylation of benzoic acid derivatives [33]. Newly designed pyridine-pyridone ligand **L7** showed exceptional reactivity in such reactions, featuring strong *ortho*-directed effect with respect to the carboxylate directing group, even in the presence of potentially stronger directing groups. Accordingly, not only a large platform of innovative phenol congeners, previously difficult to access, could be obtained in a single step, but also a LS diversification of a biologically active probenecid occurred smoothly, delivering the desired product in 60% yield (Figure 1.4a). Despite the fundamental importance of this methodology, the use of BQ oxidant together with O_2 are far from optimal. Seeking more industrially relevant transformation, the same group shortly after an additional protocol is compatible with the use of H_2O_2 as the privileged source of the hydroxyl group [34]. A modified carboxy-pyridone ligand (**L8**) generates sufficiently reactive catalytic cycle to promote direct hydroxylation of a bench of benzoic acids as well as phenylacetic acids. Remarkably, while implementing this strategy for the LSF of ibuprofen, remarkably scale up of the reaction (>200 g) was conducted efficiently using only 1 mol% of the Pd-catalyst at room temperature. Considering the diversity of phenylacetic acid motif in drug design, the optimized protocol could be rapidly evaluated against a bench of biologically active scaffolds, including loxoprofen, itanapraced, and ketoprofen (Figure 1.4b). It is also worth mentioning that high selectivity of this protocol toward carboxylate directing group warrants good functional group tolerance.

Ortho-directed Pd-catalyzed hydroxylation may also be conducted using oxime ether as a directing group, as reported by Jiao [35]. The use of simple PPh_3 or DEAD as ligand generated a catalytic system compatible with the use of oxone as the coupling reagent (Figure 1.4c). Under the optimized reaction conditions, various hydroxylated compounds, including zaltoprofen, were afforded in good yields and excellent site-selectivity.

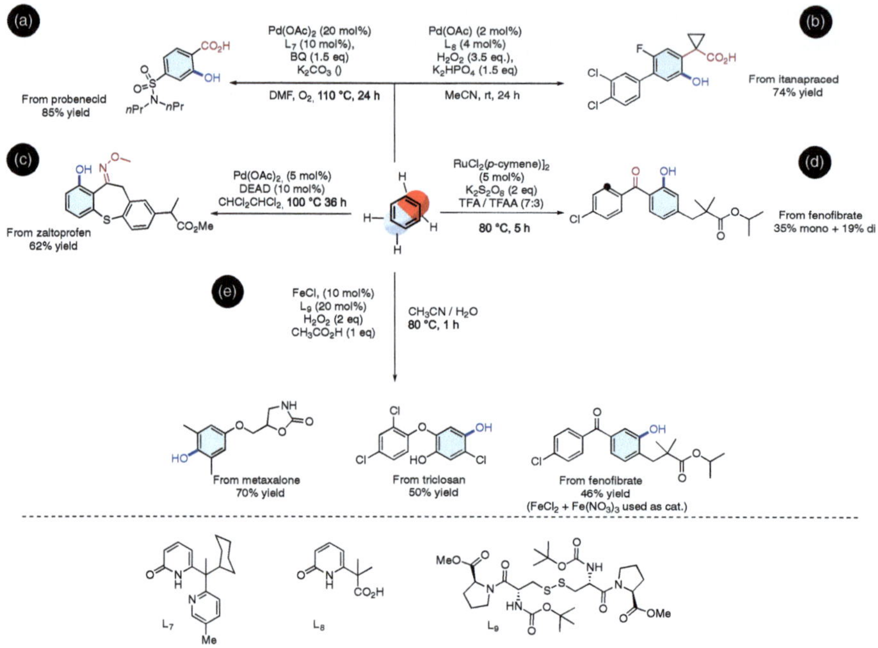

Figure 1.4 LSF of aromatics via direct C−O bond formation. a) Pd-catalyzed late-stage hydroxylation of probenecid developed by Yu et al. [33] b) Pd-catalyzed late-stage hydroxylation of itanapraced developed by Yu et al. [34] c) Pd-catalyzed late-stage hydroxylation of zaltoprofen developed by Jiao et al. [35] d) Ru-catalyzed late-stage hydroxylation of fenofibrate developed by Rao et al. [36] e) Fe-catalyzed late-stage hydroxylation of metaxalone, triclosane and fenofibrate developed by Han et al. [37].

The *ortho*-hydroxylation is not limited to Pd-catalyzed reactions. While interesting example of Ru-based catalytic system was reported for ketone-directed functionalization, the control of the monohydroxylation vs. dihydroxylation is far from trivial, as exemplified by the synthesis of fenofibrate-derivatives as a mixture of monohydroxy and dihydroxy products (Figure 1.4d) [36].

While noble metals proved their excellent reactivity in diversity of C—H activation reactions, their poor abundance, extreme costs, and often encountered high toxicity issues clearly impact the transfer of this chemistry from academic laboratories toward industry. Therefore, the design of 3d-metal-catalyzed transformations is extremely important. An attractive example of Fe-catalyzed direct hydroxylation was disclosed in 2021 by Han (Figure 1.4e) [37]. The authors discovered that this extremely cheap metal, while coordinated with cysteine-derived ligand **L9**, is prompt to promote efficient non-directed hydroxylation of a diversity of aromatics using hydrogen peroxide as the terminal oxidant. Excellent functional group tolerance and quite mild reaction conditions render this strategy a perfect solution for LSF, as illustrated by the synthesis of hydroxy metaxalone or triclosane for example, as well as a regioisomer of the previously described fenofibrate derivative.

1.2.5 LSF via C—N Bond Formation

Nitrogen atom has a pivotal role in nature, and therefore it is not surprising that the majority of drug candidates contain at least one N-atom. Indeed, nitrogen participates readily in diversity of interactions, warranting specific recognition between a drug candidate and an enzymatic pocket, for example. Accordingly, the introduction of N-atom at a late stage on a structurally complex molecule present not only a tremendous opportunity but also a great challenge. While first examples of LS nitrogenation for direct diversification of drug-like molecules were developed using mainly Cu-based catalyst, the introduction of an external directing group, such as oxazoline or oxime ether motif, was necessary [38]. In contrast, a more general strategy toward C—H amidation, compatible with several different DGs, was reported by Loh [39]. While the protocol for Rh$^{(III)}$-catalyzed direct C—N bond formation was optimized using N-heterocycle-based DG, alternative coordinating groups such as amine, amide, and ketones also showed their efficiency to direct the desired functionalization at the *ortho*-position. Interestingly, the use of hypervalent iodine reagent containing I—N bond (Zhdankin's reagent) turned out to be particularly appealing, granting the amidation event to occur under relatively mild reaction conditions (Figure 1.5a). Remarkably, thanks to the compatibility of this protocol with various DGs, it could be efficiently used for the direct nitrogenation of camptothecin *N*-oxide, delivering the desired compound in 70% yield.

Introduction of N-based functionality can also be a perfect opportunity to directly convert a desired molecule into a conjugate for target drug delivery, "PROTAC

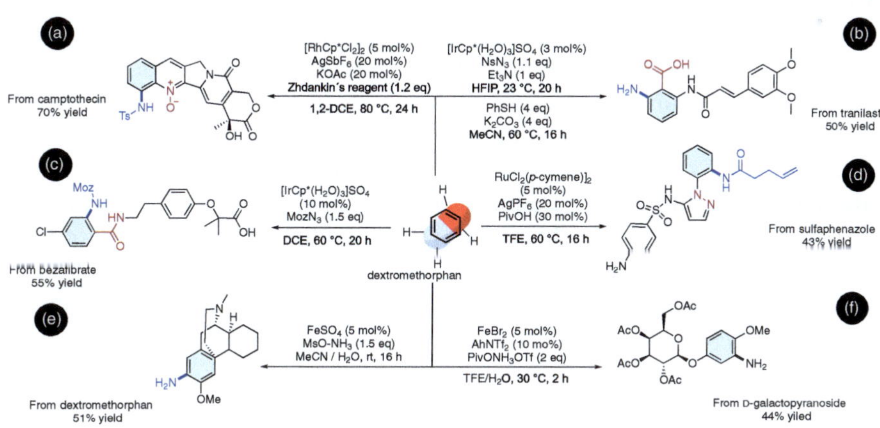

Figure 1.5 Late-stage direct C—N bond formation. a) Rh-catalyzed late-stage nitrogenation of camptothecin developed by Loh et al. [39] b) Ir-catalyzed late-stage nitrogenation of tranilast developed by Johansson, Martin-Matute et al. [40] c) Ir-catalyzed late-stage nitrogenation of bezafibrate developed by Johansson, Martin-Matute et al. [41] d) Ru-catalyzed late-stage nitrogenation of sulfaphenazole developed by Ackermann, Johansson et al. [42] e) Fe-catalyzed late-stage nitrogenation of dextromethorphan developed by Morandi et al. [43] f) Fe-catalyzed late-stage nitrogenation of D-galactopyranoside developed by Jiao et al. [44].

synthesis," or introduction of a fluorescent motif. This overall goal motivated a collaborative project between the group in AstraZeneca led by Johansson and Martin-Matute to explore direct amidation of carboxylic acid derivatives [40]. Implementation of HTE approach guaranteed rapid optimization of the reaction conditions and determination of the reaction protocol, allowing not only a high yielding reaction on a benchmark substrate but also important functional group tolerance (evaluated via the screening of various additives). IrCp*-based catalyst showed the best reactivity, thus allowing direct C—N bond formation under remarkably mild reaction conditions (Figure 1.5b). The compatibility of this protocol with various carboxylate-containing drugs was further validated, allowing the synthesis of amino-derivatives of molecules such as flavoxate and tranilast (*in situ* deprotection can be achieved). Shortly after, the same consortium further extended the generality of direct nitrogenation reactions by focusing on substrate bearing various directing groups [41]. HTE-technology was one more time the key to success to rapidly discover a modified catalytic system, able to promote the *ortho*-amidation of various building blocks, including N-heterocycles containing DGs, amides, imines, benzamides, etc. The impressive generality of this approach unlocked the door toward LSF of large panel of drug-like molecules (12 examples), including bezafibrate, levamisole, and protelivir (Figure 1.5c).

While $Ir^{(III)}$ and $Rh^{(III)}$ catalysts are recognized for their potential in direct C—N bond formation, the high cost associated with their usage may clearly hamper the industrial applications. In contrast, although Ru is a noble metal as well, its larger availability and significantly reduced costs make Ru a highly appealing alternative. Indeed, Ackermann and Johansson focused on implementing this alternative catalytic system for LSF of diversity of drug-like products [42]. In addition, dioxazolone reagents were selected as desirable aminating agents (Figure 1.5d). Under extensive high-throughput optimization, two sets of the reaction conditions were determined to promote the expected *ortho*-nitrogenations in the presence of various directing groups. The desired transformation turned out to be highly general and easily applicable to diversity of drug-like molecules. Although the yields of this LS amination are not always very high, the synthetic simplicity and rapidity of this strategy, obviating the need of the *de novo* construction of each derivative, clearly illustrate the exceptional importance of this approach. In addition, compatibility of this transformation with bifunctional dioxazolones allows LS linker installation for possible applications in PROTAC systems.

While the previous examples were based on directed approaches, benefiting for the native coordinating groups to direct the nitrogenation at the *ortho*-position, an alternative approach toward direct introduction of the nitrogen function may relay on the electrophilic activation of an aromatic ring followed by Fe-catalyzed functionalization. In 2016, Morandi disclosed an efficient and straightforward strategy for non-directed amination of electron-rich aromatics [43]. Remarkably, combination between cheap $FeSO_4$ and protonated hydroxylamine N-source warrants direct access to the primary amine derivatives without a need for additional functional group interconversions. Besides, as the aromatic substrates are used as the limiting agents, this protocol offers a unique potential for the late stage amination of drug candidates, including flurbiprofen and dextromethorphan (Figure 1.5e). Almost simultaneously, the group of Jiao reported a conceptually similar approach toward undirected, electrophilic amination of electron-rich arenes [44]. In their case, a more complex aminating agent, $PivONH_3OTf$, was finally used as the aminating agent in the presence of Fe as a catalyst and Ag-salt as a co-catalyst (Figure 1.5f). The reaction proceeded smoothly not only on a diversity of electron-rich aromatic compounds but also several applications toward LSF of complex molecules were showcased, including the synthesis of amine-derived compounds bearing D-galactopyranoside. Finally, undirected Fe-catalyzed direct amination can also be run in HFIP as solvent [45].

1.2.6 LSF via C—B Bond Formation

Direct C—H borylation, pioneered by Smith [46] and tremendously expanded by Hartwig [47, 48], is among the early and also the most powerful direct C—H activation reactions. Indeed, direct conversion of a latent C—H bond into synthetically useful C—B bond presents an extraordinary potential, paving the way toward subsequent conversion into variety of other functional groups. Therefore, selective borylation reactions warrant straightforward access to a diversity of molecular scaffolds [49].

The initial examples of direct borylations were mainly based on the use of an Ir-catalyst in combination with B_2pin_2 as the coupling partner. As heterocyclic compounds are particularly appealing in such transformations, site-selectivity was primarily controlled via combination of steric and electronic reasons, thus obviating a need for preinstallation of a directing group. Thus, non-directed direct borylation turned out to be extremely powerful and reliable strategy to afford diversity of heterocyclic building blocks [50]. One of the early elegant examples of two-step direct functionalization of complex drug-type molecules was disclosed by Hartwig in 2014 [51]. Careful evaluation of the regioselectivity of the borylation of various heterocyclic substrates allowed establishing reactivity rules, thus unlocking the door toward predictable non-directed functionalization of drug-like compounds (Figure 1.6a). The synthetic importance of this approach was beautifully illustrated while functionalizing c-Met kinase inhibitor. The site-selective C—B bond formation occurred selectively on benzoxazole moiety, thus affording an activate intermediate for a sequential Suzuki coupling. The two-step protocol of borylation followed by cross-coupling thus allowed formation of an interesting pyridine-derived drug candidate. Another attractive illustration of the synthetic importance of direct borylation of heterocycles was disclosed a few years later. Focusing on substrates bearing thiophene or pyridine-derived scaffolds, regioselective Ir(cod)-catalyzed borylation followed by copper-catalyzed methylation furnished rapidly methylated congeners of drugs such as clopidogrel or loratadine (Figure 1.6b) [52].

The excellent reactivity of the Bpin motif in various metal-catalyzed cross-coupling reactions renders this group extremely useful in synthetic chemistry and drug development. This concept was clearly demonstrated by Segawa and Itami while exploring direct borylation of strychnine (Figure 1.6c) [53]. Remarkably, the

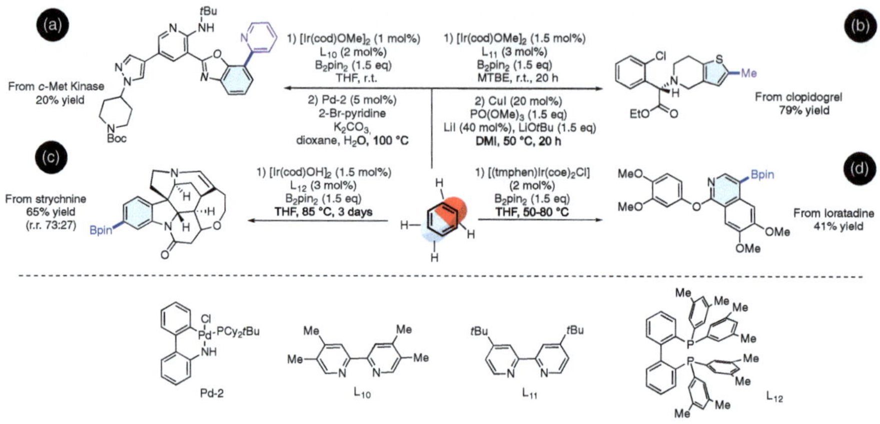

Figure 1.6 Direct borylation via LSF. a) Ir-catalyzed late-stage borylation of c-Met kinase inhibitor developed by Hartwig et al. [51] b) Ir-catalyzed late-stage borylation of clopidogrel developed by Hartwig et al. [52] c) Ir-catalyzed late-stage borylation of strychnine developed by Segawa, Itami et al. [53] d) Ir-catalyzed late-stage borylation of loratadine developed by Tu, Shekhar, Hartwig et al. [54].

Ir-based catalytic system, in combination with a finely designed diphosphine ligand **L12**, allowed direct conversion of C—H bond at C3 position of the indole motif into the desired C—B moiety. The newly accessed compound could thus be used as a starting material in a variety of reactions. Therefore, direct borylation step combined with an additional catalytic event led to the formation of a library of 15 C3-substitiuted strychnine derivatives, compounds that would be extremely difficult to access via *de novo* synthesis.

While $[Ir(cod)OMe]_2$, combined with N-based ligands, such as bipyridine- or phenantroline-derived ligands turned out to be the privileged catalysts for direct C—B bond formation, the air sensitivity of this catalytic system renders the reactions poorly reproducible and difficult to handle at larger scale. Accordingly, the design of more user-friendly catalysts has gained significant attention of the scientific community. In 2024, single-component, air-stable, and easy-to-handle Ir-catalyst has been reported by Tu, Shekhar, and Hartwig [54]. The air-stable Ir-catalyst bearing trimethylphenantroline ligand promotes the desired borylation reactions with comparable selectivities and efficiencies. Notably, its inherent stability allows direct functionalizations to be conducted, if requested, also on submicromolar scale, for applications in HTE, for example. While implementing this improved catalytic system, direct C—B functionalization of loratadine was reached in 70% yield (Figure 1.6d).

While the initial efforts on direct borylation focused mainly on transformations governed by electronic and steric properties of substrates, fine design of alternative catalytic systems to address selectively *ortho*-, *meta*-, and *para*-positions rapidly caught attention of the scientific community. Various systems have thus been designed to allow selective DG-controlled introduction of the Bpin motif [51b]. Very elegant example of the synthetic utility of the ligand-free borylation occurring under directing-group selectivity control was reported in the context of the medicinal chemistry project in 2020 by Scott et al. [55]. The authors focused on the LS diversification approach to generate fluorinated analogs of the drug-candidate AZD9833. Several conceptually different routes have been envisioned to introduce the F-atom at the different positions of the advanced molecules. In particular, direct borylation/fluorination sequence was designed to introduce the halogen atom at the C6 position of the indazole core. The presence of the pyridine motif was expected to warrant high regioselectivity via the initial pre-coordination of the Ir-complex. As expected, the desired transformation occurred smoothly under rather mild reaction conditions, delivering the expected borylated product in synthetically useful yield of 64% (Figure 1.7a). Although the further steps including amination and fluorination turned out to be more challenging, this study clearly illustrates the potential of the directed borylation using inherently present functionalities of an advanced molecule.

More systematic study of ligand-free DG-controlled *ortho*-LS borylation was reported in 2022 by Sunoj and Chattopadhyay [56]. The authors discovered that while $[Ir(cod)OMe]_2$ is used in the absence of any specific ligand, *ortho*-directing effect can be observed for several classes of substrates, including 2-phenylpyridines, 2-anilinpyridines, and benzylamines. Accordingly, this operationally simple protocol could be efficiently implemented for the LS borylation of several drug type scaffolds, including dapoxetine and sertraline (Figure 1.7b). Remarkably, the desired compounds were isolated in very high yields, thus rapidly furnishing important building blocks for further modifications.

While interesting examples can be obtained using simple $[Ir(cod)OMe]_2$, the design of significantly more general catalysts, featuring high reactivity

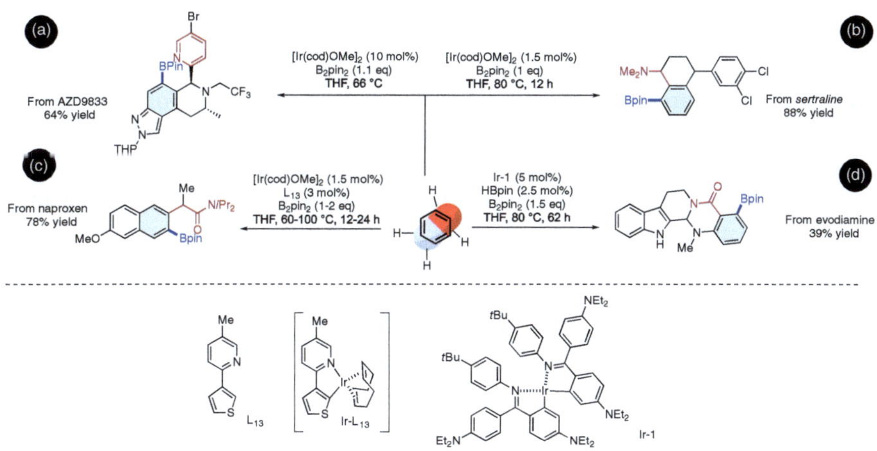

Figure 1.7 Late-stage direct C—H borylation. a) Ir-catalyzed late-stage borylation of AZD9833 developed by Scott et al. [55] b) Ir-catalyzed late-stage borylation of sertraline developed by Sunoj, Chattopadhyay et al. [56] c) Ir-catalyzed late-stage borylation of naproxen developed by Chattopadhyay et al. [58] d) Ir-catalyzed late-stage borylation of evodiamine developed by Ackermann, Smejkal, Wencel-Delord et al. [59].

and *ortho*-selectivity in the presence of structurally very different substrates, seems extremely appealing from medicinal chemistry perspective. Following this goal, Chattopadhyay proposed an original *in situ*-generated metalacyclic [57] Ir-C(thienyl) and Ir-C(furyl) anionic complexes [58]. These new catalysts feature excellent directing group compatibility, thus allowing *ortho*-selective borylation of aromatics bearing functional groups such as amides, esters, ketones, imines, ethers, amines, thioethers, and carbamates. Although relatively harsh reaction conditions are requested (at temperatures typically between 60 and 100 °C), the generality of this protocol renders it particularly suited for LS borylation, as illustrated on several examples including sertraline and naproxen (Figure 1.7c).

Following the concept of metalacyclic complexes for robust and general C—H activation, bis-cyclometallated Ir-complexes bearing imine ligands were designed by a collaboration between Smejkal from Syngenta and the groups of Ackermann and Wencel-Delord [59]. While the bis-cyclometallated complex structure is quite surprising, such Ir-species turned out to be remarkably stable and easily isolable, thus facilitating catalyst's handling. In addition, high reactivity of this complex in several different solvents, including Me-THF, octane, *i*PrOAc, CPME, and eucalyptol, under moderate reaction temperature (50 °C) and its compatibility with DGs, including amides, ketones, esters, ethers, oximes, etc., renders it highly appealing for various late stage applications, as illustrated via direct borylation of evodiamine (Figure 1.7d).

Considering the key importance of the late stage borylation for rapid expansion of the molecular complexity of advanced scaffolds, recently HT approach has attracted growing scientific interest. In a collaborative effort from researchers from Syngenta, AstraZeneca, and the group of Wencel-Delord, HT borylation of various drug-like molecules was explored, using several standard protocols, including ligand-free protocol using $[Ir(cod)OMe]_2$ precursor, $[Ir(cod)OMe]_2$ in combination with phenanthroline ligand, bipyridine ligand, and **L13**, as well as **Ir-1** [60]. Miniaturization study combined with HT technic allowed to rapidly evaluate the efficiency of such various reaction conditions for direct borylation of 45 complex drug molecules. While the reactions performed poorly or with low selectivity for 21 examples, 24 compounds furnished the functionalized structures. Even though the isolation of the borylated drugs often turned out to be highly challenging and additional derivatization steps were required, this study unambiguously illustrates the high potential of the LS borylation. The regioselectivity of the reactions appears highly substrate-dependent; in some cases, such as donepezil or tamibarotene, *ortho*-effect of the inherently present coordinating groups is evident, while in other cases, including flumazenil or olaparib, the site-selectivity seems to be controlled by steric and electronic properties of the substrate. In some cases, following the reaction conditions used, an interesting regioselectivity switch was evidenced. While this advanced study does not allow yet to predict the outcome of the direct borylation at the advanced structures, it clearly indicates the reactivity trends and selectivity/reactivity relationship depending on the nature of the catalytic system used.

Following the same objective of the rationalization of the LS borylation, the researchers from Roche, in collaboration with Konrad and Schneider, endeavored combining HTE approach for direct borylation with deep machine learning [61]. Selection of 23 drug compounds, 12 drug-like structures, and 5 frequently occurred substrates was settled as library. Standard reaction conditions, i.e., the use of $[Ir(cod)OMe]_2$ precursor in different solvents, were selected as parameters, designing a set of 956 experiments that were subsequently used as database for machine learning. The deep learning approach fructified in establishing a model able to predict reaction yield for a known reaction with an absolute error margin of 4–5%. The reactivity prediction for a new reaction was slightly less efficient but still reached the level of 92–67%, and the regioselectivity of the major product could be generally adequately foreseen.

1.2.7 LSF via C–X Bond Formation

Introduction of a Fluor atom on drug candidates can have a critical impact on the biological properties, increasing the lipophilicity, modifying pharmacokinetics and dynamic properties of a molecule, and enhancing its solubility [62]. Accordingly, approximately 20% of the commercialized pharmaceuticals and 50% of agrochemicals contain at least one fluor atom. Therefore, a late-stage fluorination of advanced drug candidates is extremely important [63].

One of the early examples of the direct fluorination of a complex molecule was described in 2015 by Xu and Xu [64]. The authors developed an efficient, Pd-based catalytic system for C—H/C—F exchange benefitting from a removable 2-pyridyloxy group. While direct fluorination is believed to be particularly challenging due to slow and unfavorable C—F bond forming reductive elimination, the use of NFSI as both the fluorinating agent and a strong oxidant warrants the desired reaction to occur smoothly for a diversity of phenol-containing substrates. Particular attention should be paid on direct fluorination of diflufenican, where high site-selectivity could be reached owing to the coordinating properties of the pyridine ring (Figure 1.8a). Use of NFSI as both, the oxidant and the fluorinating agent, was also the solution of choice to promote direct fluorination of benzamides and benzeneacetamides [65]. Slightly modified reaction protocol, using $PdCl_2$ together with $AgNO_2$, thus allowed the synthesis of diversity of fluorinated compounds, including drug derivatives such as ketoprofen and flurbiprofen, in 90 and 66% isolated yields, respectively (Figure 1.8b).

While the previous examples correspond to the traditional C—H activation scheme, occurring via insertion of a metal catalyst into a C—H bond, a conceptually very different approach toward direct fluorination of arenes was disclosed by Ritter [66]. In this case, a finely designed Pd-catalyst is used to catalytically generate, *in situ*, a highly reactive electrophilic fluorinating agent, prompting to undergo electrophilic aromatic substitution. Remarkably, this transformation offers alternative selectivity pattern, as the functionalization occurs at the electronically activated positions. The reaction features important functional group tolerance, thus paving a way toward direct diversification of a few compounds, including ciprofibrate and nateglinide derivatives (Figure 1.8c).

While direct fluorination allows to strongly impact the biological properties of drugs, the presence of other halogen atoms, in particular, chlorine, can also be an

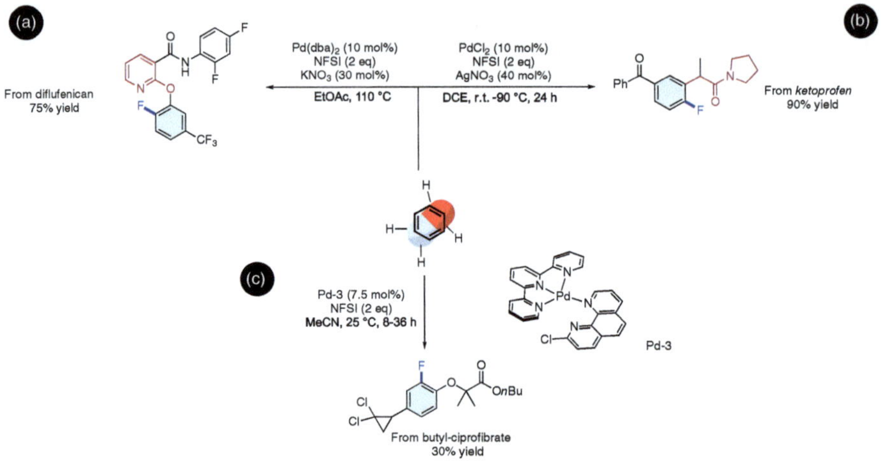

Figure 1.8 Late-stage direct fluorination. a) Pd-catalyzed late-stage fluorination of diflufenican developed by Xu, Xu et al. [64] b) Pd-catalyzed late-stage fluorination of ketoprofen developed by Lou, Xu et al. [65] c) Pd-catalyzed late-stage fluorination of ciprofibrate developed by Ritter et al. [66].

interesting tool toward improved pharmacological profile. Similarly to the late-stage fluorination directed by 2-pyridyloxy group, a direct introduction of Cl-atom can also be achieved via this Pd-catalyzed transformation [67]. The use of NCS as the chlorinating agent turned out to be optimal, allowing *in situ* generation of high oxidation state Pd(IV) intermediates, thus facilitating the challenging reductive elimination C—Cl bond formation step. This synthetic potential of this protocol toward late stage chlorination was demonstrated using diflufenican substrate, yielding the desired halogenated congener in remarkably 94% yield (Figure 1.9a).

A very interesting combination of LS functionalization and the concept of transient directing group toward direct chlorination of complex structures was reported in 2019 by Zhang [68]. While using inherently present aldehyde, combined with *p*-trifluoroaniline additive (to generate the imine directing group *in situ*), in the presence of Pd-catalyst and pyridone ligand **L14**, the *ortho*-methoxylation and chlorination occurred efficiently. When challenged with a complex structure,

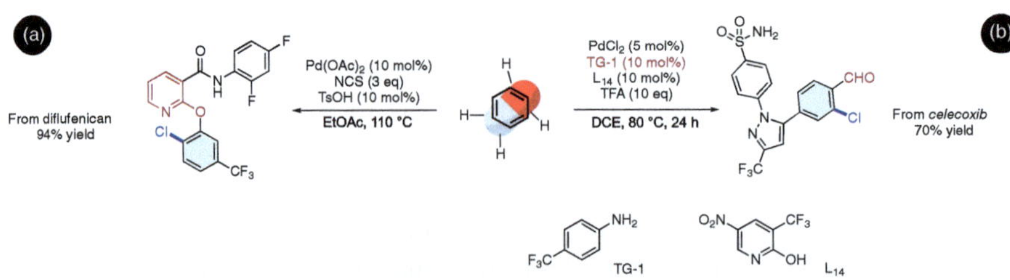

Figure 1.9 Late-stage direct chlorination. a) Pd-catalyzed late-stage chlorination of diflufenican developed by Wu, Wu et al. [67] b) Pd-catalyzed late-stage chlorination of celecoxib developed by Zhang et al. [68].

such as celecoxib, the catalytic system performed efficiently, delivering the halogenated compound in 70% yield and high 5:1 mono-functionalization vs. di-functionalization (Figure 1.9b).

1.2.8 LSF via C—D Bond Formation

Isotopically labeled compounds are of prime importance for the pharmaceutical industry [69]. Indeed, they are crucial for the elucidation of biological activity, thus being commonly used to perform ADME studies required for registration. Besides, a selective introduction of deuterium can also alter metabolic properties of a given drug candidate. Therefore, development of sustainable and straightforward hydrogen isotope exchange (HIE) reactions is highly appealing. In this context, metal-catalyzed C—H activation, in particular, LSF offers a unique route to rapidly prepare deuterated compounds while obviating the need for *de novo* synthesis with numerous synthetic steps and deuterated building blocks.

While several examples of late-stage deuteration of drug candidates have been developed using Ir-based Crabtree's catalyst [70] as well as Kerr's Ir–NHC complex [71], and other Ir-based catalysts [72], the development of alternative routes compatible with less expensive catalyst was highly appealing [73].

Elegant and sustainable strategy toward hydrogen isotope exchange was proposed by Chirik in 2016, focusing on Fe-based catalyst. Following their work on Fe-catalyzed hydrogenation of non-activated alkanes, they discovered that bis(arylimidazol-2-ylidene)pyridine iron bis(dinitrogen) complex **Fe-1** shows exceptional reactivity in hydrogen/deuterium exchange [74]. Remarkably, gaseous $^2H_2$ can be used under moderate overpressure (4 atm) and mild reaction conditions (45 °C). The selectivity of this reaction is controlled by the steric factors, with the preferential functionalization occurring at the most accessible position. The reaction is also favored in the case of electron-poor substrates and is compatible, as well, with various heterocyclic substrates. While challenging this catalytic system with complex substrates, including drug derivatives, several deuterated congeners could be directly accessed, including paroxetine, loratadine, and flumazenil (Figure 1.10a). An additional unique advantage of this protocol is its compatibility with tritiation. Indeed, under only subatmospheric pressure of $^3H_2$, tritium labeling occurred smoothly, warranting expedient access to the tritium-labeled drugs, such as souvorexant, cinacalcet, and flumazenil. While this strategy offers an interesting route toward important scaffolds, extreme water and air sensitiveness of the catalyst makes these transformations not easily applicable for less experienced practitioners.

Few years later, the same research groups focused on Ni-based catalysts for hydrogen isotope exchange. By developing nickel hydride precatalyst (**Ni-1**), they succeeded in establishing an operationally simple and highly efficient protocol for direct deuteration and tritiation of various heterocyclic substrates, using gaseous $^2H_2$ and $^3H_2$ [75, 76]. Compared to the previous Fe-based system, Ni-complex shows improved reactivity toward electron-rich heteroarenes, including pyrroles and indoles, without a need of an N-directing group. The isotope exchange occurs preferentially at positions adjacent to the nitrogen atom for N-heterocyclic

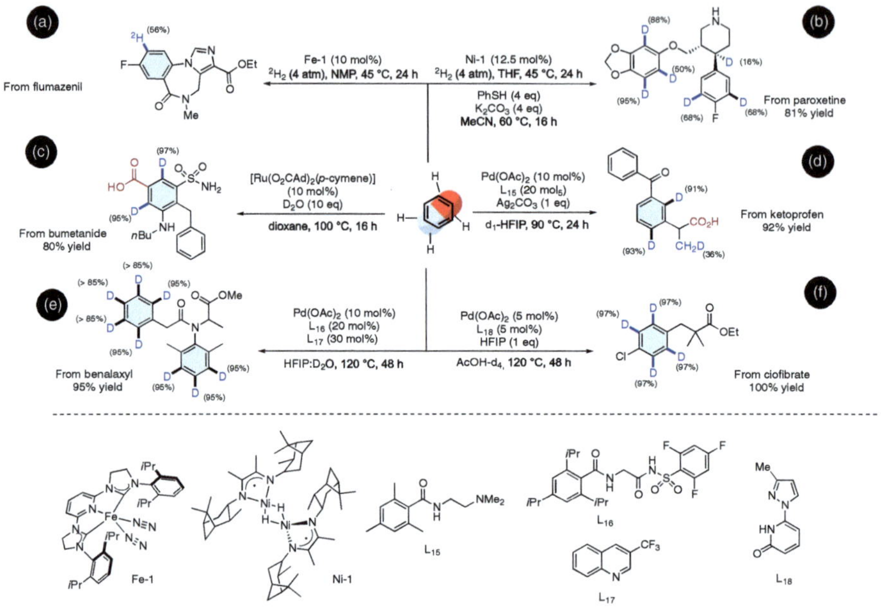

Figure 1.10 Direct hydrogen isotope exchange for the synthesis of deuterated drugs. a) Fe-catalyzed late-stage deuteration of flumazenil developed by Chirik et al. [74] b) Ni-catalyzed late-stage deuteration of paroxetine developed by Chirik et al. [75] c) Ru-catalyzed late-stage deuteration of bumetanide developed by Ackermann et al. [77] d) Pd-catalyzed late-stage deuteration of benaalaxyl developed by van Gemmeren et al. [79] e) Pd-catalyzed late-stage deuteration of clofibrate developed by Joo et al. [80] f) Pd-catalyzed late-stage deuteration of ketoprofen developed by van Gemmeren et al. [78].

substrates and at the sterically most available positions for aromatic compounds. This Ni-based catalytic system thus paved a way toward isotopically labeled entities, such as etoricoxib, papaverine, or paroxetine (Figure 1.10b).

While Chirik's Fe- and Ni-based complexes showed particular efficiency in hydrogen isotope exchange at sterically available positions, alternative regioselectivity trend was observed by Ackermann, while establishing Ru(II)-catalyzed protocol [77]. Benefiting from high reactivity of ruthenium biscarboxylate complexes and reversibility of the C—H activation step, the group succeeded in developing fully *ortho*-selective hydrogen isotope exchange protocol, using carboxylate as largely present in nature DG. The use of D_2O as deuterium source further increases the operational simplicity of this methodology, even though high reaction temperature (100 °C) is required to reach full conversion. The high potential of this hydrogen isotope exchange approach was further highlighted via LS deuteration of several pharmaceutical drugs and biological active sulfonamides, including repaglinide, bumetanide, and telmisartan (Figure 1.10c).

An alternative protocol toward deuteration of carboxylic acids was reported one year later by the group of van Gemmeren [78]. Benefiting from the coordinating properties of the carboxylate DG, combined with the addition of an amide type ligand **L15** in charge of promoting the CMD (concerted metalation-deprotonation)

C—H cleavage step, reversible C—H metalation and the following *ortho*-selective deuteration were performed successfully. While the silver salt was often necessary to reach high conversion, both d_1-HFIP and D_2O could be used as the deuterating agents. Interestingly, the reaction protocol tolerates both, aliphatic and aromatic substrates, allowing direct deuteration of structurally different compounds. As the benzyl carboxylate motif is largely present in various natural products and drugs, the compatibility of this reaction with LS deuteration was demonstrated using substrates such as ketoprofen, ciofibric acid, and ciprofibrate (Figure 1.10d). While developing this Pd-based system, the authors also discovered that under slightly modified reaction conditions undirected deuteration can be promoted [79]. Indeed, in the presence of two catalysts, i.e., bicoordinating amide **L16** combined with a pyridine-based ligand **L17**, general deuteration of the aromatic rings was observed at various aromatic positions. In case of pharmaceutically relevant scaffolds, such as benalaxyl or praziquantel, high level of D-insertion was observed at all aromatic positions using HFIP/D_2O as solvent mixture under harsh conditions (at a temperature of 120 °C) (Figure 1.10e).

Conceptually, closely related per deuteration of aromatic substrates was reported in 2023 by Joo. Capitalizing on the privileged $Pd(OAc)_2$ catalyst, they designed an alternative catalytic cycle using pyrazolopyridone (**L18**) ligand [80]. The pyridine motif acts as an internal base during the C—H cleavage step, warranting reversibility of this event and thus promoting efficient hydrogen isotope exchange. The reaction conditions remain harsh, with the reaction temperature of 120 °C and the use of acetic acid as the deuterium source, but they are still applicable for the synthesis of labeled drugs, including fenofibrate and clofibrate (Figure 1.10f).

1.3 Late-Stage Functionalization of Aliphatic Compounds

While LS functionalization of aromatic compounds has reached some level of maturity, direct introduction of functional groups on aliphatic parts of a complex molecule is infinitely more challenging. Indeed, $C(sp^3)$—H bonds are considerably less reactive in inner-sphere C—H activation type reactions. In addition, in most cases, derivatization of an advance molecule to introduce a specific DG is requested, as exemplified by a work of Chang [81]. The elegant Ir-catalyzed strategy has been reported for direct amidation of unactivated methyl group. As oxime was necessary as DG, direct functionalization of complex molecules such as friedelin or lanosterol could be achieved, but prior conversion of ketones or secondary alcohol motifs into the desired ketoxime was essential. In addition, inner-sphere metal-catalyzed protocols are generally limited to the sterically accessible positions, thus significantly limiting the diversity of possible substrates.

Alternative strategy toward metal-catalyzed direct functionalization of $C(sp^3)$—H bonds implies outer-sphere C—H activation strategies, particularly suited for C—N and C—C bond formation via nitrenoid and carbenoid intermediates. Drawing inspiration from nature, spectacular direct oxidation type transformations have been reported using finely designed Fe- and Mn-based complexes [82]. In this case, minor modifications of the ligand structure allows precise control of the regioselectivity of the direct oxidation, thus allowing direct installation of an O-atom in a precise and predictable manner [83].

1.3.1 LSF of Aliphatic Substrates via Direct C—C Bond Formation

While Pd-catalyzed direct C—H activation typically requires the use of a DG to pre-coordinate the catalyst to the substrate and promote regioselective direct arylation, carboxylic acid, often encountered motifs in natural products and drugs, can play this role and being used as an inherent directing group. This concept has been elegantly used by Yu for direct C—H arylation of activated aliphatic substrates, i.e., cyclopropanes, in 2019 [84]. The careful design of the amino acid-derived ligand **L19** allowed not only to reach high reactivity in this C—H activation reaction, but also enantioselective outcome of this arylation was achieved, delivering the arylated cyclopropanes in high ees. While challenging the catalytic system with itanapraced, cyclopropane-containing drug, the LS arylation occurred smoothly, delivering the arylated congener in excellent yield and 94% ee (Figure 1.11a). Few years later, the concept of the carboxylic acid-directed transannular C—H arylation was extended toward a large diversity of cycloalkanes [85]. Addition of the quinuclidine-pyridone ligand (**L20**) allowed transannular γ-methylene C—H activation that was successfully exploited for a diversity of cyclobutene, cyclopentane, cyclohexane, cycloheptane, and cyclooctane substrates. Remarkably, this protocol could also be exploited in the context of direct arylation of isosteviol, where the regioselectivity is perfectly controlled by the geometry of the metalacyclic intermediate (Figure 1.11b).

An alternative example of the applications of native functional groups as DG for C—H activation was reported by Gaunt et al. in 2020 [86]. The authors discovered that tertiary amines are able to coordinate the Pd catalyst, directing the C—H activation toward γ-position via generation of 5-membered metalacyclic intermediate. Addition of an external ligand, such as monoprotected amino acid, was crucial to enhance the C—H cleavage step, and simple *N*-acetyl *t*-leucine (**L21**) turned out to be the optimal additive to efficiently trigger the CMD-type metalation event. Under the optimized reaction conditions, the desired arylation could thus be conducted using diversity of propylamine derivatives. Interestingly, benefiting from the inherent structure of trimipramine, this direct arylation reaction could be implemented to directly furnish several aromatic derivatives of this antidepressant drug (Figure 1.11c). It is also worth noting an elegant work of Sanford, where transannular arylation of cyclic amines could be achieved using Pd-catalyst. However, installation of precise directing group was necessary to promote LS functionalization of varenicline derivative [87].

A conceptually and mechanistically very different example of LS functionalization of aliphatic substrates was reported by Davies and Beckwith [88]. Outer sphere Rh-catalyzed strategy was used to diversify alkaloids, including brucine. The reactivity of Rh-carbenoid insertion of donor/acceptor diazo coupling partner

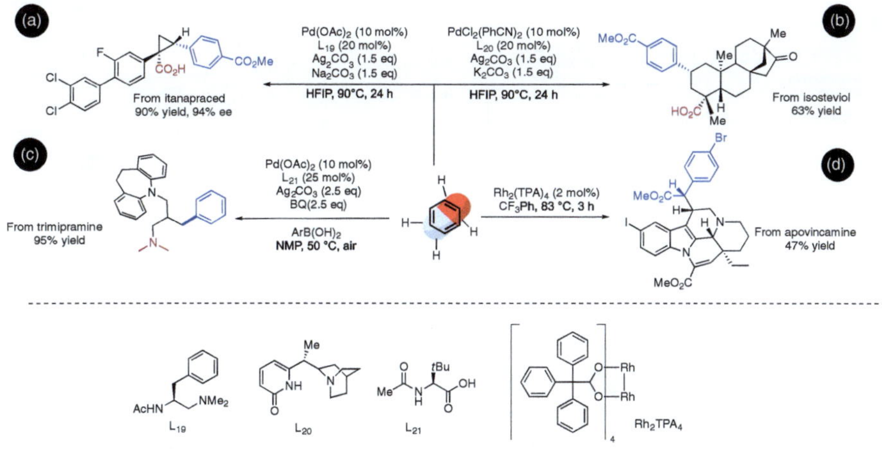

Figure 1.11 Late-stage diversification of aliphatic substrates via C–C bond formation.
a) Pd-catalyzed C(sp^3)-H late-stage arylation of itanapraced developed by Yu et al. [84]
b) Pd-catalyzed C(sp^3)-H late-stage arylation of isosteviol developed by Yu et al. [85]
c) Pd-catalyzed C(sp^3)-H late-stage arylation of trimipramine developed by Gaunt et al. [86]
d) Pd-catalyzed C(sp^3)-H late-stage alkylation of apovincamine developed by Davies, Beckwith et al. [88].

favors functionalization at electronically enriched positions, i.e., benzylic, allylic, α-atom to heteroatom, or tertiary positions. In addition, fine-tuning of the ligand allows to further control site, and if relevant, stereoselectivity of the reaction. Benefiting from such a high selectivity of the $Rh^{(II)}$-carbenoid insertion, the authors succeeded in designing three catalytic systems amenable to differentiate very similar C(sp^3)—H bonds, thus delivering selectively three different regioisomers of the functionalized compound. The reaction could be further extended toward other alkaloid-type substrates, including securinine and apovincamine (Figure 1.11d). Over the following years, additional examples of $Rh^{(II)}$-catalyzed insertion reactions have been used to allow direct synthesis of complex natural products and derivatization of drugs [89].

1.3.2 LSF via C–O Bond Formation

Direct oxidation of aliphatic compounds is one of the key processes in nature. Therefore, it is not surprising that development of similar strategies to direct installing oxygen atom on a complex molecule is a rapidly evolving research field.

Among the catalytic systems that have attracted key scientific attention, Fe-based complexes are particularly interesting. Following the excellent work of White group [83], exceptional reactivity of Fe(PDP)-catalyst was discovered, allowing non-directed but highly selective direct oxidation of aliphatic substrates. Indeed, this catalyst features a unique capability of being able to differentiate even very similar C(sp^3)—H bonds based on their electronic (activation by electron-donating groups), steric (favored functionalization of 3° site), and stereoelectronic properties (conformational effects are contributing factors to

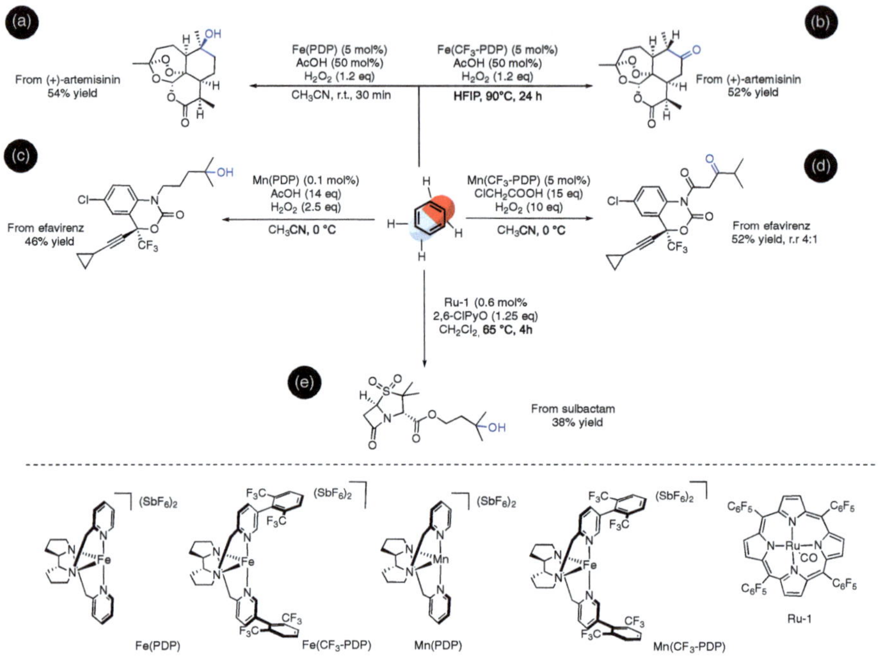

Figure 1.12 LS oxygenation of aliphatic complex molecules. a) Fe-catalyzed C(sp^3)-H late-stage oxidation of artemisinin developed by White et al. [91] b) Fe-catalyzed C(sp^3)-H late-stage oxidation of artemisinin developed by White et al. [92] c) Mn-catalyzed C(sp^3)-H late-stage oxidation of efavirenz developed by White et al. [93] d) Mn-catalyzed C(sp^3)-H late-stage oxidation of efavirenz developed by White et al. [93] e) Ru-catalyzed C(sp^3)-H late-stage oxidation of sulbactam developed by Hartwig et al. [94].

the product distribution). The detailed analysis of the combination of these parameters allows highly predictable transformations, even when using complex substrates, including dihydropleuromutilone [90] or artemisinin (Figure 1.12a) [91]. Remarkably, precise modification of the ligand's structure impacts significantly its chemical behavior, altering the selectivity of the oxidation step. Indeed, introduction on the external part of the ligand two aromatic units bearing sterically demanding and electron-withdrawing CF_3 motifs contributes to the increased sensitivity of the catalyst toward steric hindrance of a substrate. Accordingly, the newly designed Fe(CF_3-PDP) catalyst promotes oxidation of the 2°-positions, delivering regiomeric products compared to the standard Fe(PDP) system (Figure 1.12b) [92].

Extending the work of Fe(PDP), White et al. investigated a closely related Mn(PDP)-complex [93]. This new catalyst also showed important reactivity toward direct C—O bond formation at 3°-position, and it tolerates well the presence of nitrogen motifs on complex molecules. Accordingly, the LS functionalization of efavirenz was achieved selectively in a synthetically useful yield, using very low catalyst loading of 0.1 mol% (Figure 1.12c). Following the same trend as in case of the Fe-complexes, more sterically accessible Mn(PDP) promotes selective oxidation at the 3°-position, while the use of more sterically hindering (CF_3-PDP) gives access to the regiomeric products with the oxygen being installed at the 2°-position (Figure 1.12d).

Very recently, a new approach toward direct hydroxylation of various compounds, including natural products and drug derivatives, was disclosed by Hartwig [94]. Ru-porphyrine complex turned out to be extremely reactive toward direct C—O bond formation at 3° site, featuring, in addition, extensive functional group compatibility, including nitrogenated heterocycles, halogen substituents, nitro groups, and Bpin motifs. Considering the generality of this approach and its potential toward LS functionalization was rapidly evidenced, using diverse natural products and drug derivatives, including fenofibrate, sulbactam, and thalidomide (Figure 1.12e). In all cases, high yielding reaction was conducted in the presence of only 0.6 mol% of the Ru-catalyst.

1.3.3 LSF via C—N Bond Formation

While considering outer-sphere C—H functionalization of C(sp^3)—H bonds, $Rh^{(II)}$-based catalytic systems, largely explored by Du Bois [95], Dauban [96], and others [97], are frequently considered benchmark strategy toward direct amination of aliphatic substrates. However, the initial work of Du Bois showed that this approach might be poorly efficient while challenged with complex molecules, in particular containing several nitrogen atoms. The remarkably breakthrough toward more general amination approach, compatible with LSF mindset, was reported in 2018 by Du Bois [98]. Precise selection of solvent, namely, *t*BuCN, turned out to be crucial to allow, on the one hand, solubilize densely functionalized compounds and, on the other hand, stabilize the dirhodium catalyst. Under the optimized reaction conditions, this protocol could thus be used toward highly regioselective direct amination of variety of compounds, including drugs such as tadalafil. The regioselectivity is controlled by the inherent reactivity of the substrate and thus positions into N-atom or 3°-position (Figure 1.13a).

Following the same conceptual approach as Fe- and Mn-catalyzed direct oxidation, amination of aliphatic C(sp^3)—H bonds can be efficiently conducted in the presence of a Fe- and Mn-based catalyst. While H_2O_2 was commonly used in case of the oxidation reactions, direct amination can frequently benefit from the application of iminoiodinane as the aminating agents [99]. While iminoiodinane can be generated *in situ* while reacting sulfamate and $PhI(OPiv)_2$, this reaction may be slow. Therefore, addition of a preformed iminoiodinane is frequently a solution of choice. Detailed catalytic system's optimization revealed that a polychlorinated porphyrine Mn-based catalyst outcompetes other catalysts. Contrary to the direct oxidation reaction, the C—N bond formation occurs preferentially at the benzylic position. The catalyst shows, however, strong steric preferences with the functionalization occurring at more accessible position and more electron-rich positions are addressed first. As an illustration of this methodology, direct nitrogenation of sulbactamb or sertraline can be indicated (Figure 1.13b).

Outer-sphere C—H activation of C(sp^3)—H bonds is also particularly suited to promote direct azidation-type transformations. In 2015, Groves demonstrated that

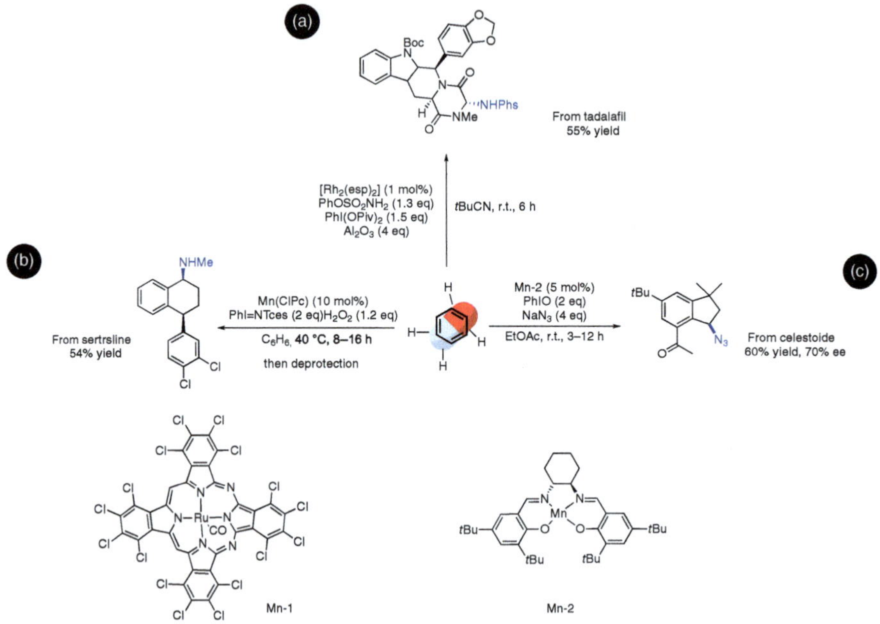

Figure 1.13 LS nitrogenation of the aliphatic substrates. a) Mn-catalyzed C(sp^3)-H late-stage nitrogenation of sertraline developed by White et al. [99] b) Mn-catalyzed C(sp^3)-H late-stage nitrogenation of celestoide developed by Groves et al. [100].

Shiff-base Mn complexes can be used successfully to install N_3-motif on various complex molecules without prior prefunctionalization [100]. The salient feature of this reaction includes the use of aqueous sodium azide solution as the coupling partner and the robustness of the catalytic system (reactions can be run under air). The azidation occurs inherently at the most reactive positions, i.e., at the benzylic or 3° site. This operationally simple protocol is thus a promising handle for the direct functionalization of a number of natural products and drugs, including artemisinin, celestoide, and papaverine (Figure 1.13c).

1.3.4 LSF via C—D Bond Formation

While metal-catalyzed direct H/D exchange can be relatively easily achieved by means of reversible C—H activation approach, the situation is significantly more complex when aliphatic C—H substrates are considered. Indeed, the metalation event, already challenging, needs to be reversible to warrant efficient H/D scrambling. One of the rare examples of LS deuteration of aliphatic carboxylic acids was reported in 2021 by van Gemmeren.[80] Combining (1) the coordinating properties of the carboxylic function inherently present on a given molecule with (2) the amine-amide ligand **L22** prone to facilitate the metalation event, efficient C(sp^3)—H activation followed by trapping with d_1-HFIP, furnished the deuterated alkanes in high efficiency. This protocol could be efficiently applied for the synthesis

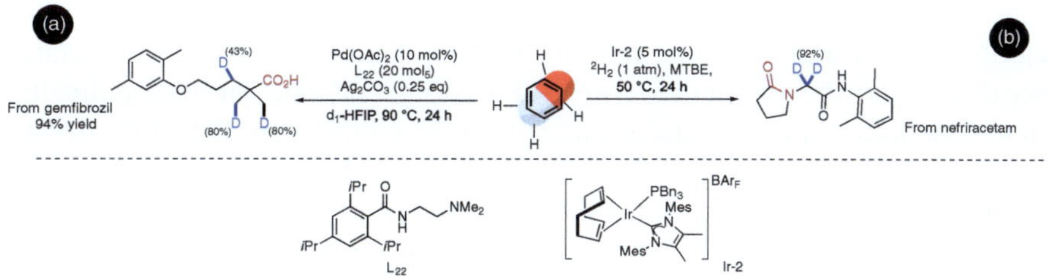

Figure 1.14 Hydrogen isotope exchange on aliphatic drug-like molecules. a) Pd-catalyzed C(*sp*^{3})-H late-stage deuteration of gemfibrozil developed by van Gemmeren et al. [78] b) Ir-catalyzed C(*sp*^{3})-H late-stage deuteration of nefriracetam developed by Kerr et al. [101].

of isotopically labeled complex molecules, including gemfibrozil and bezafibrate (Figure 1.14a).

An alternative protocol allowing direct hydrogen isotope exchange on amide-type substrates was reported in 2024 by Kerr [101]. The authors, capitalizing in their previous work on Ir-catalyzed H/D exchange, designed a catalytic system by promoting highly selective deuteration of C(*sp*^{3})—H bonds in α to nitrogen atom. Excellent levels of deuterium incorporation combined with high predictability of the regioselectivity of the reaction and interesting functional group tolerance make this strategy an interesting tool to afford labeled pharmaceuticals, including laurocapram, nefriracetam, and unifiram (Figure 1.14b).

1.4 Conclusions and Perspectives

While the pioneering and sporadic examples of the use of C—H activation reactions for LSF of drug candidates have appeared in the literature at the beginning of this century, this concept has witnessed a tremendous expansion over the last few years. The diversity of newly discovered C—H activation reactions paved the way for a large panel of systems that allow direct diversification of complex molecules. Importantly, the panel of applications of LSF is extremely broad. On the one hand, diversity of functional groups can be directly introduced, including aromatic and aliphatic units, O- and N-based motifs, as well as halogen substituents (e.g., F and Cl) and boron-based groups. On the other hand, various inherently present functional groups on a drug candidate, including carboxylate or *N*-heterocycle motifs, can be used to control the site-selectivity of the reaction and promote selective *ortho*-functionalizations. Importantly, the chemical diversity of compounds can be further expanded while implementing distal C—H activation approaches to address alternative C—H positions on a molecule. However, while reaching different selectivities is relatively well known in the case of phenol derivatives, direct *meta*- and *para*-functionalization of more complex structures, without the need for preinstallation of a complex directing group, remains extremely challenging and very case-specific.

The analysis of the state-of-the-art literature also clearly shows that the vast majority of the transformations rely on the use of expensive and toxic metals. Since this step is to be used at the very end of the drug candidate synthesis, difficulty in removing traces of metal-catalysts might thus be a limitation. Therefore, development of alternative strategies, based on the use of cheap and more abundant 3d-metals, is of high demand and will certainly progress considerably in the near future. However, the difficulty toward this challenging goal lies in the common use of bis-coordinating DGs in 3d-metal-catalyzed C—H activation, which is fundamentally incompatible with the LSF approach.

Important to notice is also a real change in the perception of the LSF by the scientific community. Still, a decade ago, challenging a newly developed methodology with complex drug-like molecules was rather seldom. Currently, demonstration of a functional group tolerance and compatibility of a new reaction with complex structures is almost a standard requirement to validate the synthetic utility of a project. In addition, clear implementation of the LSF in the R&D sector in industry further highlights the importance of this topic. Importantly, the industrially driven projects also show that even poorly regioselective LSF reactions can still be of high interest as they provide direct access to several regioisomeric products, thus further expanding the chemical space.

The last decade of research in the field of C—H activation clearly evidenced that this strategy offers a unique possibility toward direct functionalization of complex molecules. However, several limitations, including selectivity toward other than *ortho*-C—H bonds, common use of Pd-, Ru-, and Ir-complexes, and compatibility issues of a given reaction with several coordinating groups present on a drug-like molecule, are still persisting challenges. However, considering the real motivation of the academic groups to showcase the synthetic utility of their new methodologies, combined with the expanding interest of industrial partners, we are strongly convinced that the field will evolve very rapidly in the near future, bringing academic research and concrete applications even closer.

References

1 Papanikolaou, G., Centi, G., Perathoner, S., and Lanzafame, P. (2024). Green synthesis and sustainable processing routes. *Current Opinion in Green and Sustainable Chemistry* 47: 100918.

2 Sharma, S., Gallou, F., and Handa, S. (2024). Towards a sustainable tomorrow: advancing green practices in organic chemistry. *Green Chemistry* 26 (11): 6289–6317.

3 For general reviews on C—H activaton see: (a) Wencel-Delord, J., Dröge, T., Liu, F., and Glorius, F. (2011). Towards mild metal-catalyzed C—H bond activation. *Chemical Society Reviews* 40 (9): 4740–4761. (b) Dalton, T., Faber, T., and Glorius, F. (2021). C—H activation: toward sustainability and applications. *ACS Central Science* 7: 245–261. (c) De Jesus, R., Hiesinger, K., and Van Gemmeren, M. (2023). Preparative scale applications of C—H activation in medicinal chemistry. *Angewandte Chemie, International Edition* 62 (45): e202306659.

(d) Gensch, T., Hopkinson, M.N., Glorius, F., and Wencel-Delord, J. (2016). Mild metal-catalyzed C—H activation: examples and concepts. *Chemical Society Reviews* 45 (10): 2900–2936. (e) Hartwig, J.F. (2016). Evolution of C—H bond functionalization from methane to methodology. *Journal of the American Chemical Society* 138 (1): 2–24. (f) Kuhl, N., Hopkinson, M.N., Wencel-Delord, J., and Glorius, F. (2012). Beyond directing groups: transition-metal-catalyzed C—H activation of simple arenes. *Angewandte Chemie, International Edition* 51 (41): 10236–10254.

4 (a) Ali, W., Oliver, G.A., Werz, D.B., and Maiti, D. (2024). Pd-catalyzed regioselective activation of C(sp^2)—H and C(sp^3)—H bonds. *Chemical Society Reviews* 53 (19): 9904–9953. (b) Ma, W., Gandeepan, P., Li, J., and Ackermann, L. (2017). Recent advances in positional-selective alkenylations: removable guidance for twofold C—H activation. *Organic Chemistry Frontiers* 4 (7): 1435–1467.

5 Gandeepan, P., Müller, T., Zell, D. et al. (2019). 3d transition metals for C—H activation. *Chemical Reviews* 119 (4): 2192–2452.

6 Ackermann, L. (2015). Robust Ruthenium(II)-catalyzed C—H arylations: carboxylate assistance for the efficient synthesis of angiotensin-II-receptor blockers. *OPRD* 19 (1): 260–269.

7 (a) Caro-Diaz, E.J.E., Urbano, M., Buzard, D.J., and Jones, R.M. (2016). C—H activation reactions as useful tools for medicinal chemists. *Bioorganic & Medicinal Chemistry Letters* 26 (22): 5378–5383. (b) Cernak, T., Dykstra, K.D., Tyagarajan, S. et al. (2016). The medicinal chemist's toolbox for late stage functionalization of drug-like molecules. *Chemical Society Reviews* 45 (3): 546–576. (c) Lam, N.Y.S., Wu, K., and Yu, J. (2021). Advancing the logic of chemical synthesis: C—H activation as strategic and tactical disconnections for C—C bond construction. *Angewandte Chemie, International Edition* 60: 15767–15790.

8 Dai, H.-X., Stepan, A.F., Plummer, M.S. et al. (2011). Divergent C—H functionalizations directed by sulfonamide pharmacophores: late-stage diversification as a tool for drug discovery. *Journal of the American Chemical Society* 133 (18): 7222–7228.

9 (a) Shang, W., Sun, H., Chen, W., and Liu, J. (2023). Diversification of pharmaceutical molecules via late-stage C(sp^2)—H functionalization. *Green Synthesis and Catalysis* 4 (2): 104–123. (b) Shim, S.Y. (2023). Late-stage C—H activation of drug (derivative) molecules with Pd(II) catalysis. *Chemistry – A European Journal* 29 (71): e202302620. (c) Huo, T., Zhao, X., Cheng, Z. et al. (2024). Late-stage modification of bioactive compounds: improving druggability through efficient molecular editing. *Acta Pharmaceutica Sinica B* 14 (3): 1030–1076. (d) Guillemard, L., Kaplaneris, N., Ackermann, L., and Johansson, M.J. (2021). Late-stage C—H functionalization offers new opportunities in drug discovery. *Nature Reviews Chemistry* 5 (8): 522–545. (e) Castellino, N.J., Montgomery, A.P., Danon, J.J., and Kassiou, M. (2023). Late-stage functionalization for improving drug-like molecular properties. *Chemical Reviews* 123 (13): 8127–8153.

10 (a) Zhan, B.-B., Jiang, M.-X., and Shi, B.-F. (2020). Late-stage functionalization of peptides *via* a palladium-catalyzed C(sp^3)—H activation strategy. *Chemical Communications* 56 (90): 13950–13958. (b) Barahdia, A.S., Thakare, K.L.,

Kaur, L., and Jain, R. (2024). Endogenous group-directed late-stage C—H functionalization of peptides. *Advanced Synthesis and Catalysis* 366 (13): 2844–2858. (c) Wang, W., Lorion, M.M., Shah, J. et al. (2018). Late-stage peptide diversification by position-selective C—H activation. *Angewandte Chemie, International Edition* 57 (45): 14700–14717. (d) Tang, X., Van Der Eycken, E.V., and Song, L. (2025). Functionalization of amino acids and peptides *via* rhodium-catalyzed C—H activation. *Chemical Communications* 61 (63): 11746–11757. (e) Sengupta, S. and Mehta, G. (2017). Late stage modification of peptides via C H activation reactions. *Tetrahedron Letters* 58 (14): 1357–1372.

11 Thuy-Boun, P.S., Villa, G., Dang, D. et al. (2013). Ligand-accelerated *ortho*-C—H alkylation of arylcarboxylic acids using alkyl boron reagents. *Journal of the American Chemical Society* 135 (46): 17508–17513.

12 Ni, S., Hribersek, M., Baddigam, S.K. et al. (2021). Mechanochemical solvent-free catalytic C—H methylation. *Angewandte Chemie, International Edition* 60 (12): 6660–6666.

13 Hogg, A., Wheatley, M., Domingo-Legarda, P. et al. (2022). Ruthenium-catalyzed monoselective C—H methylation and d_3-methylation of arenes. *JACS Au* 2 (11): 2529–2538.

14 Friis, S.D., Johansson, M.J., and Ackermann, L. (2020). Cobalt-catalysed C—H methylation for late-stage drug diversification. *Nature Chemistry* 12 (6): 511–519.

15 Kaplaneris, N., Rogge, T., Yin, R. et al. (2019). Late-stage diversification through manganese-catalyzed C—H activation: access to acyclic, hybrid, and stapled peptides. *Angewandte Chemie, International Edition* 58 (11): 3476–3480.

16 Guillemard, L., Ackermann, L., and Johansson, M.J. (2024). Late-stage meta-C—H alkylation of pharmaceuticals to modulate biological properties and expedite molecular optimisation in a single step. *Nature Communications* 15 (1): 3349.

17 Yuan, C., Zhu, L., Chen, C. et al. (2018). Ruthenium(II)-enabled para-selective C–H difluoromethylation of anilides and their derivatives. *Nature Communications* 9 (1): 1189.

18 Dou, Y., Kenry, Liu, J. et al. (2019). Late-stage direct *o*-alkenylation of phenols by Pd^{II}-catalyzed C—H functionalization. *Chemistry – A European Journal* 25 (28): 6896–6901.

19 Lim, Y., Tang, G., Ye, Z. et al. (2023). A late-stage aryl C—H olefination strategy and its application towards global proteome profiling of Δ^8-tetrahydrocannabinol. *Chemistry – A European Journal* 29 (29): e202300531.

20 Naksomboon, K., Valderas, C., Gómez-Martínez, M. et al. (2017). *S,O*-Ligand-promoted palladium-catalyzed C—H functionalization reactions of nondirected arenes. *ACS Catalysis* 7 (9): 6342–6346.

21 Chen, X.-Y., Wu, Y., Zhou, J. et al. (2019). Synthesis of β-arylethenesulfonyl fluoride via Pd-catalyzed nondirected C—H alkenylation. *Organic Letters* 21 (5): 1426–1429.

22 Zhang, Z., Tanaka, K., and Yu, J.-Q. (2017). Remote site-selective C—H activation directed by a catalytic bifunctional template. *Nature* 543 (7646): 538–542.

23 Wang, P., Verma, P., Xia, G. et al. (2017). Ligand-accelerated non-directed C—H functionalization of arenes. *Nature* 551 (7681): 489–493.

24 Fan, Z., Chen, X., Tanaka, K. et al. (2022). Molecular editing of aza-arene C—H bonds by distance, geometry and chirality. *Nature* 610 (7930): 87–93.

25 Simonetti, M., Cannas, D.M., Just-Baringo, X. et al. (2018). Cyclometallated ruthenium catalyst enables late-stage directed arylation of pharmaceuticals. *Nature Chemistry* 10 (7): 724–731.

26 Deng, K., Sukowski, V., and Fernández-Ibáñez, M.Á. (2024). Non-directed C—H arylation of anisole derivatives via Pd/*S,O*-ligand catalysis. *Angewandte Chemie, International Edition* 63 (19): e202400689.

27 Liu, L.-Y., Qiao, J.X., Yeung, K.-S. et al. (2019). *Meta* C—H arylation of electron-rich arenes: reversing the conventional site selectivity. *Journal of the American Chemical Society* 141 (37): 14870–14877.

28 Catellani, M., Frignani, F., and Rangoni, A. (1997). A complex catalytic cycle leading to a regioselective synthesis of *o,o′*-disubstituted vinylarenes. *Angewandte Chemie (International Ed. in English)* 36 (1–2): 119–122.

29 Shi, H., Lu, Y., Weng, J. et al. (2020). Differentiation and functionalization of remote C—H bonds in adjacent positions. *Nature Chemistry* 12 (4): 399–404.

30 Dhankhar, J., González-Fernández, E., Dong, C.-C. et al. (2020). Spatial anion control on palladium for mild C—H arylation of arenes. *Journal of the American Chemical Society* 142 (45): 19040–19046.

31 Liu, C., Ji, C., Zhou, T. et al. (2021). Bimetallic cooperative catalysis for decarbonylative heteroarylation of carboxylic acids via C—O/C—H coupling. *Angewandte Chemie, International Edition* 60 (19): 10690–10699.

32 Feng, R., Zhen, Y.-Q., Wu, D. et al. (2024). Late-stage modification of complex drug: base-controlled Pd-catalyzed regioselective synthesis and bioactivity of arylated osimertinibs. *Science Advances* 10 (10): eadl0026.

33 Li, Z., Wang, Z., Chekshin, N. et al. (2021). A tautomeric ligand enables directed C—H hydroxylation with molecular oxygen. *Science* 372 (6549): 1452–1457.

34 Li, Z., Park, H.S., Qiao, J.X. et al. (2022). Ligand-enabled C—H hydroxylation with aqueous H_2O_2 at room temperature. *Journal of the American Chemical Society* 144 (39): 18109–18116.

35 Liang, Y.-F., Wang, X., Yuan, Y. et al. (2015). Ligand-promoted Pd-catalyzed oxime ether directed C—H hydroxylation of arenes. *ACS Catalysis* 5 (10): 6148–6152.

36 Shan, G., Han, X., Lin, Y. et al. (2013). Broadening the catalyst and reaction scope of regio- and chemoselective C—H oxygenation: a convenient and scalable approach to 2-acylphenols by intriguing Rh(II) and Ru(II) catalysis. *Organic & Biomolecular Chemistry* 11 (14): 2318.

37 Cheng, L., Wang, H., Cai, H. et al. (2021). Iron-catalyzed arene C—H hydroxylation. *Science* 374 (6563): 77–81.

38 (a) Shang, M., Wang, M., Saint-Denis, T.G. et al. (2017). Copper-mediated late-stage functionalization of heterocycle-containing molecules. *Angewandte Chemie, International Edition* 56 (19): 5317–5321. (b) Xu, L.-L., Wang, X.,

Ma, B. et al. (2018). Copper mediated C—H amination with oximes: en route to primary anilines. *Chemical Science* 9 (23): 5160–5164. (c) Kim, H., Heo, J., Kim, J. et al. (2018). Copper-mediated amination of aryl C—H bonds with the direct use of aqueous ammonia via a disproportionation pathway. *Journal of the American Chemical Society* 140 (43): 14350–14356.

39 Hu, X.-H., Yang, X.-F., and Loh, T.-P. (2016). Chelation-assisted rhodium-catalyzed direct amidation with amidobenziodoxolones: C(sp^2)—H, C(sp^3)—H, and late-stage functionalizations. *ACS Catalysis* 6 (9): 5930–5934.

40 Weis, E., Johansson, M.J., and Martín-Matute, B. (2021). Late-stage amination of drug-like benzoic acids: access to anilines and drug conjugates through directed iridium-catalyzed C—H activation. *Chemistry – A European Journal* 27 (72): 18188–18200.

41 Weis, E., Johansson, M., Korsgren, P. et al. (2022). Merging directed C—H activations with high-throughput experimentation: development of iridium-catalyzed c–h aminations applicable to late-stage functionalization. *JACS Au* 2 (4): 906–916.

42 Antermite, D., Friis, S.D., Johansson, J.R. et al. (2023). Late-stage synthesis of heterobifunctional molecules for PROTAC applications via ruthenium-catalysed C—H amidation. *Nature Communications* 14 (1): 8222.

43 Legnani, L., Prina Cerai, G., and Morandi, B. (2016). Direct and practical synthesis of primary anilines through iron-catalyzed C—H bond amination. *ACS Catalysis* 6 (12): 8162–8165.

44 Liu, J., Wu, K., Shen, T. et al. (2017). Fe-catalyzed amination of (hetero)arenes with a redox-active aminating reagent under mild conditions. *Chemistry – A European Journal* 23 (3): 563–567.

45 D'Amato, E.M., Börgel, J., and Ritter, T. (2019). Aromatic C—H amination in hexafluoroisopropanol. *Chemical Science* 10 (8): 2424–2428.

46 Iverson, C.N. and Smith, M.R. (1999). Stoichiometric and catalytic B—C bond formation from unactivated hydrocarbons and boranes. *Journal of the American Chemical Society* 121 (33): 7696–7697.

47 Ishiyama, T., Takagi, J., Ishida, K. et al. (2002). Mild iridium-catalyzed borylation of arenes. high turnover numbers, room temperature reactions, and isolation of a potential intermediate. *Journal of the American Chemical Society* 124 (3): 390–391.

48 Mkhalid, I.A.I., Barnard, J.H., Marder, T.B. et al. (2010). C—H activation for the construction of C—B bonds. *Chemical Reviews* 110 (2): 890–931.

49 Guria, S., Hassan, M.M.M., and Chattopadhyay, B. (2024). C—H borylation: a tool for molecular diversification. *Organic Chemistry Frontiers* 11 (3): 929–953.

50 Hassan, M.M.M., Guria, S., Dey, S. et al. (2023). Transition metal-catalyzed remote C—H borylation: an emerging synthetic tool. *Science Advances* 9 (16): eadg3311.

51 Larsen, M.A. and Hartwig, J.F. (2014). Iridium-catalyzed C—H borylation of heteroarenes: scope, regioselectivity, application to late-stage functionalization, and mechanism. *Journal of the American Chemical Society* 136 (11): 4287–4299.

52 He, Z.-T., Li, H., Haydl, A.M. et al. (2018). Trimethylphosphate as a methylating agent for cross coupling: a slow-release mechanism for the methylation of arylboronic esters. *Journal of the American Chemical Society* 140 (49): 17197–17202.

53 Saito, Y., Yamanoue, K., Segawa, Y., and Itami, K. (2020). Selective transformation of strychnine and 1,2-disubstituted benzenes by C—H borylation. *Chem* 6 (4): 985–993.

54 D'Angelo, K.A., La, C., Kotecki, B. et al. (2024). An air-stable, single-component iridium precatalyst for the borylation of C—H bonds on large to miniaturized scales. *Journal of the American Chemical Society* 146 (47): 32717–32729.

55 Scott, J.S., Moss, T.A., Barlaam, B. et al. (2020). Addition of fluorine and a late-stage functionalization (LSF) of the oral SERD AZD9833. *ACS Medicinal Chemistry Letters* 11 (12): 2519–2525.

56 Mahamudul Hassan, M.M., Mondal, B., Singh, S. et al. (2022). Ir-catalyzed ligand-free directed C—H borylation of arenes and pharmaceuticals: detailed mechanistic understanding. *The Journal of Organic Chemistry* 87 (6): 4360–4375.

57 For a review on potential of metallacyclic complexes in C—H activation see: Zakis, J.M., Smejkal, T., and Wencel-Delord, J. (2022). Cyclometallated complexes as catalysts for C—H activation and functionalization. *Chemical Communications* 58 (4): 483–490.

58 Hoque, M.E., Hassan, M.M.M., and Chattopadhyay, B. (2021). Remarkably efficient iridium catalysts for directed C(sp^2)—H and C(sp^3)—H borylation of diverse classes of substrates. *Journal of the American Chemical Society* 143 (13): 5022–5037.

59 Zakis, J.M., Messinis, A.M., Ackermann, L. et al. (2024). Air-stable bis-cyclometallated iridium catalysts for ortho-directed C(sp^2)—H borylation. *Advanced Synthesis and Catalysis* 366 (10): 2292–2304.

60 Zakis, J.M., Lipina, R.A., Bell, S. et al. (2025). High-throughput enabled iridium-catalyzed C—H borylation platform for late-stage functionalization. *ACS Catalysis* 15 (4): 3525–3534.

61 Nippa, D.F., Atz, K., Hohler, R. et al. (2024). Enabling late-stage drug diversification by high-throughput experimentation with geometric deep learning. *Nature Chemistry* 16 (2): 239–248.

62 For selected reviews see: (a) Besset, T., Poisson, T., and Pannecoucke, X. (2014). Recent progress in direct introduction of fluorinated groups on alkenes and alkynes by means of C—H bond functionalization. *Chemistry – A European Journal* 20 (51): 16830–16845. (b) Cheng, Q. and Ritter, T. (2019). New directions in C—H fluorination. *Trends in Chemistry* 1 (5): 461–470. (c) Monsigny, L., Doche, F., and Besset, T. (2023). Transition-metal-catalyzed C—H bond activation as a sustainable strategy for the synthesis of fluorinated molecules: an overview. *Beilstein Journal of Organic Chemistry* 19: 448–473.

63 Wang, M., Ruskin, J., Marques, J. et al. (2025). Selective fluorination of complex molecules: late-stage functionalization. *Chemical Reviews* 125 (19): 9382–9428.

64 Lou, S.-J., Chen, Q., Wang, Y.-F. et al. (2015). Selective C—H bond fluorination of phenols with a removable directing group: late-stage fluorination of 2-phenoxyl nicotinate derivatives. *ACS Catalysis* 5 (5): 2846–2849.

65 Ning, X.-Q., Lou, S.-J., Mao, Y.-J. et al. (2018). Nitrate-promoted selective C—H fluorination of benzamides and benzeneacetamides. *Organic Letters* 20 (8): 2445–2448.

66 Yamamoto, K., Li, J., Garber, J.A.O. et al. (2018). Palladium-catalysed electrophilic aromatic C—H fluorination. *Nature* 554 (7693): 511–514.

67 Gao, C., Li, H., Liu, M. et al. (2017). Regioselective C—H chlorination: towards the sequential difunctionalization of phenol derivatives and late-stage chlorination of bioactive compounds. *RSC Advances* 7 (74): 46636–46643.

68 Li, F., Zhou, Y., Yang, H. et al. (2019). Monodentate transient directing group enabled Pd-catalyzed *Ortho*-C—H methoxylation and chlorination of benzaldehydes. *Organic Letters* 21 (10): 3692–3695.

69 Kopf, S., Bourriquen, F., Li, W. et al. (2022). Recent developments for the deuterium and tritium labeling of organic molecules. *Chemical Reviews* 122 (6): 6634–6718.

70 Hesk, D. (2020). Highlights of C(sp^2)—H hydrogen isotope exchange reactions. *Labelled Comp Radiopharmac* 63 (6): 247–265.

71 (a) Brown, J.A., Cochrane, A.R., Irvine, S. et al. (2014). The synthesis of highly active iridium(I) complexes and their application in catalytic hydrogen isotope exchange. *Advanced Synthesis and Catalysis* 356 (17): 3551–3562. (b) Kerr, W.J., Knox, G.J., and Paterson, L.C. (2020). Recent advances in iridium(I) catalysis towards directed hydrogen isotope exchange. *Journal of Labelled Compounds and Radiopharmaceuticals* 63 (6): 281–295.

72 Jess, K., Derdau, V., Weck, R. et al. (2017). Hydrogen isotope exchange with iridium(I) complexes supported by phosphine-imidazolin-2-imine P,N ligands. *Advanced Synthesis and Catalysis* 359 (4): 629–638.

73 For a general review see: Atzrodt, J., Derdau, V., Kerr, W.J., and Reid, M. (2018). C—H functionalisation for hydrogen isotope exchange. *Angewandte Chemie, International Edition* 57 (12): 3022–3047.

74 Pony Yu, R., Hesk, D., Rivera, N. et al. (2016). Iron-catalysed tritiation of pharmaceuticals. *Nature* 529 (7585): 195–199.

75 Zarate, C., Yang, H., Bezdek, M.J. et al. (2019). Ni(I)—X complexes bearing a bulky α-diimine ligand: synthesis, structure, and superior catalytic performance in the hydrogen isotope exchange in pharmaceuticals. *Journal of the American Chemical Society* 141 (12): 5034–5044.

76 For an initial work on Ni-catalyzed HIE see: Yang, H., Zarate, C., Palmer, W.N. et al. (2018). Site-selective nickel-catalyzed hydrogen isotope exchange in *N*-heterocycles and its application to the tritiation of pharmaceuticals. *ACS Catalysis* 8 (11): 10210–10218.

77 Müller, V., Weck, R., Derdau, V., and Ackermann, L. (2020). Ruthenium(II)-catalyzed hydrogen isotope exchange of pharmaceutical drugs by C—H deuteration and C—H tritiation. *ChemCatChem* 12 (1): 100–104.

78 Uttry, A., Mal, S., and Van Gemmeren, M. (2021). Late-stage β-C(sp^3)—H deuteration of carboxylic acids. *Journal of the American Chemical Society* 143 (29): 10895–10901.

79 Farizyan, M., Mondal, A., Mal, S. et al. (2021). Palladium-catalyzed nondirected late-stage C—H deuteration of arenes. *Journal of the American Chemical Society* 143 (40): 16370–16376.

80 Yun, S.J., Kim, J., Kang, E. et al. (2023). Nondirected Pd-catalyzed C—H perdeuteration and *meta*-selective alkenylation of arenes enabled by pyrazolopyridone ligands. *ACS Catalysis* 13 (7): 4042–4052.

81 Kang, T., Kim, Y., Lee, D. et al. (2014). Iridium-catalyzed intermolecular amidation of sp^3 C—H bonds: late-stage functionalization of an unactivated methyl group. *Journal of the American Chemical Society* 136 (11): 4141–4144.

82 (a) Stone, K.L. and Borovik, A. (2009). Lessons from nature: unraveling biological C H bond activation. *Current Opinion in Chemical Biology* 13 (1): 114–118. (b) Bakanas, I., Lusi, R.F., Wiesler, S. et al. (2023). Strategic application of C—H oxidation in natural product total synthesis. *Nature Reviews Chemistry* 7 (11): 783–799. (c) Costas, M. (2011). Selective C–H oxidation catalyzed by metalloporphyrins. *Coordination Chemistry Reviews* 255 (23–24): 2912–2932.

83 White, M.C. and Zhao, J. (2018). Aliphatic C—H oxidations for late-stage functionalization. *Journal of the American Chemical Society* 140 (43): 13988–14009.

84 Shen, P.-X., Hu, L., Shao, Q. et al. (2018). Pd(II)-catalyzed enantioselective C(sp^3)—H arylation of free carboxylic acids. *Journal of the American Chemical Society* 140 (21): 6545–6549.

85 Kang, G., Strassfeld, D.A., Sheng, T. et al. (2023). Transannular C—H functionalization of cycloalkane carboxylic acids. *Nature* 618 (7965): 519–525.

86 Rodrigalvarez, J., Nappi, M., Azuma, H. et al. (2020). Catalytic C(sp^3)—H bond activation in tertiary alkylamines. *Nature Chemistry* 12 (1): 76–81.

87 Topczewski, J.J., Cabrera, P.J., Saper, N.I., and Sanford, M.S. (2016). Palladium-catalysed transannular C—H functionalization of alicyclic amines. *Nature* 531 (7593): 220–224.

88 He, J., Hamann, L.G., Davies, H.M.L., and Beckwith, R.E.J. (2015). Late-stage C H functionalization of complex alkaloids and drug molecules via intermolecular rhodium-carbenoid insertion. *Nature Communications* 6 (1): 5943.

89 Begum, T., Prakash, R., Duarah, G. and Gogoi, S. (2020). Utilization of Rh-carbenoid C—H insertion reactions for the synthesis of bioactive natural products. In: *Studies in Natural Products Chemistry*, vol. 65, 349–380. Elsevier.

90 Chen, M.S. and White, M.C. (2010). Combined effects on selectivity in Fe-catalyzed methylene oxidation. *Science* 327 (5965): 566–571.

91 Chen, M.S. and White, M.C. (2007). A predictably selective aliphatic C—H oxidation reaction for complex molecule synthesis. *Science* 318 (5851): 783–787.

92 Gormisky, P.E. and White, M.C. (2013). Catalyst-controlled aliphatic C—H oxidations with a predictive model for site-selectivity. *Journal of the American Chemical Society* 135 (38): 14052–14055.

93 Chambers, R.K., Zhao, J., Delaney, C.P., and White, M.C. (2020). Chemoselective tertiary C—H hydroxylation for late-stage functionalization with Mn(PDP)/chloroacetic acid catalysis. *Advanced Synthesis and Catalysis* 362 (2): 417–423.

94 Kang, Y.C., Volochnyuk, D.M., Ryabukhin, S.V., and Hartwig, J.F. (2025). Late-stage oxidation of C(sp^3)—H bonds with high efficiency and alkyl group dihydroxylation of complex molecules. *Angewandte Chemie, International Edition* 64: e202513418.

95 (a) Du Bois, J. (2011). Rhodium-catalyzed C—H amination. An enabling method for chemical synthesis. *Organic Process Research and Development* 15 (4): 758–762. (b) Roizen, J.L., Zalatan, D.N., and Du Bois, J. (2013). Selective intermolecular amination of C—H bonds at tertiary carbon centers. *Angewandte Chemie, International Edition* 52 (43): 11343–11346. (c) Zalatan, D.N. and Du Bois, J. (2008). A chiral rhodium carboxamidate catalyst for enantioselective C—H amination. *Journal of the American Chemical Society* 130 (29): 9220–9221.

96 (a) Liang, C., Robert-Peillard, F., Fruit, C. et al. (2006). Efficient diastereoselective intermolecular rhodium-catalyzed C—H amination. *Angewandte Chemie, International Edition* 45 (28): 4641–4644. (b) Brunard, E., Boquet, V., Van Elslande, E. et al. (2021). Catalytic intermolecular C(sp^3)—H amination: selective functionalization of tertiary C—H bonds vs activated benzylic C—H bonds. *Journal of the American Chemical Society* 143 (17): 6407–6412. (c) Brunard, E., Boquet, V., Saget, T. et al. (2024). Catalyst-controlled intermolecular homobenzylic C(sp^3)—H amination for the synthesis of β-arylethylamines. *Journal of the American Chemical Society* 146 (9): 5843–5854.

97 For general reviews see: (a) Li, H.-H., Chen, X., and Kramer, S. (2023). Recent developments for intermolecular enantioselective amination of non-acidic C(sp^3)—H bonds. *Chemical Science* 14 (46): 13278–13289. (b) Davies, H.M.L. and Manning, J.R. (2008). Catalytic C—H functionalization by metal carbenoid and nitrenoid insertion. *Nature* 451 (7177): 417–424. (c) Collet, F., Lescot, C., and Dauban, P. (2011). Catalytic C—H amination: the stereoselectivity issue. *Chemical Society Reviews* 40 (4): 1926.

98 Chiappini, N.D., Mack, J.B.C., and Du Bois, J. (2018). Intermolecular C(sp^3)—H amination of complex molecules. *Angewandte Chemie, International Edition* 57 (18): 4956–4959.

99 Clark, J.R., Feng, K., Sookezian, A., and White, M.C. (2018). Manganese-catalysed benzylic C(sp^3)—H amination for late-stage functionalization. *Nature Chemistry* 10 (6): 583–591.

100 Huang, X., Bergsten, T.M., and Groves, J.T. (2015). Manganese-catalyzed late-stage aliphatic C—H azidation. *Journal of the American Chemical Society* 137 (16): 5300–5303.

101 Knight, N.M.L., Thompson, J.D.F., Parkinson, J.A. et al. (2024). Iridium-catalysed C(sp^3)—H activation and hydrogen isotope exchange via nitrogen-based carbonyl directing groups. *Advanced Synthesis and Catalysis* 366 (11): 2577–2586.

2

Light-Triggered Methods for Late-Stage Functionalization

Camilla Russo[1], *Gian Cesare Tron*[2], *and Mariateresa Giustiniano*[1]

[1] *University of Naples Federico II, Department of Pharmacy, 80131 Napoli, Italy*
[2] *University of Piemonte Orientale, Department of Drug Science, 28100 Novara, Italy*

2.1 Introduction

2.1.1 Definition of Late-Stage Functionalization

The term *late-stage functionalization* (LSF) has been associated with ambiguity until 2020, when T. Ritter defined LSF reactions as chemoselective transformations on complex molecules, which do not require the pre-installation of a functional group but rather exploit the innate reactivity of the substrates [1]. The LSF provides at least one analog in sufficient quantity and purity. Accordingly, a reaction can be classified as LSF even if the regioselectivity is not excellent (according to T. Ritter, site-selectivity is indeed *optional* given that C(sp^2)—H vs. C(sp^3)—H is key), thus leading to mixture of mono-, di-, and multi-functionalized adducts [2]. Albeit the availability of different constitutional isomers can be desirable in structure–activity relationship (SAR) studies, such mixtures can lead to purity concerns due to troubleshooting separations and can decrease significantly the quantity of the desired product. As for chemical complexity, the authors refer to an intuitive understanding of the term, which shall rely on the synthetic efforts needed to access a molecule. For instance, natural compounds are complex molecules by definition, whereas bioactive scaffolds, such as *hit* and *lead* compounds, often become even more complex as their optimization advances in order to improve potency, efficacy, on- and off-target selectivity, and pharmacokinetic features (absorption–distribution–metabolism–excretion–toxicity [ADMET]) including solubility, bioavailability, metabolic stability, and toxicity [3]. As a matter of example, indomethacin can undergo an LSF if its carboxylic acid functional group is involved in a photoredox-catalyzed oxidative decarboxylation (Scheme 2.1a), whereas the use of its bromine analog as a substrate in a light-triggered difluoromethylation (Scheme 2.1b) cannot be defined as a strict LSF (since the bromine functional group is not native to indomethacin) [4, 5]. However, from a medicinal chemist perspective, the latter transformation can be more useful than the former as it does

Late-Stage Functionalization and Diversification in Organic Synthesis: Methods and Applications, First Edition. Edited by Tatiana Besset.

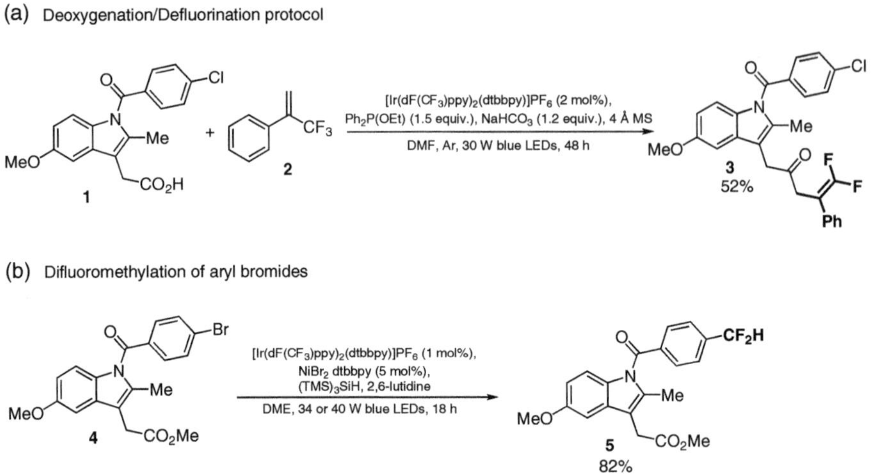

Scheme 2.1 (a) LSF of indomethacin to derivative **3** and (b) synthesis of difluromethylated derivative of indomethacin **5**, which cannot define LSF according to T. Ritter's definition. *Source:* (a) Adapted from Wang et al. [4]. (b) Adapted from MacMillan et al. [5].

not necessarily modify a pharmacophoric structural element (the carboxylic acid provides indeed an ionic interaction with the Cox enzymes, the biological targets of indomethacin).

Furthermore, albeit LSF strives to meet many goals of the Green Chemistry worldwide agenda, such as either minimizing or eliminating waste and reducing unnecessary derivatizations, T. Ritter's definition includes transformations where protecting groups are introduced to the substrates in order to overcome poor orthogonality of the LSF method with respect to innate functionalities of the drug/natural compounds, such as amines and carboxylic acids. The latter need indeed to be previously converted into their carbamate or ester derivatives and, after the LSF, one or more synthetic steps will enable the cleavage to free amines and carboxylic acids, respectively, thus producing waste and decreasing green chemistry metrics (atom economy, process mass intensity, E-factor, etc.). More recently, F. Glorius has revised literature photocatalytic LSF approaches by including them as the only exception to T. Ritter's definition on one-pot transformations relying on sulfonium and pyridinium salts (whose formation is indeed a pre-functionalization necessary to enable the LSF) [6]. On the other hand, it is possible to define an LSF as the introduction of a linchpin (i.e., a versatile functional group readily convertible into a variety of derivatives). In other words, an LSF step can also afford a synthetic intermediate and must not necessarily lead to the final target compound. Hence, in LSF, a pre-functionalization step must not be required while the need for post-functionalization, e.g., of a chemical platform (bromine, boryl, and thianthrenyl derivatives) does not preclude such synthetic intermediates to be considered LSF products.

2.1.2 Photo(Redox) Catalysis in LSF

Given the strict requirements in terms of chemoselectivity, in the last decade, the availability of chemical methods to generate reactive species, such as radicals, under mild, light-induced conditions, has provided a significant contribution to advance the field of LSF [3c, 6]. As for the C—H functionalization, every chemoselective transformation can be classified as an LSF process, thus resulting in the most populated group of reactions with a plethora of methods to achieve either C(sp^2)—H or C(sp^3)—H derivatization [7]. Alternatively, photochemical approaches can capitalize on the reactivity of specific functional groups, which can afford open shell species via either oxidation or reduction, such as carboxylic acids, amines, and halogens (chloro, bromo, and iodo derivatives). Alcohols and thiols, on the other hand, can either undergo addition to electrophiles generated *in situ* by means of light-activating strategies or get oxidized under photoredox catalytic conditions. Albeit chemoselective methods are also available for, but not limited to, other functional groups, such as carbonyls, imines, aryl iodonium and aryl diazonium salts, carbonates, and boronic acids and esters, these will not be herein considered not or rarely present on biorelevant scaffolds. Figure 2.1 reports the structures of common metal-based and organic photoredox catalysts cited in the present chapter.

2.2 C(sp^3)−H Functionalization

The direct functionalization of aliphatic C—H bonds is as challenging as desirable. In many cases, indeed, the site-selectivity is hardly achieved due to the lack of unique features of a given aliphatic C—H bond over others present on the same scaffold. This issue has been addressed by using either activating or directing groups, while in photocatalysis the homolytic cleavage of an aliphatic C—H group *via* a hydrogen atom transfer event (HAT) represents one of the most active research areas [8]. A HAT event can be considered a concerted transfer of an electron and a proton from a hydrogen donor to a suitable acceptor, which can be referred to as the hydrogen abstractor. The control of the site-selectivity, albeit challenging, can be achieved by considering the following key features: (1) the relative stability of the generated alkyl radical (tertiary ones are favored with respect to secondary and primary); (2) hyperconjugative effects (orbitals' overlapping in cyclic substrates); (3) strain release upon C-H cleavage (e.g., with cyclohexane); (4) steric hindrance; (5) the presence of heteroatoms (e.g., —*N*, —*O*, and —*S*), which can donate nonbonding electrons thus activating the α-position *via* hyperconjugation; (6) polarity matching/mismatching between the bond to be cleaved and the hydrogen abstractor (very important to be considered in the selection of the reaction solvent, whose polarity shall be mismatched with respect to the hydrogen abstractor); (8) stereoelectronic

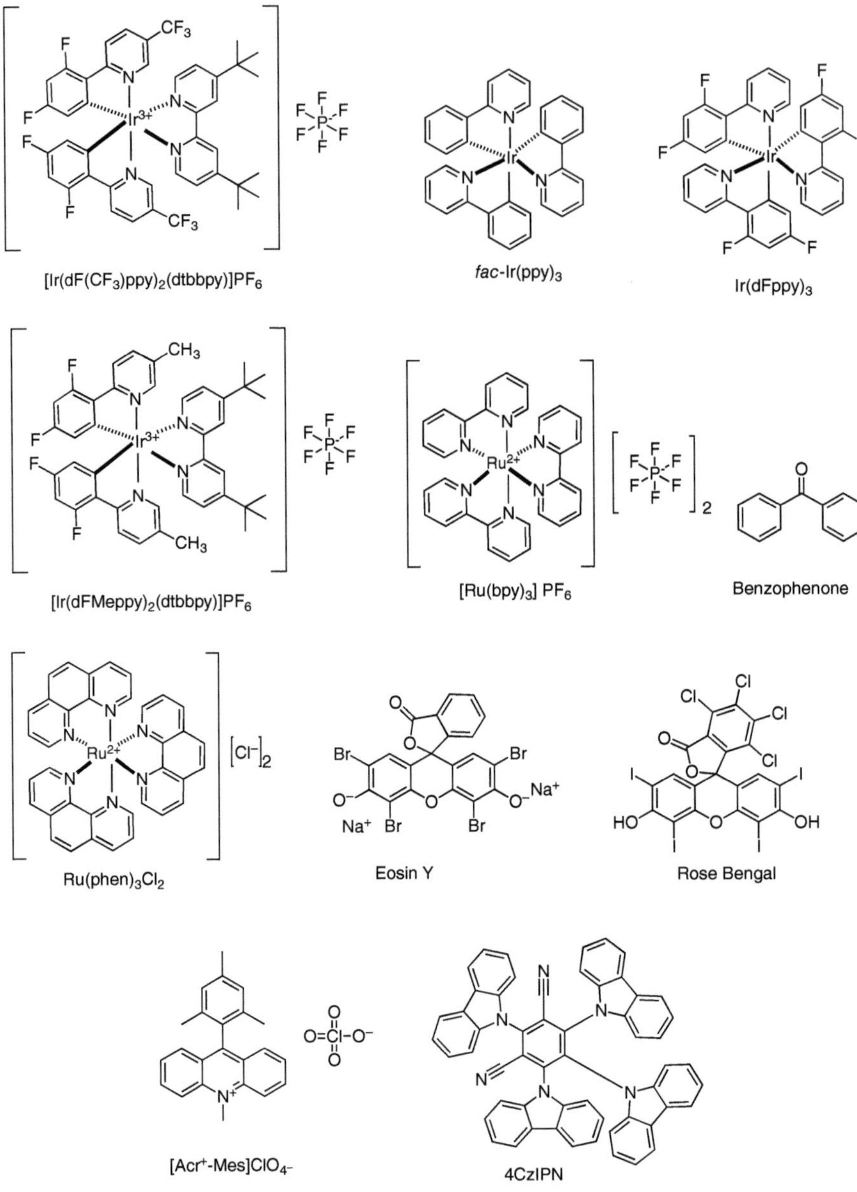

Figure 2.1 Structures of common photoredox catalysts cited in the present chapter.

effects, which can drive selectivity toward linear over cyclic substrates. Furthermore, unproductive back-hydrogen transfer events can be avoided by considering the bond dissociation energies (BDEs) for both the substrate (i.e., the hydrogen donor) and the product [8c]. Hence, the newly formed bond shall be stronger with respect to the C—H bond to cleave, whereas the latter are typically around 100 kcal/mol or 85–95 kcal/mol in the presence of a heteroatom at the α-position.

From a mechanistic point of view, the HAT event can be considered direct (*d*-HAT), when an electronically excited photocatalyst PC_{HAT}* (PC = photocatalyst) directly abstracts a hydrogen atom, or indirect (*i*-HAT) when an electronically excited photocatalyst PC_{SET}* generates the hydrogen abstractor X$^{\bullet}$ via a single electron transfer (SET) with a suitable electron acceptor/donor. For a comprehensive outlook on the *d*-HAT and on both *d*- and *i*-HAT in the LSF of biorelevant scaffolds, the reader can refer to recent authoritative review articles [3d, 8c]. To the aim of the current chapter, a selection of complex substrates will be presented in order to delineate a general but not comprehensive overview of the synthetic methods for LSF currently available in literature.

2.2.1 LSF of Hydrocarbon C—H Bonds

Sclareolide can be considered one of the most representative substrates undergoing site-selective LSF. Herein, the site-selectivity in the cleavage of an aliphatic C—H bond is driven by both the distance from the electron-withdrawing lactone and steric features, which make other C—H bonds with innate comparable reactivity inaccessible (i.e., the most electron-rich).

As a matter of example, sclareolide **6** (Scheme 2.2) can undergo either chlorination or fluorination in the presence of tetrabutylammonium decatungstate (TBADT) as a polyoxometalate PC_{HAT} and *N*-chlorosuccinimide (NCS) or *N*-fluorobenzenesulfonimide (NFSI), respectively [9, 10]. While the PC_{HAT} loading required is different (10 mol% for the chlorination vs. 2 mol% in the case of fluorination), both the reactions were performed in acetonitrile (MeCN) as a solvent, under an inert atmosphere (either argon or nitrogen). The wavelength required is 365 nm and was generated either with UV light-emitting diodes (UV LEDs) or a 15 W black light bulb (BLB). Interestingly, the chlorination was accomplished with a Schlenk-in-flow technique relying on the use of argon to drive the reagents through the flow tubular reactors. Mechanistically, the excited photocatalyst PC_{HAT}* abstracts a hydrogen atom from **6**, which undergoes either homolytic chlorination or homolytic fluorination with NCS or NFSI, respectively, while the N-centered

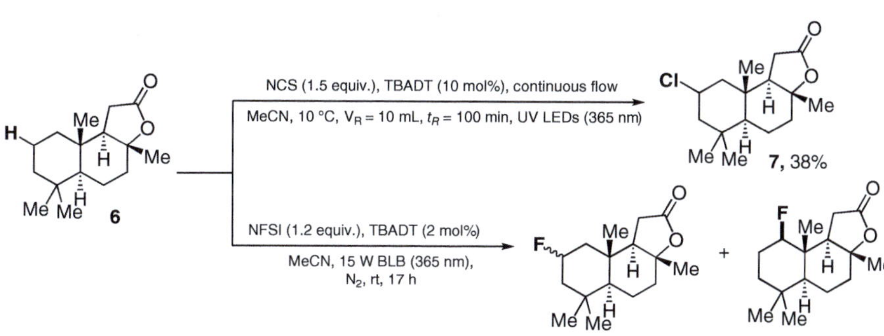

Scheme 2.2 Representative LSFs of sclareolide to chloro and fluoro derivatives **7–9** (BLB: black light bulb).

radicals formed upon halogen-atom transfer (XAT) can behave as a back-hydrogen abstractor to regenerate the ground-state PC_{HAT}.

Fluorination of sclareolide can also be promoted by visible light as reported by E. J. Sorensen et al. [11]. The authors exploited the uranyl cation UO_2^{2+} as the PC_{HAT} and NFSI (1.5 equivalents) as the fluorine source in a mechanism which is very similar to that occurring with TBADT, i.e., *d*-HAT (Scheme 2.3). The reaction, performed in deuterated acetonitrile at room temperature under irradiation with blue LEDs for 16 hours, led to late-stage fluorinated adduct **10** in a 26% NMR yield as a 1.6:1 mixture of C2/C3 regioisomers.

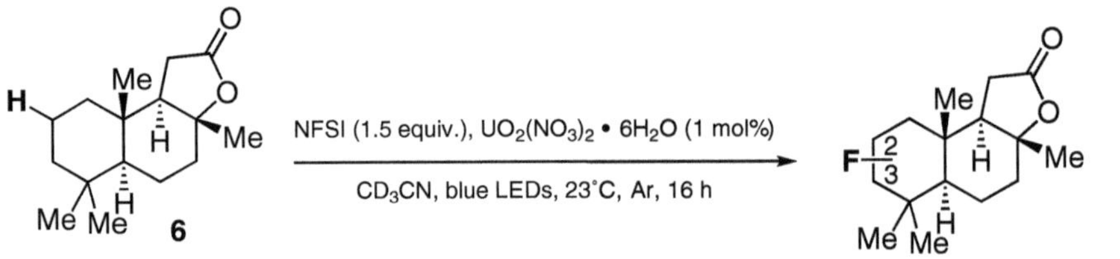

10, 26% (*C2/C3*: 1.6:1)

Scheme 2.3 Late-stage fluorination of sclareolide **6** to a regioisomeric mixture **10**.

Continuous-flow conditions were also set up to perform a late-stage oxidation of sclareolide **6** to derivative **11** as depicted in Scheme 2.4 [12]. The reaction was promoted by TBADT (5 mol%) as a PC_{HAT} under irradiation with 365 nm LEDs and oxygen was added with a T-mixer. A 2.5:1 mixture of MeCN/HCl 1 M was key as a solvent system, and a residence time of 1.5 hours afforded **11** in 44% yield as a 4.8:1 C_2:C_1 mixture of regioisomers.

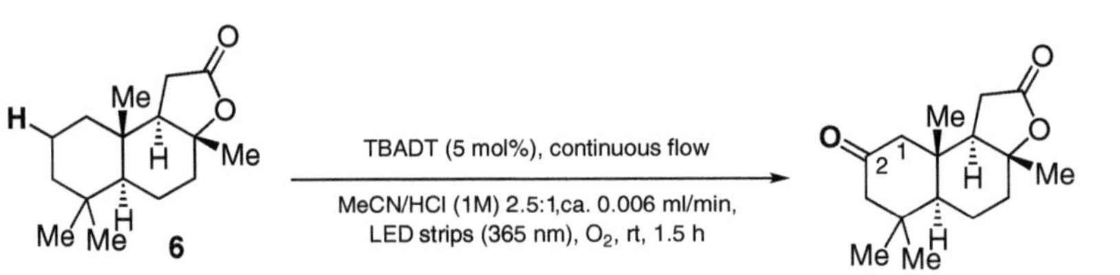

11, 44% (*C2/C1*: 4.8:1)

Scheme 2.4 Late-stage oxidation of sclareolide **6** to a mixture of regioisomers **11**.

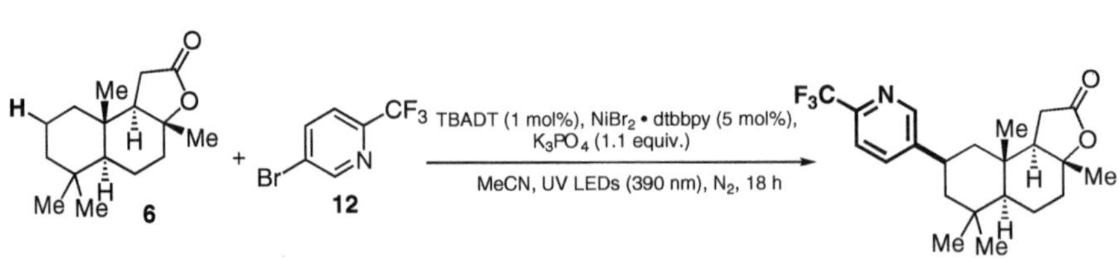

13, 41%

Scheme 2.5 Late-stage heteroarylation of sclareolide **6** to derivative **13**.

Sclareolide can be heteroarylated by harnessing a dual TBADT/nickel catalytic approach [13]. The PC_{HAT} was responsible for a direct hydrogen abstraction forming an alkyl radical, while the nickel catalyst activated an aryl bromide via oxidative addition. The formation of a nickel complex promoted the cross-coupling process leading to product **13** in 41% yield (Scheme 2.5).

2.2.2 LSF of Aliphatic C–H Bonds with Heteroatoms at the α-Position

2.2.2.1 LSF via α-Amino Alkyl Radicals

Fluoxetine has been herein selected as a representative substrate for the functionalization of C(sp^3)—H bonds with an amine at the α-position. Both the synthetic methods reported in Scheme 2.6 are *i*-HAT processes, where quinuclidine **16** or an organic phosphate salt (formed *in situ* from **18**) has been used as a hydrogen abstraction mediator to achieve either alkylation or cyanation of amino-CH_3 group [14]. As for the alkylation, the reaction mechanism relies on a SET event from **16** to the excited iridium-photocatalyst to afford **16**$^{\bullet+}$. The latter is the effective hydrogen abstractor and able to afford the α-aminoalkyl radical **14**$^{\bullet}$ with the concomitant formation of

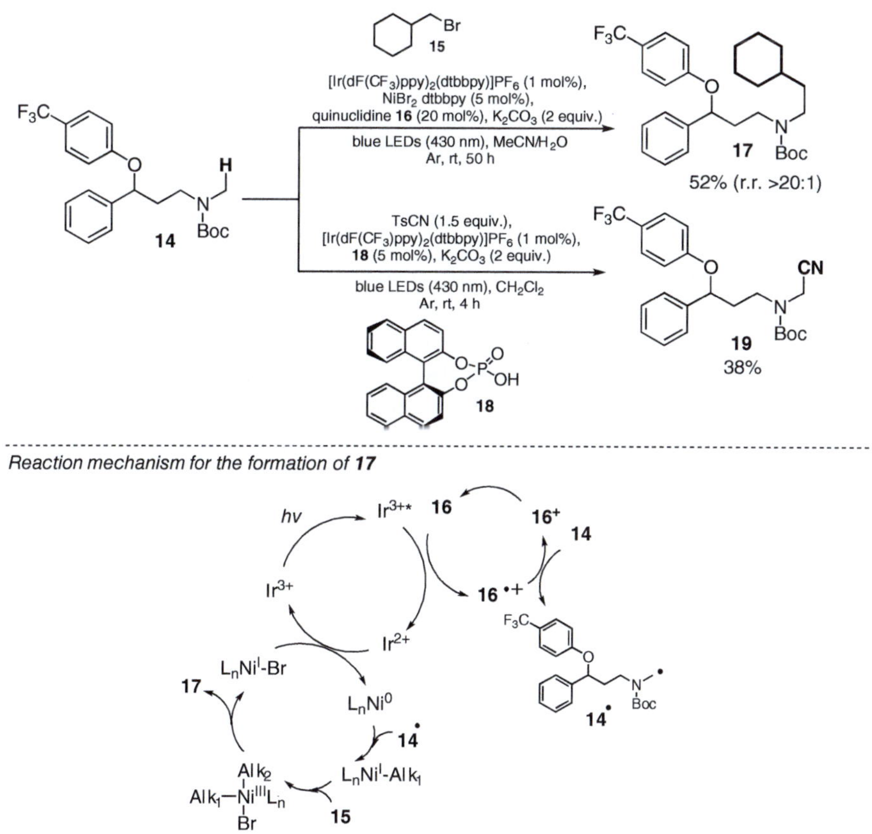

Scheme 2.6 LSF of fluoxetine **14** to derivatives **17** and **19**.

a quinuclidinium ion **16**$^+$. **14**$^\cdot$ is then trapped by a nickel-based co-catalyst to give a nickel complex undergoing oxidative addition onto the alkyl bromide **15**. Reductive elimination finally leads to the alkylated product **17** in 52% yield (regioisomeric ratio > 20:1). The cyanation of fluoxetine occurs with the same iridium photocatalyst (1 mol% loading in both cases), but with a different mediator **18** (Scheme 2.6) [15].

One more *i*-HAT mechanistic pathway relying on the generation of a *tert*-butoxy radical as the hydrogen abstractor upon [Ir(dF[CF_3]ppy)$_2$(dtbbpy)]PF_6-mediated SET reduction of *tert*-butyl peracetate (*t*-BPA) has been exploited to achieve a direct cross-dehydrogenative coupling of amines and heteroarenes, as shown in Scheme 2.7 [16]. Once an α-aminoalkyl radical is generated via *i*-HAT, it can be added to a protonated electron-deficient heteroarene via a Minisci-type pathway to give a radical cation. Oxidation of the latter to regenerate the Ir^{3+} ground-state catalyst, followed by deprotonation, affords heteroarylated fluoxetine **20** in 30% yield (Scheme 2.7).

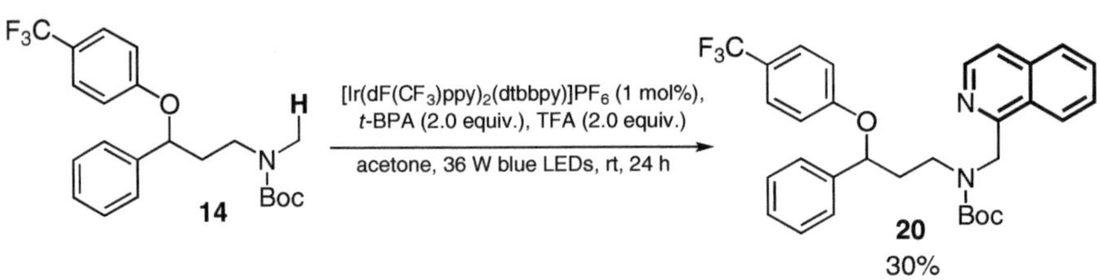

Scheme 2.7 Late-stage heteroarylation of fluoxetine **14** to derivative **20** (*t*-BPA: *tert*-butyl peracetate).

2.2.2.2 LSF Via α-Oxyalkyl Radicals

The versatility of α-oxyalkyl radicals in LSF processes can be represented by ambroxide, a natural terpenoid used in perfumery to create ambergris notes and as a flavoring agent in food. Photo(redox) chemical methods have been reported to promote (1) cyanation [17]; (2) formal formylation [18]; (3) C(sp^3)—H to C(sp^3)—C(sp^3) bond conversion [19]; and (iv) trifluoromethyl thiolation [20] (Scheme 2.8). More in detail, cyanation was reported in 2011 by M. Inoue et al. by using benzophenone as a photoactive hydrogen abstractor (1 equivalent required) under irradiation with a 100 W medium-pressure mercury lamp (Hg lamp) and tosyl cyanide (4 equivalents). Mechanistically, an α-oxyalkyl radical, generated upon direct hydrogen-atom abstraction from ambroxide **21** mediated by benzophenone (BP) in its excited state, is trapped by tosyl cyanide to provide a tosyl radical and the late-stage cyanation product **22** in 71% yield as a 4:3 mixture of diastereoisomers (Scheme 2.8). The tosyl radical, in turn, should be able to regenerate BP, albeit the stoichiometric amount required could be due to an unwanted radical addition of the tosyl radical to BP.

Similarly, the formal formylation can be achieved by using 4-benzoylpyridine (BzPy) as a photoactive promoter (1 equivalent required) under near-UV light irradiation (100 W Hg lamp). BzPy* abstracts the hydrogen atom at the α-position with respect to the oxygen atom, thus forming the α-oxyalkyl radical of ambroxide (**21**$^\cdot$). The latter is trapped by a sulfonyl oxime **23**, to give a radical intermediate

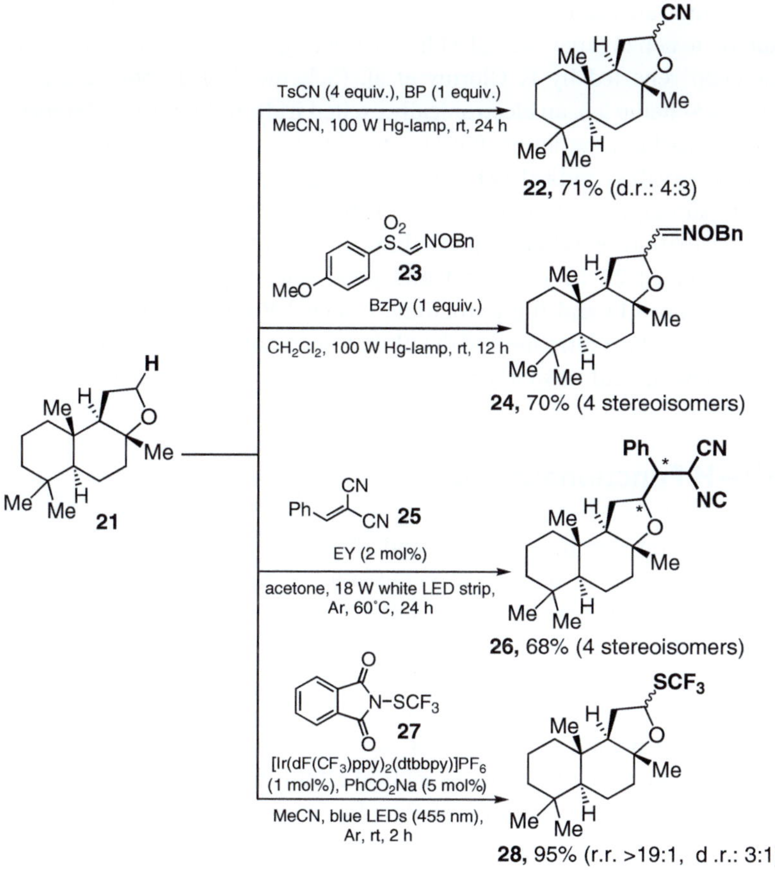

Scheme 2.8 LSF of ambroxide **21** to derivatives **22**, **24**, **26**, and **28** (BP: benzophenone; EY: Eosin Y).

further undergoing fragmentation to (*p*-methoxy)benzene sulfonyl radical and product **24** in 70% yield and as a mixture of four isomers (*E*/*Z* and *syn*/*anti*). Finally, the (*p*-methoxy)benzene sulfonyl radical shall promote the regeneration of the ground-state photocatalyst albeit the formation of a sulfonamide, upon trapping of (*p*-methoxy)benzene sulfonyl radical by BzPy, can explain the stoichiometric amount required for the latter.

Interestingly, in 2018, J. Wu et al. demonstrated that Eosin Y could be harnessed as a PC_{HAT} to get alkylation of the α-oxy-position of ambroxide (Scheme 2.8). The excited PC_{HAT}^* promotes a *d*-HAT event to form the α-oxyalkyl radical of ambroxide (**21**$^\bullet$), which undergoes addition to an electron poor alkene such as benzylidenemalononitrile **25**. The resulting open shell species regenerates the PC_{HAT}, thus affording the LSF adduct **26** in 68% yield as a mixture of four stereoisomers (Scheme 2.8). Optimal reaction conditions require a 2 mol% of catalyst loading, acetone as a solvent, and irradiation with an 18 W white LED strip for 24 hours and by heating at 60 °C.

On the other hand, an example of *i*-HAT using sodium benzoate ($PhCO_2Na$) as a precursor of a benzoyloxy radical ($PhCO_2^{\cdot}$) acting as the hydrogen atom abstractor has been reported by F. Glorius et al. (Scheme 2.8). In the reaction manifold shown in Scheme 2.8, an electronically excited iridium catalyst undergoes SET to oxidize the carboxylate anion $PhCO_2^-$ to $PhCO_2^{\cdot}$. The latter triggers the formation of an α-oxyalkyl radical (**21**˙) upon hydrogen atom abstraction from ambroxide, with subsequent trapping of the trifluoromethylthiolating reagent Phth-SCF_3 (**27**, *N*-[trifluoromethylthio]phthalimide, 1.5 equivalents) to give the late-stage functionalized **28** in 95% yield and good selectivities (r.r. > 19:1 and diastereoisomeric ratio 3:1) and the phthalimido radical Phth˙, which closes the catalytic cycle. Concomitant formation of the phthalimido anion $Phth^-$ enables the deprotonation of benzoic acid, so that the reaction does not require any further base.

2.3 C(sp^2)—H Functionalization

Unsaturated systems involve both alkenes and (hetero)arenes, with the former showing a higher occurrence in natural compounds with respect to synthetic drugs [3c]. Alkenes are considered the most suitable radical acceptors undergoing radical conjugate addition reactions with both carbon- (Giese-type reactions) and heteroatom-centered radicals [21, 22]. The direct functionalization of aromatic C—H bonds via light-triggered methods can occur via two different patterns: (1) generation of an open-shell species (either carbon or heteroatom-centered) via photoredox catalytic processes followed by radical addition to a(n) (hetero)arene (i.e., a closed-shell π-system) and (2) photoredox catalytic SET oxidation of electron-rich arenes to arene radical cations, which can be added to closed-shell nucleophiles. Over the time, the application of both approaches has provided an increasing number of procedures to achieve regioselective alkylation, arylation, and the introduction of heteroatoms, even of unstable isotopes. As for the latter, a late-stage introduction enables to overcome issues associated with a short lifetime, which is not always compatible with a *de novo* synthesis. Recently, T. Ritter et al. provided a critical analysis of selected examples of late-stage aromatic C-H bond functionalization (but not limited to photoredox catalytic methods) by considering reactivity, chemoselectivity, site-selectivity, and substrate scope. The evaluation of these four criteria provided a kind of semi-quantitative analysis, albeit they did not involve other key features such as practicality, scalability, sustainability, and costs [23]. For a general overview on C—C bond formation from C(sp^2)—H bonds, the readers can refer to a recent review article by D. J. F. Soul é et al. [22].

2.3.1 LSF of Alkene Functional Groups

To the aim of this chapter, two different natural compounds, namely limonene and nootkatone, have been selected as representative complex substrates undergoing chemo- and site-selective functionalization of a native double bond. Noteworthy, there are many examples in literature which can witness the orthogonality of alkenes functionalization with respect to other functional groups [3c], but as they have been performed on analogs of drugs where the alkene functional group is introduced to enable the given transformation, these methods cannot be considered LSF as they violate the strict definition provided by T. Ritter et al. [1].

Limonene, a cyclic monoterpene, is the main component in the essential oil of citrus peels responsible for the orange flavor, and hence it is used as a flavoring agent in food. The (+)-isomer (corresponding to the [*R*]-enantiomer) is exploited as a synthetic precursor of carvone and as a renewable solvent in cleaning products. While in common Giese-type addition reactions a nucleophilic radical is trapped by an electron-poor alkene, limonene represents an "unactivated" (i.e., electron-rich) olefin. Accordingly, the LSF processes often rely on "polarity umpolung" reactions with electrophilic radicals. As a matter of example, electrophilic C-centered radicals have been exploited to achieve haloalkylation and azidotrifluoromethylation of limonene, the latter being a three-component reaction [24, 25]. The haloalkylation proceeds via an atom-transfer radical addition (ATRA) reaction with diethylbromomalonate **30** by using *p*-anisaldehyde **31** as a photosensitizer (Scheme 2.9).

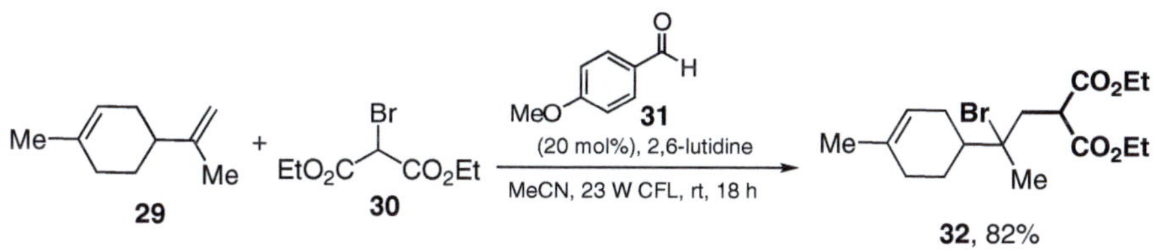

Scheme 2.9 Late-stage haloalkylation of limonene **29**.

As for the azidotrifluoromethylation, the Umemoto's reagent **33** affords a trifluoromethyl radical $CF_3^{\cdot}$ upon reductive fragmentation induced by SET with a ruthenium photocatalyst (Scheme 2.10). The addition of $CF_3^{\cdot}$ to the double bond leads to an open-shell species, which regenerates the ground-state $[Ru(bpy)_3]PF_6$, thus getting oxidized to a carbocation, finally quenched by $TMSN_3$ (azidotrimethylsilane). LSF of limonene is accomplished in 51% yield (diastereomeric ratio 1:1) by using dichloromethane as a solvent, at room temperature, and under irradiation with blue LEDs for 2.5 hours.

(Alkoxy)-carbonyl methyl radicals represent one more electrophilic open-shell species undergoing addition to limonene (Scheme 2.11) [26]. They can be formed from ester-stabilized phosphorous ylides, such as **35**, upon SET reduction promoted by *fac*-$Ir(ppy)_3$ in its excited state. The elongated ester **37** of limonene was obtained in 43% yield in 1,2-dimethoxyethane (DME)/H_2O 1:1 mixture as the solvent system, under blue LEDs irradiation, and at room temperature for 40 hours.

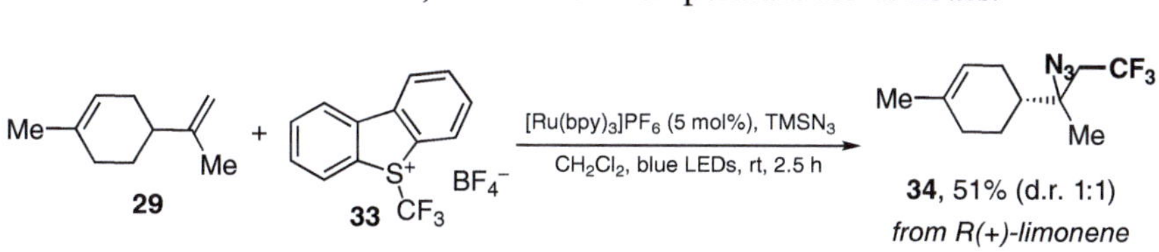

Scheme 2.10 Azidotrifluoromethylation of limonene **29**.

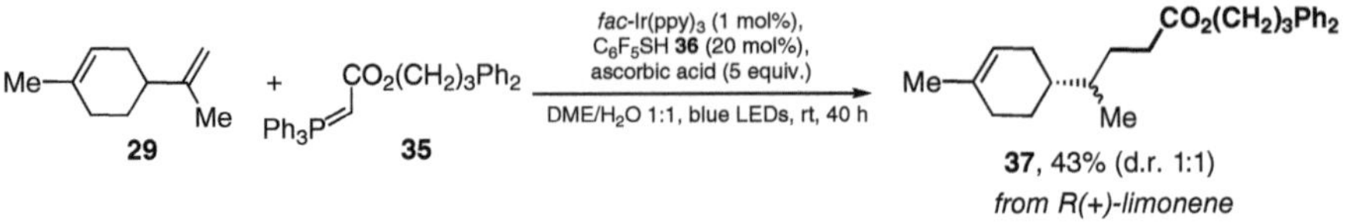

Scheme 2.11 Synthesis of elongated ester **37** from limonene **29**.

Later on, limonene **29** has been reported to undergo alkylation with sulfone-bearing radical precursors such as **38** (Scheme 2.12), where the presence of the —SO_2 functional group confer electrophilicity to an alkyl radical **38**$^{\cdot}$ generated upon SET from an excited iridium catalyst followed by mesolytic fragmentation with a loss of a bromide anion [27]. After addition of the alkyl radical **38**$^{\cdot}$ to the terminal alkene, the heteroaryl group undergoes an intramolecular migration extruding SO_2 (Scheme 2.12). Optimal reaction conditions affording **40** in 93% yield required *tert*-dodecylthiol **39** as both a hydrogen source and a final reductant, potassium carbonate (K_2CO_3) as a base, DMF (*N*,*N*-dimethylformamide) as the solvent, and irradiation with 15 W blue LEDs at room temperature.

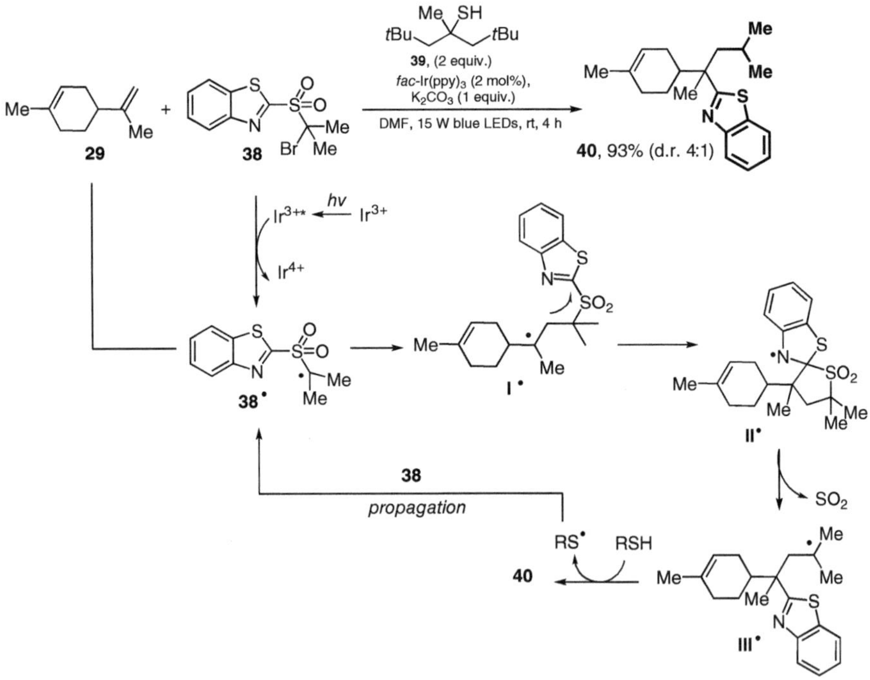

Scheme 2.12 Late-stage alkylation/heteroarylation of limonene **29**.

As for the site-selectivity, the abovementioned LSFs performed on limonene highlight how the terminal alkene can be exclusively functionalized with respect to an endocyclic one. This is also true for nootkatone, where the presence of an α-carbonyl group leads to a polarity mismatch with electrophilic radicals further contributing to the site-selectivity observed.

Nootkatone **41** (Scheme 2.13), a sesquiterpenoid contributing to most of the grapefruit aroma, is used as an environmentally friendly repellant/insecticide. It is non-toxic for humans, and its use is approved as an additive in food, cosmetics, and pharmaceuticals.

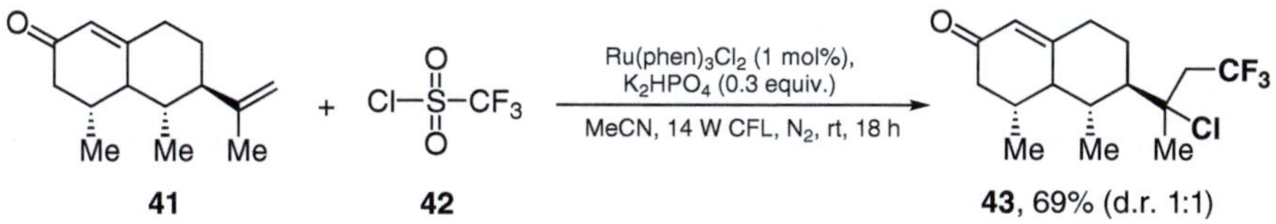

Scheme 2.13 Chlorotrifluoromethylation of nootkatone **41**.

Nootkatone has been reported to undergo chlorotrifluoromethylation with CF_3SO_2Cl (triflyl chloride, TfCl) as a source of both trifluoromethyl radical $CF_3^{\cdot}$ and a chloride anion Cl^-, in the presence of $Ru(phen)_3Cl_2$ as a photoredox catalyst, K_2HPO_4 (dipotassium phosphate) as a base, and MeCN as the solvent, under irradiation with a 14 W Compact Fluorescent Lamp (CFL) and inert atmosphere, at room temperature, for 18 hours (Scheme 2.13) [28].

Organic azides can lead to aziridination of nootkatone **41**, when used in combination with an iron porphyrin actually acting both as a photosensitizer and as a catalyst by forming a reactive iron-nitrene intermediate [29]. The latter is responsible for the observed chemo- and regioselectivities, which are typically poor in light-driven nitrene transfer and insertion reactions. The nootkatone aziridine derivative **45** was obtained in moderate yield (30%) by using $Fe(TF_4DMAP)Cl$ (TF4DMAP = *meso*-tetrakis(*o*,*o*,*m*,*m*-tetrafluoro-*p*-[dimethylamino]phenyl)-porphyrinato dianion) in DCE as the solvent, at room temperature, and under blue LEDs irradiation (Scheme 2.14).

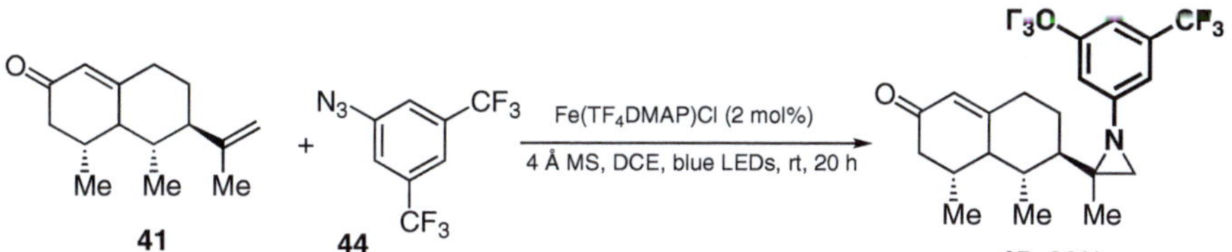

Scheme 2.14 Aziridination of nootkatone **41**.

Finally, concomitant iododifluoromethylation-benzothiazolylation of nootkatone is achievable via a "docking-migration" approach similar to that previously cited for limonene (Scheme 2.15) [30]. Nootkatone derivative **47** was obtained in 72% yield and 2:1 d.r. in acetone, with tetrabutylammonium iodide as the iodine source and under irradiation with 14 W blue LEDs at room temperature for 48 hours (Scheme 2.15).

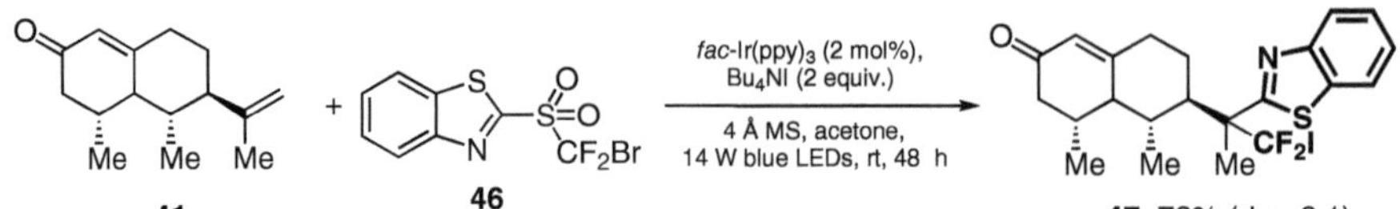

Scheme 2.15 Iododifluoromethylation-benzothiazolylation of nootkatone **41**.

2.3.2 Arenes and Heteroarenes LSF

Ibuprofen is a nonsteroidal anti-inflammatory drug (NSAID) discovered in 1961 and currently listed on the World Health Organization's List of Essential Medicines and available under a number of brand names as generic medication. Its commercial availability at reasonable prices makes it one of the most common drugs used in the examples of LSF methods. Other NSAIDs frequently occurring as representative complex scaffolds include fenoprofen, flurbiprofen, naproxen, indomethacin, and celecoxib [3c]. As for heteroarenes, caffeine, a central nervous system (CNS) stimulant, and fasudil, a Rho-kinase inhibitor and vasodilator used in the treatment of cerebral vasospasm, can be considered versatile and readily available bioactive chemical platforms for the applications of LSF methods.

The introduction of a trifluoromethyl group into ibuprofen has been pioneered by D. W. MacMillan et al. [31]. In this early report, trifluoromethanesulfonyl chloride **42** is used as a source of $CF_3^{\cdot}$, SO_2, and a chloride anion Cl^-, which are released via mesolytic cleavage upon SET with an excited photoredox catalyst, i.e., $Ir(dF^{\cdot}ppy)_3$ (Scheme 2.16). A 1.4:1 mixture of regioisomers **49** was obtained in 78% under optimal conditions, featuring MeCN as a solvent, K_2HPO_4 as a base, and a 26 W CFL as the light source under an inert atmosphere at room temperature for 24 hours (Scheme 2.16).

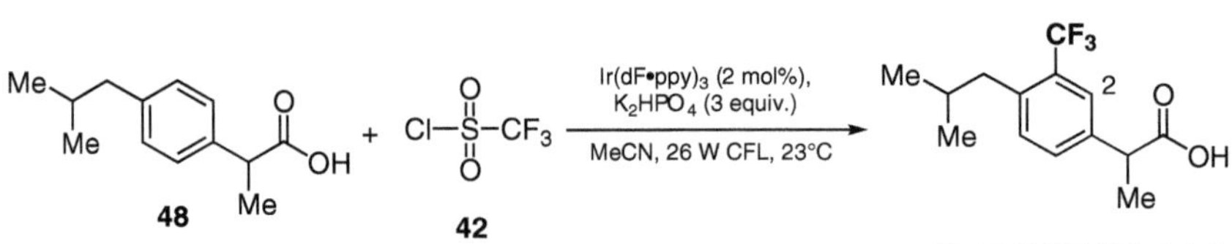

Scheme 2.16 Trifluoromethylation of ibuprofen **48**.

Later on, increased regioselectivity in the trifluoromethylation of ibuprofen has been reported by using dye-incorporated coordination polymers, such as Zn-TCTA (Tris[4-(2-thienyl)phenyl]amine [TCTA]), composed of large π-systems of thiophene-substituted triphenylamine (Scheme 2.17) [32]. Noteworthy, regioselectivity is achieved through docking of the substrates into the photoactive centers. The trifluoromethyl radical precursor is still triflyl chloride **42**, while 2,6-collidine is used as a base, in MeCN under irradiation with a 23 W CFL at room temperature (Scheme 2.17).

Among NSAIDs, fenoprofen is representative of LSF reactions relying on the oxidation of an arene moiety, followed by nucleophilic interception of the latter,

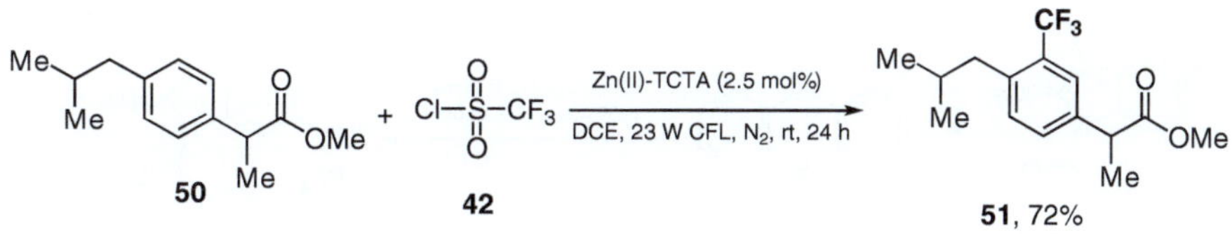

Scheme 2.17 Regioselective trifluoromethylation of ibuprofen methyl ester **50** (TCTA: Tris[4-(2-thienyl)phenyl]amine).

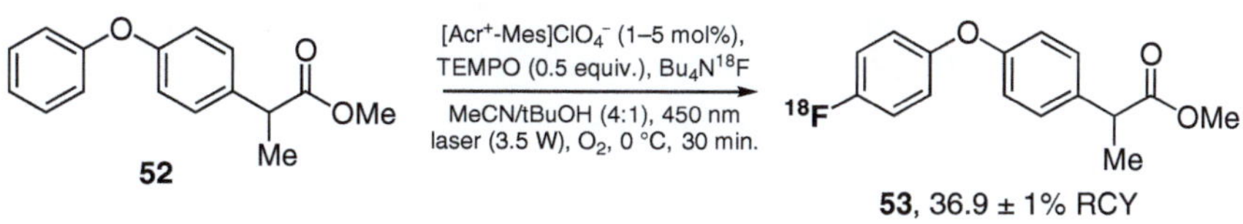

Scheme 2.18 ^{18}F-fluorination of fenoprofen methyl ester **52**.

as reported for ^{18}F-fluorination with $[^{18}F]F^-$ salts (Scheme 2.18) [33, 34]. The latter were formed *in situ* via proton bombardment of [^{18}O] water and elution of ^{18}F-fluoride with tetrabutylammonium bicarbonate. An acridinium perchlorate was required as strong photo-oxidant, TEMPO (2,2,6,6-tetramethylpiperidine-1-oxyl) as an additive, and a 4:1 mixture of MeCN/*t*-BuOH (*tert*-butanol) as the solvent (Scheme 2.18). The reaction was carried out under blue LEDs light irradiation for 30 minutes at 0 °C and under an O_2 atmosphere.

Finally, indomethacin has been herein selected as representative of the introduction of linchpins enabling further manipulation, ideally, to an array of functional groups [35].

In this context, T. Ritter, a few years ago, introduced thianthrenium salts as versatile chemical platforms, which can be obtained in high chemo- and regioselectivities from a wide range of complex substrates [36]. As a matter of example, indomethacin thianthrenium salt **55** can be obtained in 85% yield and further involved in a photoredox process in order to get amination as for derivative **57** (44% yield, Scheme 2.19) [37].

Caffeine is one of the most recurrent bioactive scaffolds in literature reported LSFs including, but not limited to, fluoromethylation [38], alkylation [39], oxyfluoroalkylation [40], (hetero)arylation [41], carbamoylation [42], olefination [43], and introduction of heteroatoms in thiolation [44] and halogenation [45]. Caffeine has also been reported in DNA-encoded libraries [46].

These reactions show a typical chemoselectivity in the functionalization of the imidazole C—H bond. As for the introduction of trifluoromethyl functional groups, there are different reports featuring the use of sodium trifluoromethanesulfonate (CF_3SO_2Na) [38a, f, g], trifluoroacetic acid (CF_3COOH) [38d, e], or trifluoroacetic anhydride [$(CF_3CO)_2O$] [38b] (Scheme 2.20). Noteworthy, the robustness of the LSF approach has been demonstrated on kilogram scale [38c]. As for the introduction of alkyl moieties, the radical precursors investigated include, among others, alkyl halides, boronic acids, alkyltrifluoroborate salts, and C(sp^3)—H such as in

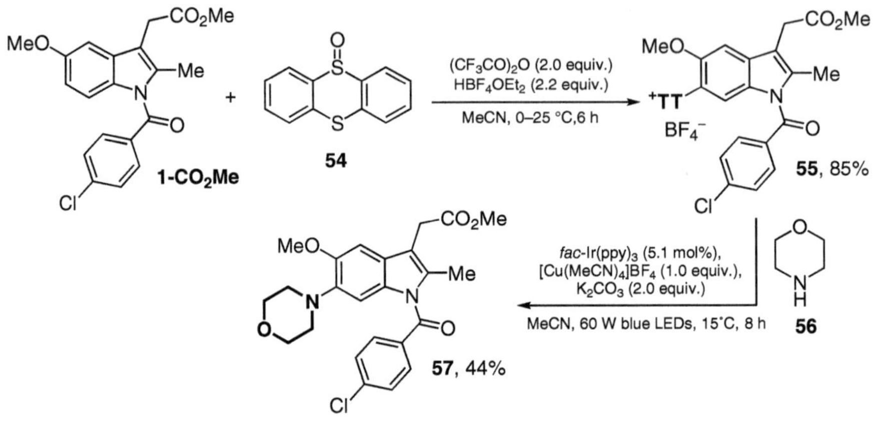

Scheme 2.19 Formation of indomethacin thianthrenium salt **55** and further amination to derivative **57**.

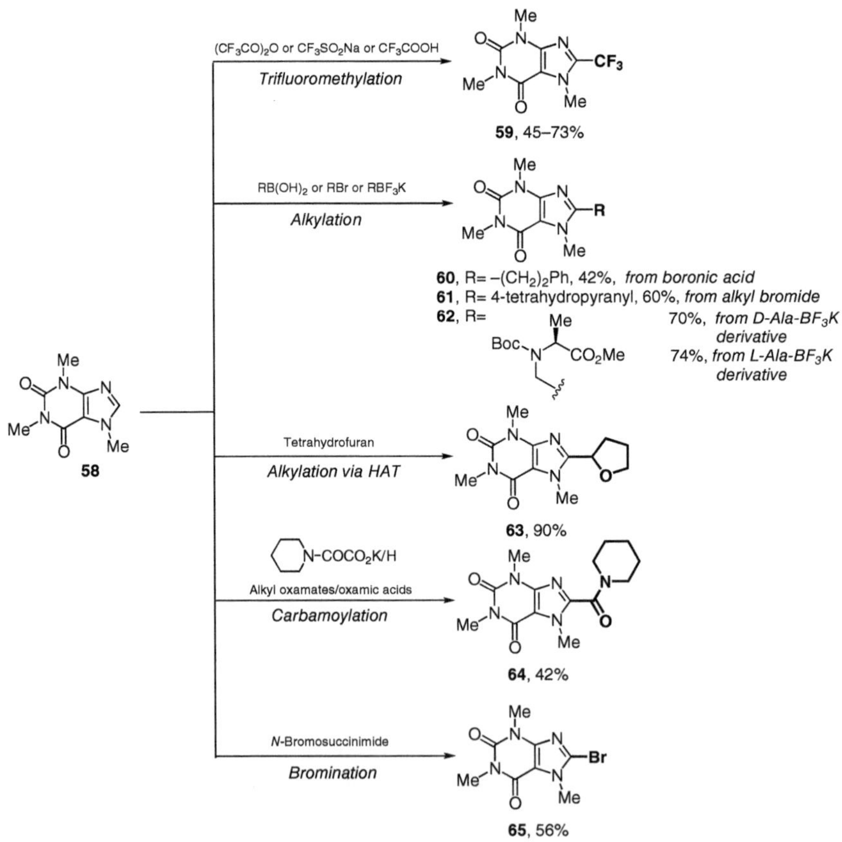

Scheme 2.20 Representative LSFs for caffeine **58**.

tetrahydrofuran (Scheme 2.20). While carbamoylation is achieved using alkyl oxamates, typical halogen donor for the introduction of a bromine atom is the *N*-bromosuccinimide (Scheme 2.20).

2.4 Functional Group-Enabled LSF

The manipulation of carboxylic acids, amines, and halogens represents the standard alternative to the direct C—H functionalization, with site-selectivity being driven by these innate functionalities. Albeit the state-of-the-art offers procedures to derivatize a wide range of functional groups, the current chapter focuses on the three above-cited ones given their higher occurrence in drugs and biorelevant scaffolds.

2.4.1 Carboxylic Acids in LSF

Carboxylic acids, both synthetic and naturally occurring ones, are considered readily available substrates, which are often bench-stable, nontoxic, and inexpensive. These features favored the investigation of their reactivities in visible light-triggered reactions [46]. Typically, carboxylic acid salts can be oxidized via a SET event with an excited state photocatalyst to undergo a homolytic cleavage to alkyl radicals and carbon dioxide, whose extrusion hampers a HAT event to give back a carboxylic acid. Alternatively, they can afford acyl radicals either via SET reduction of *in situ*-generated anhydrides or via the formation of a PPh_3-derived acyl radical [47]. Furthermore, they can be used as nucleophiles in photo(redox) catalytic reactions, occurring via radical polar crossover mechanistic pathways and involving the generation of carbon electrophiles.

As a matter of example, telmisartan **66** (Scheme 2.21), an angiotensin II receptor blocker used to treat high blood pressure, heart failure, and diabetic kidney disease, can undergo deoxygenative conversion to aldehyde **68** in 80% yield under very mild conditions by using $[Ir(dFMeppy)_2(dtbbpy)]PF_6$ as a photoredox catalyst, disulfide **67** to promote the HAT step, and PPh_3 which enables the formation of a phospho ranyl radical, a key intermediate toward the formation of an acyl radical [47]. Noteworthy, telmisartan **66** can be converted to deuterium-labeled aldehyde derivative **70** in 67% yield and 95% deuterium incorporation by using D_2O as an inexpensive

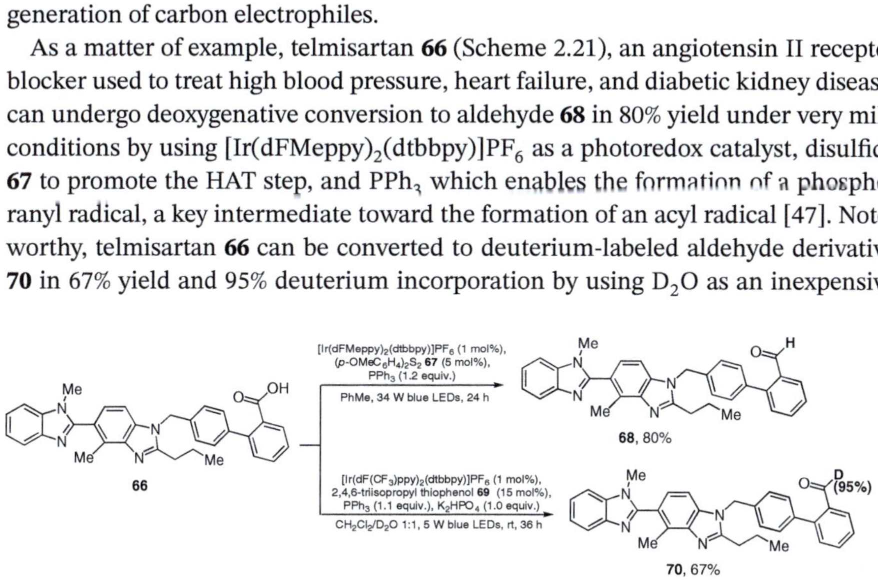

Scheme 2.21 Conversion of telmisartan **66** to aldehyde **68** and deuterium-labeled aldehyde **70**.

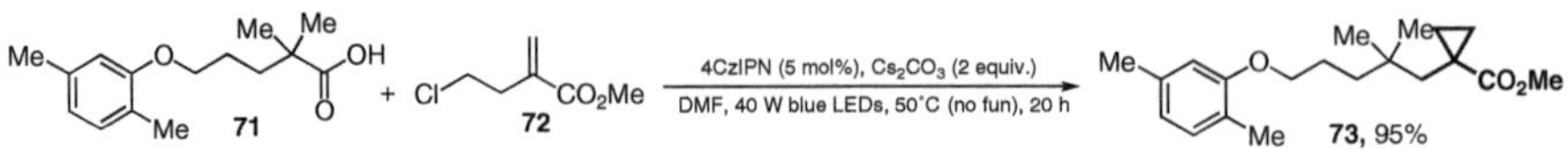

Scheme 2.22 Decarboxylative cyclopropanation of gemfibrozil **71**.

and readily available deuterium source under very similar conditions, as depicted in Scheme 2.21 [48].

As representative for decarboxylative reactions, gemfibrozil, a lipid-lowering drug, has been reported to undergo cyclopropanation to derivative **73** in 95% yield, by using 4CzIPN (1,2,3,5-tetrakis[carbazol-9-yl]-4,6-dicyanobenzene) as an organic photoredox catalyst and cesium carbonate as a base, in DMF, under irradiation with 40 W blue LEDs (Scheme 2.22) [49].

Finally, carboxylic acid drugs can be used as inputs in multicomponent reaction promoted by visible light as shown in Scheme 2.23 for indomethacine **1**. Here, an aromatic tertiary amine such as *N*,*N*-dimethylaniline **74** forms a photoactive electron-acceptor donor (EDA) complex with an aliphatic isocyanide, promoting the formation of an iminium ion via SET followed by hydrogen atom abstraction. After this radical polar crossover event, cyclohexyl isocyanide **75** and the carboxylic acid component, i.e., indomethacin **1**, can react in an Ugi-type multicomponent reaction to afford **76** in 85% yields (Scheme 2.23) [50].

Scheme 2.23 Indomethacin **1** as carboxylic acid component in an Ugi-like light-triggered reaction.

2.4.2 Amines in LSF

Similar to carboxylic acids, aliphatic amine can afford alkyl radical intermediates, which can then be intercepted by a range of acceptors [51]. Usually, this is accomplished by forming pyridinium salts known as Katritzsky salts, which are endowed with lower redox potential (~ −0.92 V *vs.* SCE) in comparison to the redox active NHPI (*N*-hydroxyphthalimide) ester derivatives (~−1.28 V *vs.* SCE) [51]. According to F. Glorius, even if in this case, a pre-functionalization is required in order to enable the LSF step, these reactions do not violate T. Ritter's definition of LSF if the formation of such intermediates is performed *in situ* as exemplified in Scheme 2.24 [6]. Herein, F. Glorius et al. performed a deaminative borylation of complex natural compounds such as dopamine, tryptamine, leucinol, and myrtanylamine. It is worth noting that this transformation was also accomplished on the Lipitor intermediate

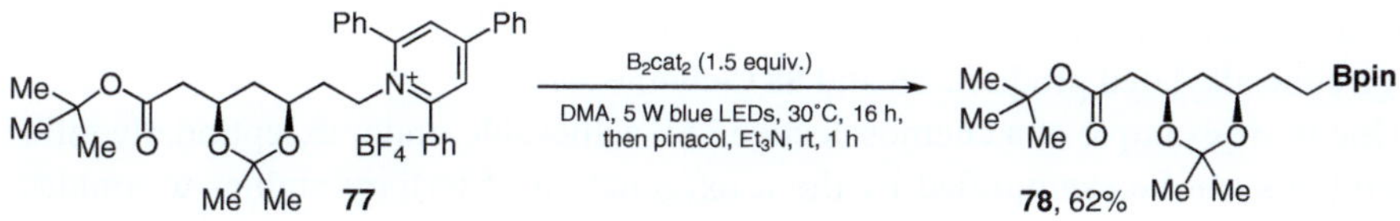

Scheme 2.24 Deaminative borylation of Lipitor intermediate **77**.

77 in 62% yield, proving a high tolerance toward functional groups such as acetal and ester (Scheme 2.24) [52].

As for the nucleophilic reactivity of amines in light-promoted LSF, desloratadine, an antihistamine drug, has been selected as representative of three-component alkylation reactions to afford **82**, **85**, and **88** in 32%, 39%, and 75% yields, respectively (Scheme 2.25) [53]. Mechanistically, as for the formation of **85** and **88**, a photo(redox) catalyst triggers the formation of a carbon-centered radical, which undergoes Giese-type addition onto a double bond, affording an open shell intermediate. The latter is then oxidized to carbocation *via* a SET event to regenerate the

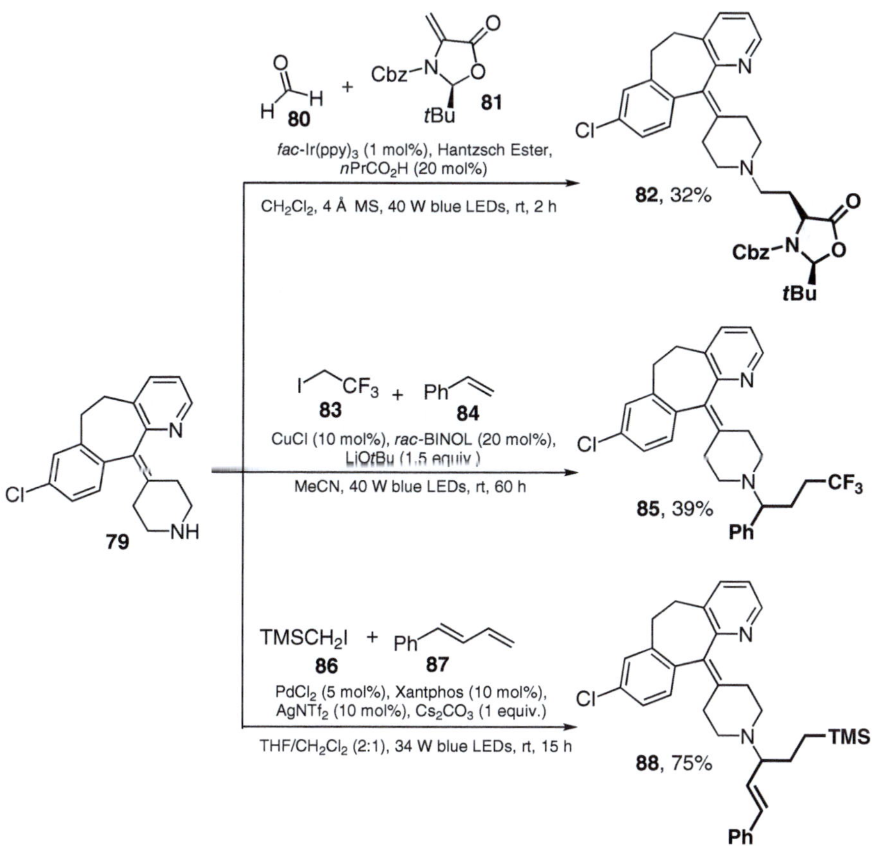

Scheme 2.25 Diversification of desloratadine **79** through light-triggered multicomponent reactions.

ground-state photo(redox) catalyst. Finally, the carbocation is trapped by the amine to give the alkylated products **85** and **88** (Scheme 2.25).

One more example of a chemoselective LSF achievable under exceptionally mild conditions can be exemplified by the α-oxygenation of tertiary amines to amides as shown in Scheme 2.26 for meclizine **89**, an antihistamine drug. The amide **90** can be obtained in 61% yield by using Rose Bengal as an organic photocatalyst, 1,5-diazabicyclo[4.3.0]non-5-ene (DBN) as a base, in DMF, and under 12 W blue LED irradiation at room temperature for 16–48 hours (Scheme 2.26) [54].

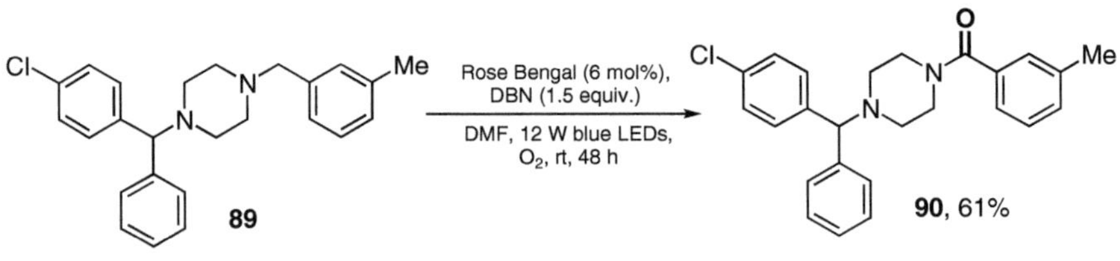

Scheme 2.26 α-Oxygenation of meclizine **89**.

2.4.3 Halides in LSF

Aliphatic halides such as chlorides, bromides, and iodides have been extensively investigated in photo(redox) catalytic methods as radical precursors of alkyl radicals [55]. However, due to their poor drug-likeness and toxicity issues, most of examples in literature rely on LSF of aromatic and heteroaromatic halides. The LSF of the latter is typically accomplished under dual catalysis, i.e., by employing a photoredox catalyst to generate an open-shell species, whose insertion into the aryl halide is promoted via the formation of a complex with a metal (very often nickel) catalyst, followed by a reductive elimination step to give the final product [56].

As an example of C(sp^2)—Cl bond manipulation, loratadine, an antihistamine drug, can be converted into its methyl analog in 51% yield via a metallaphotoredox protocol, requiring both an iridium-based photocatalyst and a nickel catalyst and by using trimethyl orthoformate as a source of methyl radical (Scheme 2.27) [57].

The formylation of aryl chlorides can be seen as a valuable transformation to introduce an aldehyde functional group that is suitable for further elaboration. M. K. Nielsen et al. exploited 1,3-dioxolane as alkyl radical precursor to introduce an acetal functional group *in situ* hydrolyzed to aldehyde derivative as shown

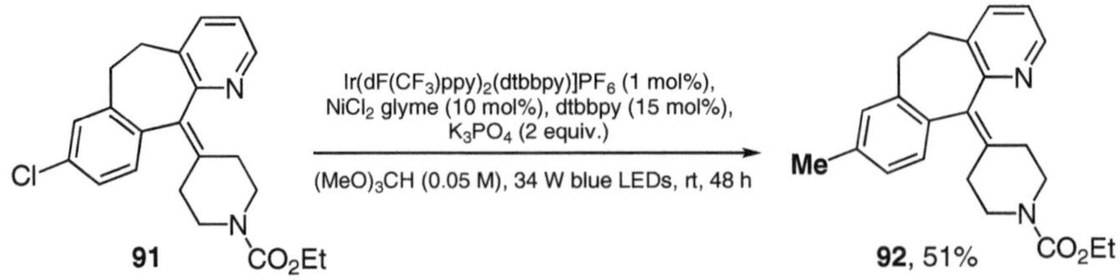

Scheme 2.27 Methylation of loratadine **91**.

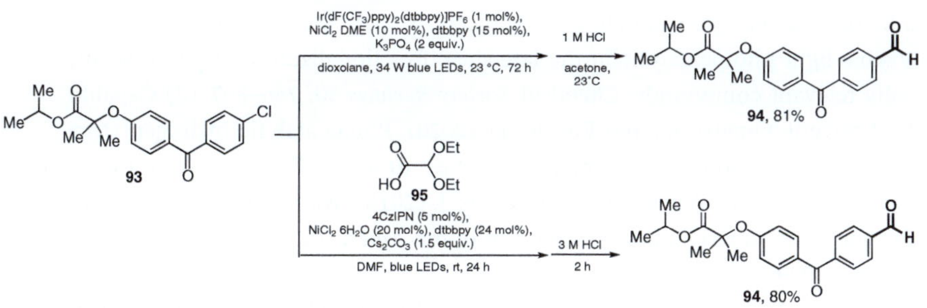

Scheme 2.28 Formylation of fenofibrate **93**.

for the LSF of fenofibrate affording derivative **94** in 80% yield (Scheme 2.28) [58]. Alternatively, a formylation reaction can be achieved by using diethoxyacetic acid **95** as a radical precursor and 4CzIPN as an organic photoredox catalyst (Scheme 2.28). In both cases, a nickel catalyst is also required to promote the insertion of the light-mediated generated radical species onto the C—Cl bond [59].

2.5 Conclusion

Starting with a critical analysis of what could and should be defined as LSF, the present chapter highlights a selection of visible light-promoted methods. It is worth noting that the selection of the examples herein presented is a personal choice of the authors, and many other valuable LSF protocols are available in literature. The selected LSFs enabled to discuss site-selectivity issues associated, in particular, with the direct C—H functionalization, showing future directions for improvement of these issues. On the other hand, the chemoselectivities (a requisite for LSF) associated with the derivatization of aliphatic and aromatic C—H bonds as well as carboxylic acids, amines, and halides pointed out the synthetic value of visible light-triggered methodologies to forge the drugs of the future.

References

1 Börgel, J. and Ritter, T. (2020). Late-stage functionalization. *Chem* 6: 1877–1887.

2 For relevant examples: (a) DiRocco, D.A., Dykstra, K., Krska, S. et al. (2014). Late-stage functionalization of biologically active heterocycles through photoredox catalysis. *Angewandte Chemie, International Edition* 53 (19): 4802–4806. (b) Loh, Y.Y., Nagao, A.J., Hesk, D. et al. (2017). Photoredox-catalyzed deuteration and tritiation of pharmaceutical compounds. *Science* 358: 1182–1187.

3 (a) Castellino, N.J., Montgomery, A.P., Danon, J.J., and Kassiou, M. (2023). Late-stage functionalization for improving drug-like molecular properties. *Chemical Reviews* 123: 8127–8153. (b) Lasso, J.D., Castillo-Pazos, D.J., and Li, C.-J. (2021). Green chemistry meets medicinal chemistry: a perspective on modern metal-free late-stage functionalization reactions. *Chemical Society Reviews*

50: 10955–10982. (c) Cannalire, R., Pelliccia, S., Sancineto, L. et al. (2021). Visible light photocatalysis in the late-stage functionalization of pharmaceutically relevant compounds. *Chemical Society Reviews* 50: 766–897. (d) Capaldo, L., Lafayette Quadri, L., and Ravelli, D. (2020). Photocatalytic hydrogen atom transfer: the philosophers stone for late-stage functionalization? *Green Chemistry* 22: 3376–3396. (e) White, M.C. and Zhao, J. (2018). Aliphatic C—H oxidations for late-stage functionalization. *Journal of the American Chemical Society* 140: 13988–14009.

4 Guo, Y.-Q., Wang, R., Song, H. et al. (2020). Visible-light-induced deoxygenation/defluorination protocol for synthesis of γ,γ-difluoroallylic ketones. *Organic Letters* 2020: 709–713.

5 Bacauanu, V., Cardinal, S., Yamauchi, M. et al. (2018). Metallaphotoredox difluoromethylation of aryl bromides. *Angewandte Chemie, International Edition* 57: 12543–12548.

6 Bellotti, P., Huang, H.-M., Faber, T., and Glorius, F. (2023). Photocatalytic late-stage C—H functionalization. *Chemical Reviews* 123: 4237–4352.

7 (a) Holmberg-Douglas, N. and Nicewicz, D.A. (2022). Photoredox-catalyzed C—H functionalization reactions. *Chemical Reviews* 122: 1925–2016. (b) Prier, C.K., Rankic, D.A., and MacMillan, D.W.C. (2013). Visible light photoredox catalysis with transition metal complexes: applications in organic synthesis. *Chemical Reviews* 113: 5322–5363.

8 (a) Sambiagio, C., Schönbauer, D., Blieck, R. et al. (2018). A comprehensive overview of directing groups applied in metal-catalysed C—H functionalisation chemistry. *Chemical Society Reviews* 47: 6603–6743. (b) Rasheed, O.K. and Sun, B. (2018). Advances in development of C—H activation/functionalization using a catalytic directing group. *ChemistrySelect* 3: 5689–5708. (c) Capaldo, L., Ravelli, D., and Fagnoni, M. (2022). Direct photocatalyzed hydrogen atom transfer (HAT) for aliphatic C—H bonds elaboration. *Chemical Reviews* 122: 1875–1924.

9 Halperin, S.D., Fan, H., Chang, S. et al. (2014). A convenient photocatalytic fluorination of unactivated C—H bonds. *Angewandte Chemie, International Edition* 53: 4690–4693.

10 Kleoff, M., Schwan, J., Christmann, M., and Heretsch, P. (2021). A modular, argon-driven flow platform for natural product synthesis and late-stage transformations. *Organic Letters* 23: 2370–2374.

11 West, J.G., Bedell, T.A., and Sorensen, E.J. (2016). The uranyl cation as a visible-light photocatalyst for C(sp^3)—H fluorination. *Angewandte Chemie, International Edition* 55: 8923–8927.

12 Govaerts, S., Nyuchev, A., and Noël, T. (2020). Pushing the boundaries of C—H bond functionalization chemistry using flow technology. *Journal of Flow Chemistry* 10: 13–71.

13 Perry, I.B., Brewer, T.F., Sarver, P.J. et al. (2018). Direct arylation of strong aliphatic C—H bonds. *Nature* 560: 70–75.

14 Le, C., Liang, Y., Evans, R.W. et al. (2017). Selective sp^3 C—H alkylation via polarity-match-based cross-coupling. *Nature* 547: 79–83.

15 Wakaki, T., Sakai, K., Enomoto, T. et al. (2018). C(sp^3)—H cyanation promoted by visible-light photoredox/phosphate hybrid catalysis. *Chemistry - A European Journal* 24: 8051–8055.

16 Dong, J., Xia, Q., Lv, X. et al. (2018). Photoredox-mediated direct cross-dehydrogenative coupling of heteroarenes and amines. *Organic Letters* 20: 5661–5665.

17 Kamijo, S., Hoshikawa, T., and Inoue, M. (2011). Photochemically induced radical transformation of C(sp^3)—H bonds to C(sp^3)—CN bonds. *Organic Letters* 13: 5928–5931.

18 Kamijo, S., Takao, G., Kamijo, K. et al. (2016). Photo-induced substitutive introduction of the aldoxime functional group to carbon chains: a formal formylation of non-acidic C(sp^3)—H bonds. *Angewandte Chemie, International Edition* 55: 9695–9699.

19 Fan, X.-Z., Rong, J.-W., Wu, H.-L. et al. (2018). Eosin Y as a direct hydrogen-atom transfer photocatalyst for the functionalization of C—H bonds. *Angewandte Chemie, International Edition* 57: 8514–8518.

20 Mukherjee, S., Maji, B., Tlahuext-Aca, A., and Glorius, F. (2016). Visible-light-promoted activation of unactivated C(sp^3)—H bonds and their selective trifluoromethylthiolation. *Journal of the American Chemical Society* 138: 16200–16203.

21 Gant Kanegusuku, A.L. and Roizen, J.L. (2021). Recent advances in photoredox-mediated radical conjugate addition reactions: an expanding toolkit for the Giese reaction. *Angewandte Chemie, International Edition* 60: 21116–21149.

22 Wang, C.-S., Dixneuf, P.H., and Soulé, j.-F. (2018). Photoredox catalysis for building C—C bonds from C(sp^2)—H bonds. *Chemical Reviews* 118: 7532–7585.

23 Zhang, L. and Ritter, T. (2022). A perspective on late-stage aromatic C—H bond functionalization. *Journal of the American Chemical Society* 144: 2399–2414.

24 Arceo, E., Montroni, E., and Melchiorre, P. (2014). Photo-organocatalysis of atom-transfer radical additions to alkenes. *Angewandte Chemie, International Edition* 53: 12064–12068.

25 Dagousset, G., Carboni, A., Magnier, E., and Masson, G. (2014). Photoredox-induced three-component azido- and aminotrifluoromethylation of alkenes. *Organic Letters* 16: 4340–4343.

26 Miura, T., Funakoshi, Y., Nakahashi, J. et al. (2018). Synthesis of elongated esters from alkenes. *Angewandte Chemie, International Edition* 57: 15455–15459.

27 Liu, J., Wu, S., Yu, J. et al. (2020). Polarity umpolung strategy for the radical alkylation of alkenes. *Angewandte Chemie, International Edition* 59: 8195–8202.

28 Oh, S.H., Malpani, Y.R., Ha, N. et al. (2014). Vicinal difunctionalization of alkenes: chlorotrifluoromethylation with CF_3SO_2Cl by photoredox catalysis. *Organic Letters* 16: 1310–1313.

29 Du, Y. D., Zhou, C. Y., To, W. P.To, W. P., Wang, H. X., Che, C. M. (2020). Iron porphyrin catalysed light driven C—H bond amination and alkene aziridination with organic azides. *Chemical Science* 11: 4680–4686.

30 Yu, J., Wu, Z., and Zhu, C. (2018). Efficient docking–migration strategy for selective radical difluoromethylation of alkenes. *Angewandte Chemie, International Edition* 57: 17156–17160.

31 Nagib, D.A. and Macmillan, D.W.C. (2011). Trifluoromethylation of arenes and heteroarenes by means of photoredox catalysis. *Nature* 480: 224–228.

32 Zhang, T., Guo, X., Shi, Y. et al. (2018). Dye-incorporated coordination polymers for direct photocatalytic trifluoromethylation of aromatics at metabolically susceptible positions. *Nature Communications* 9: 4024.

33 Chen, W., Huang, Z., Tay, N.E.S. et al. (2019). Direct arene C—H fluorination with $^{18}F^-$ via organic photoredox catalysis. *Science* 364: 1170–1174.

34 For the introduction of chlorine and iodine atoms into naproxen:
(a) Petzold, D. and König, B. (2018). Photocatalytic oxidative bromination of electron-rich arenes and heteroarenes by anthraquinone. *Advanced Synthesis and Catalysis* 360: 626–630. (b) Narobe, R., Düsel, S.J.S., Iskra, J., and König, B. (2019). Photocatalytic oxidative iodination of electron-rich arenes. *Advanced Synthesis and Catalysis* 361: 3998–4004. (c) Düsel, S.J.S. and König, B. (2020). Oxidative photochlorination of electron-rich arenes via *in situ* bromination. *European Journal of Organic Chemistry* 2020: 1491–1495.

35 For the introduction of a diazomethyl linchpin (can undergo diazodiversification) into ibuprofen:Wang, Z., Herraiz, A.G., Del Hoyo, A.M., and Suero, M.G. (2018). Generating carbyne equivalents with photoredox catalysis. *Nature* 554: 86–91.

36 Berger, F., Plutschack, M.B., Riegger, J. et al. (2019). Site-selective and versatile aromatic C—H functionalization by thianthrenation. *Nature* 567: 223–228.

37 (a) Engl, P.S., Häring, A.P., Berger, F. et al. (2019). C—N cross-couplings for site-selective late-stage diversification via aryl sulfonium salts. *Journal of the American Chemical Society* 141: 13346–13351. (b) Juliá, F., Shao, Q., Duan, M. et al. (2021). High site selectivity in electrophilic aromatic substitutions: mechanism of C—H thianthrenation. *Journal of the American Chemical Society* 143: 16041–16054.

38 (a) Abdiaj, I., Bottecchia, C., Alcazar, J., and Noël, T. (2017). Visible-light-induced trifluoromethylation of highly functionalized arenes and heteroarenes in continuous flow. *Synthesis* 49: 4978–4985. (b) Beatty, J.W., Douglas, J.J., Cole, K.P., and Stephenson, C.R.J. (2015). A scalable and operationally simple radical trifluoromethylation. *Nature Communications* 6: 7919. (c) Beatty, J.W., Douglas, J.J., Miller, R. et al. (2016). Photochemical perfluoroalkylation with pyridine *N*-oxides: mechanistic insights and performance on a kilogram scale. *Chem* 1: 456–472. (d) Lin, J., Li, Z., Kan, J. et al. (2017). Photo-driven redox-neutral decarboxylative carbon-hydrogen trifluoromethylation of (hetero)arenes with trifluoroacetic acid. *Nature Communications* 8: 14353. (e) Qiu, Y., Scheremetjew, A., Finger, L.H., and Ackermann, L. (2020). Electrophotocatalytic undirected C—H trifluoromethylations of (het)arenes. *Chemistry - A European Journal* 26: 3241–3246. (f) Ghosh, I., Khamrai, J., Savateev, A. et al. (2019). Organic semiconductor photocatalyst can bifunctionalize arenes and heteroarenes. *Science* 365: 360–366. (g) Zhang, W., Xiang, X.X., Chen, J. et al. (2020). Direct C—H difluoromethylation of heterocycles via organic photoredox catalysis. *Nature Communications* 11: 638.

39 (a) Li, G.X., Morales-Rivera, C.A., Wang, Y. et al. (2016). Photoredox-mediated Minisci C—H alkylation of *N*-heteroarenes using boronic acids and hypervalent iodine. *Chemical Science* 7: 6407–6412. (b) Quan, Y., Lan, G., Fan, Y. et al. (2020). Metal–organic layers for synergistic lewis acid and photoredox catalysis. *Journal of the American Chemical Society* 142: 1746–1751. (c) Cheng, W.M., Shang, R., and Fu, Y. (2017). Photoredox/Brønsted acid co-catalysis enabling decarboxylative coupling of amino acid and peptide redox-active esters with *N*-heteroarenes. *ACS Catalysis* 7: 907–911. (d) El Khatib, M., Serafim, R.A.M., and Molander, G.A. (2016). α-Arylation/heteroarylation of chiral α-aminomethyltrifluoroborates by synergistic iridium photoredox/nickel cross-coupling catalysis. *Angewandte Chemie, International Edition* 55: 254–258. (e) Constantin, T., Zanini, M., Regni, A. et al. (2020). Aminoalkyl radicals as halogen-atom transfer agents for activation of alkyl and aryl halides. *Science* 367: 1021–1026. (f) Matsui, J.K., Primer, D.N., and Molander, G.A. (2017). Metal-free C—H alkylation of heteroarenes with alkyltrifluoroborates: a general protocol for 1°, 2° and 3° alkylation. *Chemical Science* 8: 3512–3522. (g) Matsui, J.K. and Molander, G.A. (2017). Organocatalyzed, photoredox heteroarylation of 2-trifluoroboratochromanones via C—H functionalization. *Organic Letters* 19: 950–953.

40 Park, G.R., Moon, J., and Cho, E.J. (2017). Dual role of Cu_2O nanocubes as templates and networking catalysts for hollow and microporous Fe-porphyrin networks. *Chemical Communications* 53: 12786–12789.

41 Chen, H., Yi, H., Tang, Z. et al. (2018). External oxidant-free regioselective cross dehydrogenative coupling of 2-arylimidazoheterocycles and azoles with H_2 evolution via photoredox catalysis. *Advanced Synthesis and Catalysis* 360: 3220–3227.

42 Jouffroy, M. and Kong, J. (2019). Direct C—H carbamoylation of nitrogen-containing heterocycles. *Chemistry - A European Journal* 25: 2217–2221.

43 Sato, S. and Nakamura, H. (2013). Ligand-directed selective protein modification based on local single-electron-transfer catalysis. *Angewandte Chemie, International Edition* 52: 8681–8684.

44 Jouffroy, M., Kelly, C.B., and Molander, G.A. (2016). Thioetherification via photoredox/nickel dual catalysis. *Organic Letters* 18: 876–879.

45 (a) Trump, L., Lemos, A., Lallemand, B. et al. (2019). Late-stage ^{18}F-difluoromethyl labeling of *N*-heteroaromatics with high molar activity for PET imaging. *Angewandte Chemie, International Edition* 58: 13149–13154. (b) Rogers, D.A., Brown, R.G., Brandeburg, Z.C. et al. (2018). Organic dye-catalyzed, visible-light photoredox bromination of arenes and heteroarenes using *N*-bromosuccinimide. *ACS Omega* 3: 12868–12877. (c) Rogers, D.A., Bensalah, A.T., Espinosa, A.T. et al. (2019). Amplification of trichloroisocyanuric acid (TCCA) reactivity for chlorination of arenes and heteroarenes via catalytic organic dye activation. *Organic Letters* 21: 4229–4233. (d) Rogers, D.A., Hopkins, M.D., Rajagopal, N. et al. (2020). U.S. Food and Drug Administration-certified

food dyes as organocatalysts in the visible light-promoted chlorination of aromatics and heteroaromatics. *ACS Omega* 5: 7693–7704.
46 (a) Schwarz, J. and König, B. (2018). Decarboxylative reactions with and without light – a comparison. *Green Chemistry* 20: 323–361. (b) Beil, S.B., Chen, T.Q., Intermaggio, N.E., and MacMillan, D.W.C. (2022). Carboxylic acids as adaptive functional groups in metallaphotoredox catalysis. *Accounts of Chemical Research* 55: 3481–3494. (c) Hu, X.-Q., Liu, Z.-K., Hou, Y.-X., and Gao, Y. (2020). Single electron activation of aryl carboxylic acids. *iScience* 23: 101266. (d) Li, L., Yao, Y., and Fu, N. (2023). Free carboxylic acids: the trend of radical decarboxylative functionalization. *European Journal of Organic Chemistry* 26: e202300166.
47 Stache, E.E., Ertel, A.B., Rovis, T., and Doyle, A.G. (2018). Generation of phosphoranyl radicals via photoredox catalysis enables voltage–independent activation of strong C—O bonds. *ACS Catalysis* 8: 11134–11139.
48 Zhang, M., Yuan, X.-A., Zhu, C., and Xie, J. (2019). Deoxygenative deuteration of carboxylic acids with D_2O. *Angewandte Chemie, International Edition* 58: 312–316.
49 Shu, C., Mega, R.S., Andreassen, B.J. et al. (2018). Synthesis of functionalized cyclopropanes from carboxylic acids by a radical addition–polar cyclization cascade. *Angewandte Chemie, International Edition* 57: 15430–15434.
50 Russo, C., Graziani, G., Cannalire, R. et al. (2022). Visible-light photocatalytic metal-free multicomponent Ugi-like chemistry. *Green Chemistry* 24: 3993–4003.
51 (a) Correia, J.T., Fernandes, V.A., Matsuo, B.T. et al. (2020). Photoinduced deaminative strategies: Katritzky salts as alkyl radical precursors. *Chemical Communications* 56: 503–514. (b) Gao, Y., Jiang, N.-D., Xiang, H. et al. (2022). Recent progress in fragmentation of Katritzky Salts enabling formation of C—C, C—B, and C—S bonds. *Topics in Current Chemistry* 380: 25. (c) Yousif, A.M., Colarusso, S., and Bianchi, E. (2022). Katritzky Salts for the synthesis of unnatural amino acids and late-stage functionalization of peptides. *European Journal of Organic Chemistry* 26: e202201274.
52 Sandfort, F., Strieth-Kalthoff, F., Klauck, F.J.R. et al. (2018). Deaminative borylation of aliphatic amines enabled by visible light excitation of an electron donor–acceptor complex. *Chemistry - A European Journal* 24: 17210–17214.
53 (a) Trowbridge, A., Reich, D., and Gaunt, M.J. (2018). Multicomponent synthesis of tertiary alkylamines by photocatalytic olefin-hydroaminoalkylation. *Nature* 561: 522–527. (b) Xiong, Y., Ma, X., and Zhang, G. (2019). Copper-catalyzed intermolecular carboamination of alkenes induced by visible light. *Organic Letters* 21: 1699–1703. (c) Shing Cheung, K.P., Kurandina, D., Yata, T., and Gevorgyan, V. (2020). Photoinduced palladium-catalyzed carbofunctionalization of conjugated dienes proceeding via radical-polar crossover scenario: 1,2-aminoalkylation and beyond. *Journal of the American Chemical Society* 142: 9932–9937.
54 Zhang, Y., Riemer, D., Schilling, W. et al. (2018). Visible-light-mediated efficient metal-free catalyst for α-oxygenation of tertiary amines to amides. *ACS Catalysis* 8: 6659–6664.

55 Protti, S. and Fagnoni, M. (2020). Generation of alkyl radicals: from the tyranny of tin to the photon democracy. *Chemical Reviews* 120: 9790–9833.

56 (a) Chan, A.Y., Perry, I.B., Bissonnette, N.B. et al. (2022). Metallaphotoredox: the merger of photoredox and transition metal catalysis. *Chemical Reviews* 122: 1485–1542. (b) Zhang, J. and Rueping, M. (2023). Metallaphotoredox catalysis for sp^3 C—H functionalizations through hydrogen atom transfer (HAT). *Chemical Society Reviews* 52: 4099–4120.

57 Kariofillis, S.K., Shields, B.J., Tekle-Smith, M.A. et al. (2020). Nickel/photoredox-catalyzed methylation of (hetero)aryl chlorides using trimethyl orthoformate as a methyl radical source. *Journal of the American Chemical Society* 142: 7683–7689.

58 Nielsen, M.K., Shields, B.J., Liu, J. et al. (2017). Mild, redox-neutral formylation of aryl chlorides through the photocatalytic generation of chlorine radicals. *Angewandte Chemie, International Edition* 56: 7191–7194.

59 Huang, H., Li, X., Yu, C. et al. (2017). Visible-light-promoted nickel- and organic-dye-cocatalyzed formylation reaction of aryl halides and triflates and vinyl bromides with diethoxyacetic acid as a formyl equivalent. *Angewandte Chemie, International Edition* 56: 1500–1505.

3

Electrochemical Late-Stage Functionalization

Yanjun Li, Zhipeng Lin, Hao Long, Yang Xu, and Lutz Ackermann

Georg-August-Universität Göttingen, Institut für Organische und Biomolekulare Chemie, Tammannstraße 2, 37077 Göttingen, Germany

3.1 Introduction

Electrosynthesis offers an environmentally benign and sustainable alternative for molecular transformations, with mild reaction conditions, high selectivities, and scalability. The history of organic electrochemistry dates back nearly 200 years, beginning with Faraday's work on the electrolysis of acetic acid [1] and later with Kolbe's decarboxylative dimerization [2]. Significant industrial processes, such as the Simons fluorination [3], the Monsanto adiponitrile method [4], and the Shono oxidation [5], substantiate the practical utility of electrosynthesis. Modern developments include the Yoshida's introduction of electroauxiliaries and the cation pool concept [6–8], and Steckhan's indirect electrolysis-mediated processes [9, 10]. Direct electrolysis techniques, advanced by Schäfer [11], Lund [12], Little [13–15], Moeller [16], Jutand [17], and Amatore [18] in the 21st century, have boosted the resurgence of electro-organic synthesis. Notable contributions by Ackermann [19], Baran [20], Xu [21], Lei [22], Lambert [23], Lin [24], Gouin [25], Waldvogel [26], Stahl [27], Malins [28], and Mei [29], among others [30–37], have led to significant progresses in C—H activation, cross-electrophile coupling, alkene difunctionalization, and the late-stage functionalization of natural products and medicinal molecules.

This chapter focuses on electrochemical late-stage functionalization, which enables direct, site-selective, and chemoselective modification of C—H bonds or functional groups in bio-relevant molecules – such as natural products, pharmaceuticals, and complex molecules – to access their analogs (Scheme 3.1).

Late-Stage Functionalization and Diversification in Organic Synthesis: Methods and Applications, First Edition. Edited by Tatiana Besset.

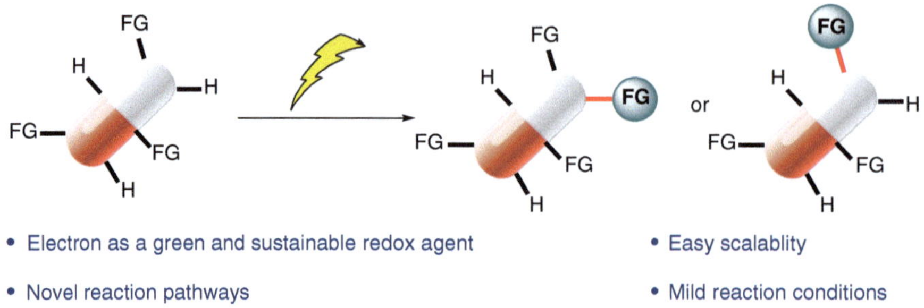

Scheme 3.1 Electrochemical late-stage functionalization of bio-relevant molecules.

3.2 Electrochemical Late-Stage Functionalization of C–H Bonds

Over the past decade, the merger of electrolysis and C—H functionalization has led to a transformative shift in the field of molecular synthesis. A considerable number of these electrochemical C—H functionalization techniques have been effectively employed in the late-stage modification of natural products and pharmaceuticals, thereby opening up new avenues in drug discovery.

3.2.1 Electrochemical Late-Stage Functionalization of C(sp^2)–H Bonds

The methylation reaction is a prominent example of a late-stage functionalization strategies for improving bioactive molecule properties, such as metabolic stability, lipophilicity, and binding interactions – commonly known as "magic methyl effect" [38]. Mei group developed an anodic C(sp^2)—H methylation with $MeBF_3K$, circumventing the use of strong oxidants [39, 40]. In contrast, Ackermann enabled the eLSF of *N*-heteroarenes in the presence of a rhodium catalyst (Scheme 3.2a) [41], thus achieving selective methylation of purines, diazepam, and amino acids. Later, Guo group further diversified the methyl donors to MeBPin and $MeB(OH)_2$, allowing the functionalization of estrone and nucleotides (Scheme 3.2b) [42].

Trifluoromethylation reaction has raised significant interest in recent years, as this approach allows the synthesis of a variety of trifluoromethylated compounds that play an important role in medicinal chemistry [43]. In 2014, Blackmond and Baran developed the C(sp^2)—H trifluoromethylation of bio-relevant heterocycles with $Zn(CF_3SO_2)_2$ under electrochemical conditions (Scheme 3.3) [44]. This electrochemical approach was subsequently extended to difluoromethylation using $Zn(CF_2HSO_2)_2$, albeit with reduced efficiency.

C—H alkynylation is a practical strategy for incorporating alkynes into molecules [45–47]. In 2020, Shi and Xie presented an electrochemical iridium-catalyzed directed C(sp^2)—H alkynylation using terminal alkynes (Scheme 3.4) [48]. The electrocatalysis operates in an undivided cell, requires no external oxidants, and gives excellent yields. Its success in modifying bioactive molecules such as estrone and purine derivatives highlights its pharmaceutical relevance.

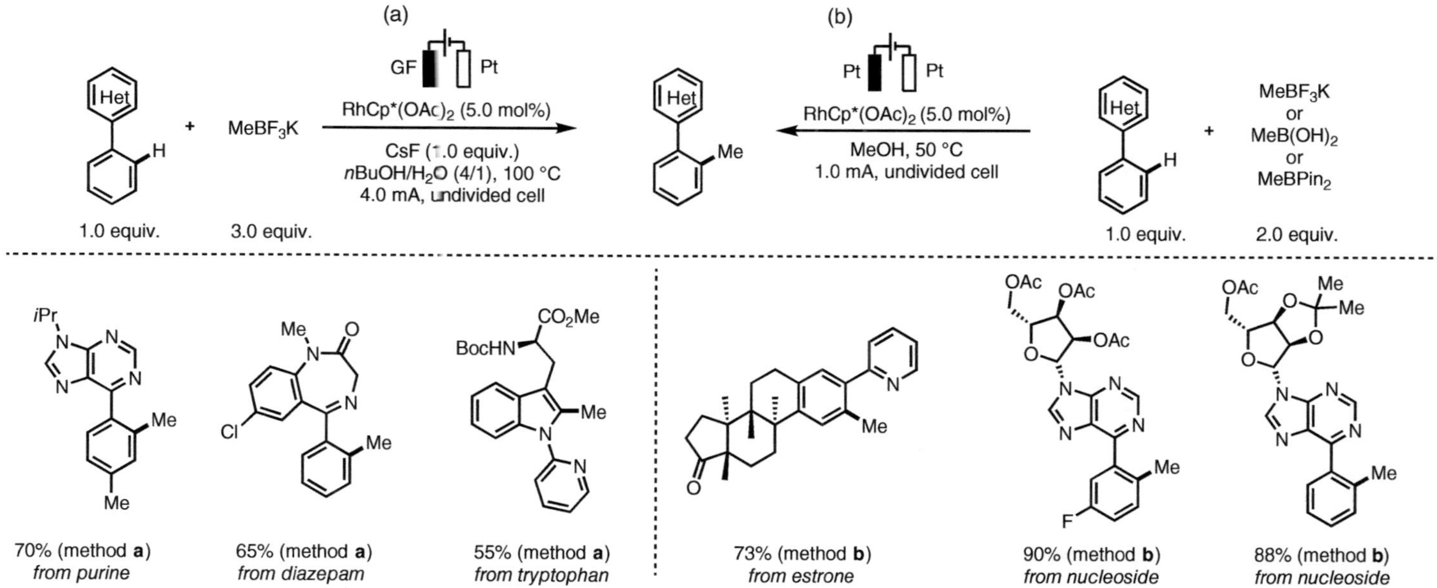

Scheme 3.2 Electrochemical rhodium-catalyzed late-stage C(sp^2)–H methylation. (a) Ackerman. (b) Yang and Gao.

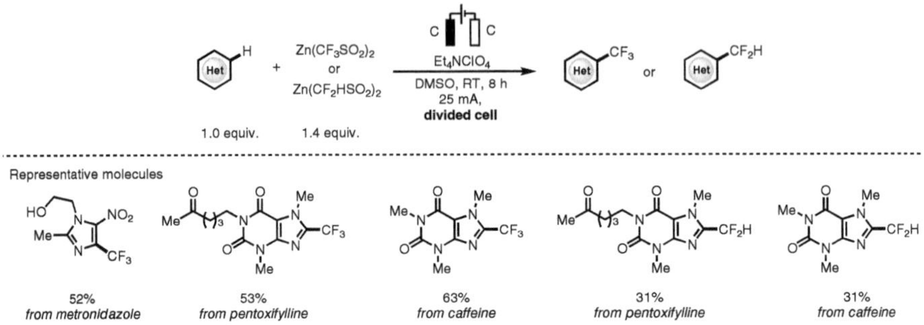

Scheme 3.3 Electrochemical late-stage trifluoromethylation and difluoromethylation.

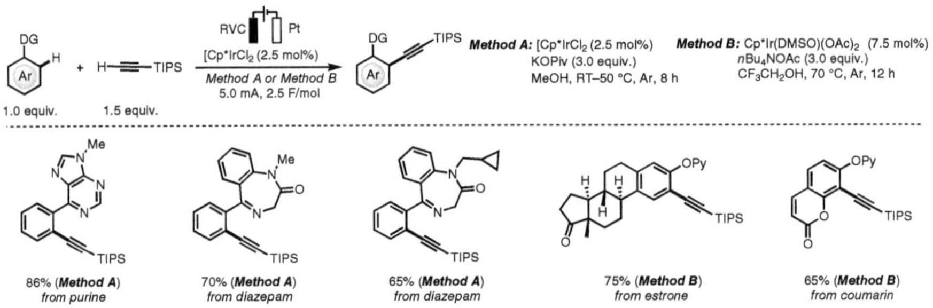

Scheme 3.4 Electrochemical late-stage C(sp^2)—H alkynylation.

Hydroxylation of arene C(sp^2)—H bonds is essential to improve drug solubility in aqueous solution. Nevertheless, the controlled electrochemical hydroxylation of the aromatic C—H bond is challenging due to the high propensity of phenol to overoxidized state [49]. To mitigate the overoxidation problem, trifluoroacetic acid (TFA)-mediated strategies that generate aryl trifluoroacetates to control oxidation are proved viable, albeit with effectiveness limited to certain arenes [50, 51]. The Xu group established an efficient solution using continuous-flow electrochemical microreactors, allowing selective hydroxylation of a wide range of arenes without external oxidants or metals (Scheme 3.5) [52]. The arenes undergo anodic oxidation to form radical cations, which subsequently interact with trifluoroacetate, yielding a cyclohexadienyl radical. Another electron transfer event, coupled with proton elimination, leads to forming an aryl trifluoroacetate intermediate. This intermediate undergoes hydrolysis in the presence of water to provide the phenol. This mild approach suppresses overoxidation and has successfully modified natural products and drug derivatives.

Electrooxidative C—H/N—H cross-coupling is a highly efficient approach to incorporate nitrogen functionalities into aromatic compounds. The Lei group reported an intermolecular cross-coupling between sulfonimides and aromatic arenes (Scheme 3.6) [53]. This transformation proceeds via a nitrogen radical addition [54] mechanism and successfully functionalized various substrates, such as flavone, caffeine, and fenofibrate. Aryl sulfonamides or aniline derivatives can be obtained after a deprotection step. Concurrently, the Ackermann group achieved

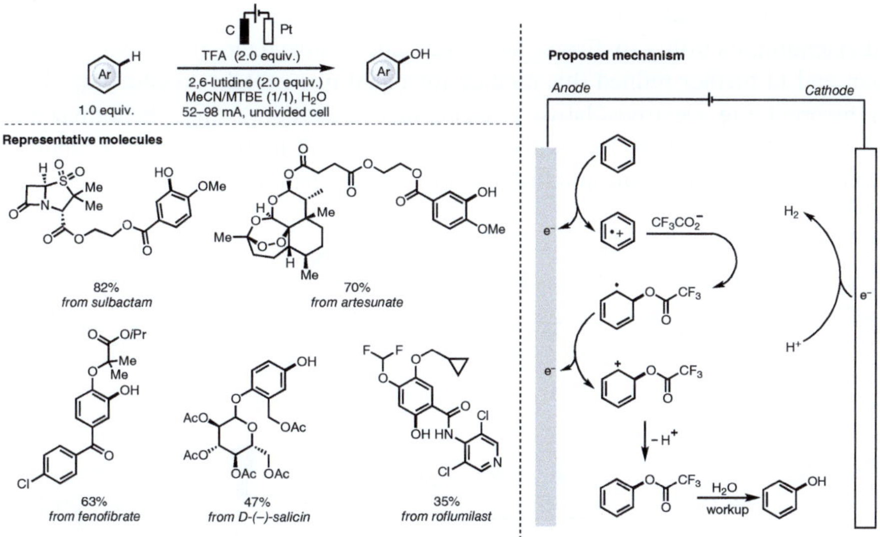

Scheme 3.5 Electrochemical late-stage hydroxylation.

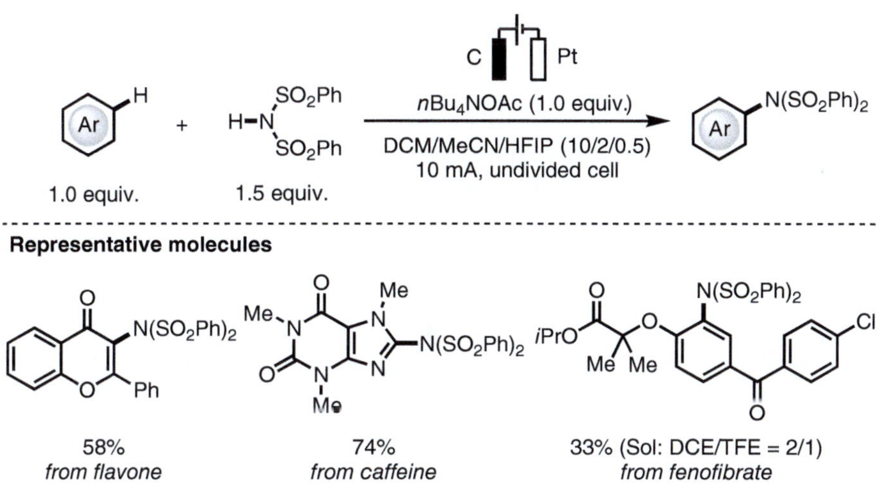

Scheme 3.6 Electrochemical late-stage aminations.

electrochemical oxidation-induced C(sp^2)—H nitrogenation of heteroarenes such as pyrroles and indoles [55]. Moreover, metallaelectro-catalyzed C(sp^2)—H aminations have been demonstrated with various 3d transition metals such as Ni, Co, and Cu [56–59].

Protein conjugation techniques are crucial for the advancement of chemical biology and therapeutic applications. In 2010, Barbas group reported on an innovative PTAD-mediated reaction with tyrosine residues, while the requirement of oxidants and co-solvents limited its practical use [60]. In 2018, Gouin achieved the Y-click reaction under alternative mild electrochemical conditions

(+0.36 V, Scheme 3.7a) [25], facilitating selective tyrosine labeling in aqueous media and extending its utility to diverse substrates, such as oxytocin and epratuzumab. Huan and Li further refined this method for metal-free protein crosslinking [61]. Lei reported the electrooxidative tyrosine bioconjugation using phenothiazine derivatives (Scheme 3.7b) [62] with high efficiency and high selectivity, enabling key biomolecules modifying and fluorescent peptide labeling.

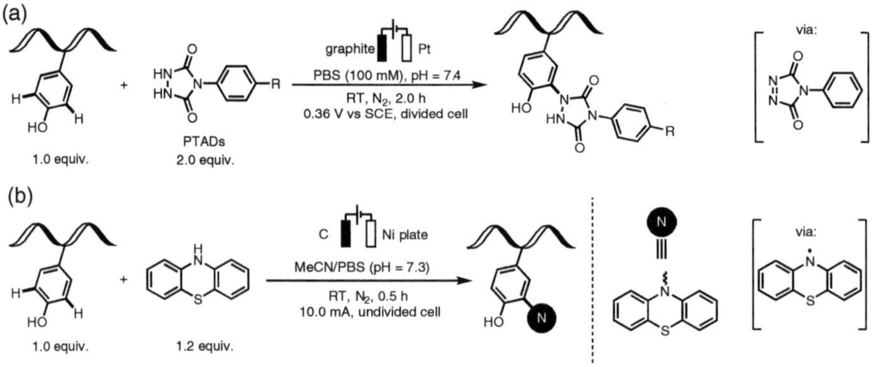

Scheme 3.7 Electrochemical peptide and protein modification on tyrosine. (a) Gouin. (b) Lei.

Late-stage phosphorylation methods are essential in medicinal chemistry, given the wide applications of organophosphorus compounds [63]. Budnikova developed a metallaelectro-catalyzed coupling of caffeine with dialkylphosphites to produce phosphorylated caffeine (Scheme 3.8) [64].

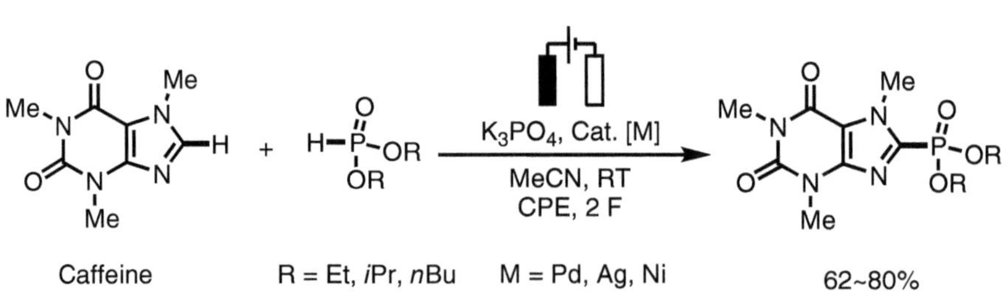

Scheme 3.8 Electrocatalytic coupling of caffeine with dialkylphosphites.

In 2019, Xu and coworkers reported an electrochemical rhodium-catalyzed aryl C(sp^2)—H phosphorylation using various phosphine oxides (Scheme 3.9) [65]. This method, notable for broad substrate scope and high functional group tolerance, has been demonstrated effective for the late-stage functionalization of bioactive molecules, such as diazepam and purine derivatives. It is noteworthy that the reaction was scalable, producing a phosphonate product in 87.7 g. A possible catalytic cycle is illustrated in Scheme 3.9. The rhodacycle **Int-1** is formed through carboxylate-assisted C—H activation. Subsequent ligand exchange between **Int-1**

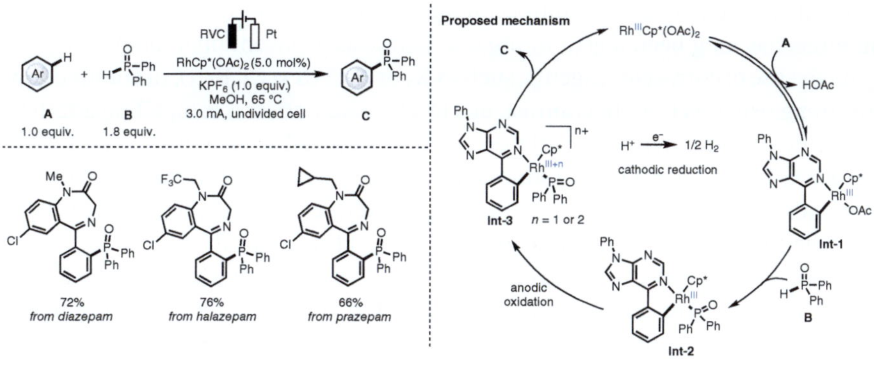

Scheme 3.9 Electrochemical rhodium-catalyzed C(sp^2)–H phosphorylation.

and **B** generates the more readily oxidized species **Int-2**, which then undergoes oxidation-induced reductive elimination, affording product **C** and a Rh^{III} or Rh^{II} complex, depending on the oxidation state of **Int-3**. If a Rh^{II} complex is formed, it will undergo anodic oxidation to generate a Rh^{III} complex. At the cathode, proton reduction produces H_2.

In 2021, the Xu group reported an electrochemical C(sp^2)—H phosphorylation reaction using triethyl phosphite in a continuous flow cell, thereby circumventing the use of external oxidants and transition metal catalysts (Scheme 3.10) [66]. The method proved effective for both electron-rich and electron-deficient arenes, as demonstrated by the continuous production of 55.0 g of a phosphonate product. The C—P bond was formed via the reaction of arenes with anodically generated P-radical cations, thereby enabling the late-stage functionalization of bioactive compounds and natural products.

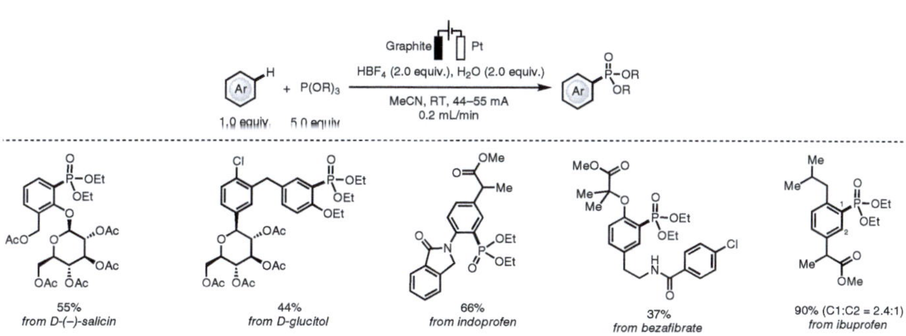

Scheme 3.10 Electrochemical late-stage C(sp^2)–H phosphorylation in continuous flow.

The incorporation of a halide atom into drug molecules has been demonstrated to significantly enhance their biological properties [67]. Halogenated arenes are also of value for the preparation of radio-labeled compounds, which are useful in the study of metabolism [68]. Furthermore, halogenated arenes and heterocycles serve as versatile intermediates in a number of organic transformations [69]. Consequently, the development of efficient and atom-economical halogenation methods under mild conditions has long been a goal in organic synthesis. Conventional methodologies rely on the use of corrosive reagents, such as X_2 or NXS (X = Cl, Br, I), in conjunction with strong oxidants [70]. In contrast, electrochemistry offers a compelling alternative by employing simple halide salts or aqueous HX as halogenating agents [71–73].

In 2017, Rivera and Liu introduced an electrochemical bromination method using NaBr in a water mixture with acetonitrile/methanol (Scheme 3.11) [74]. This reaction, carried out in a microflow electrochemical cell under mild conditions, successfully brominated a range of drug molecules such as Cytidine, Sch 48793, Tenofovir, and MK-4618, yielding the corresponding aryl bromides. The brominated derivatives of Tenofovir and MK-4618 were subsequently converted into tritium-labeled products.

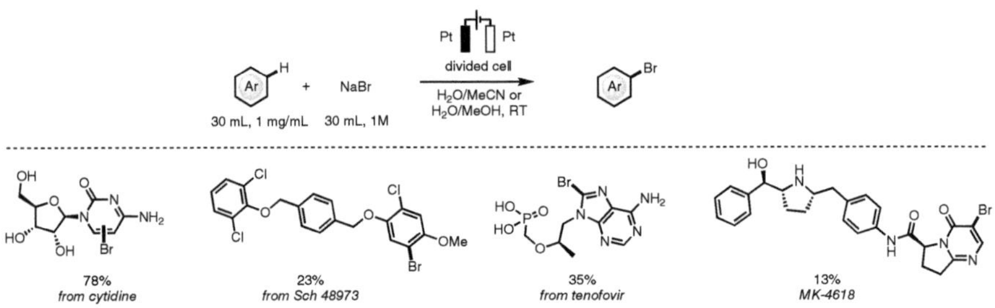

Scheme 3.11 Electrochemical late-stage bromination of drug molecules.

Jiao and coworkers developed an elegant electrochemical aromatic chlorination method using DCE as the chlorine source, additionally producing vinyl chloride as a highly valuable byproduct (Scheme 3.12) [75]. The method's mild nature and practicality were demonstrated by its scalability and compatibility with bioactive molecules such as (*S*)-naproxen methyl ester, leflunomide, and acetaminophen.

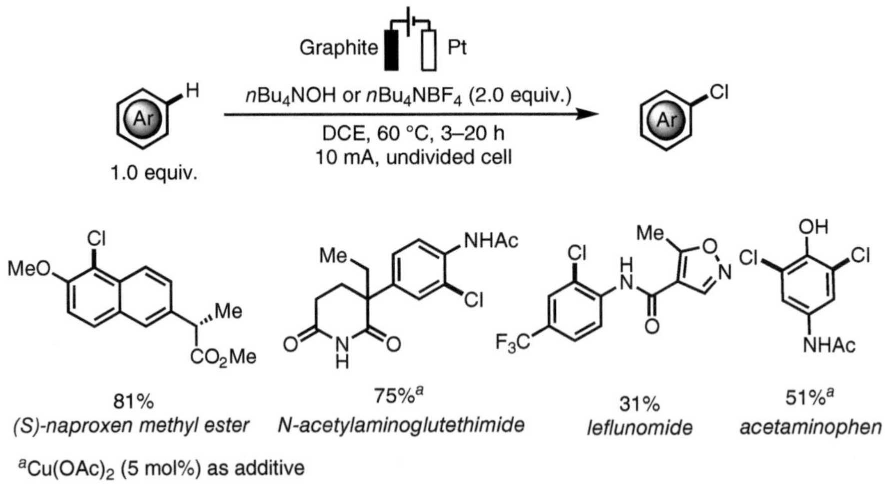

Scheme 3.12 Paired electrocatalysis for the preparation of aryl chlorides.

Similarly, McNeil and coworkers recently achieved chlorination using waste poly(vinyl chloride) as the chlorine source [35].

C—H activation-based annulations have been a significant advancement in the synthesis of cyclic compounds over the past two decades [76–78]. Electrochemical annulation methods have gained prominence for their ability to avoid sacrificial oxidants such as $Cu(OAc)_2$ and AgOAc, thereby reducing byproducts and enhancing atom efficiency [79–84]. These methods have been extensively employed to form five-, six-, and seven-membered rings. Nevertheless, their use in chemoselective electrochemical late-stage functionalization remains constrained.

In 2021, the Ackermann group presented a rhodaelectro-catalyzed formyl C—H activation approach for the annulation of 2-hydroxybenzaldehydes with alkynes (Scheme 3.13) [85]. This approach enabled the selective functionalization of tyrosine derivatives, facilitating the electro-labeling of fluorescent amino acids and peptides. A wide range of peptides, including those with oxidation-sensitive residues, were efficiently labeled.

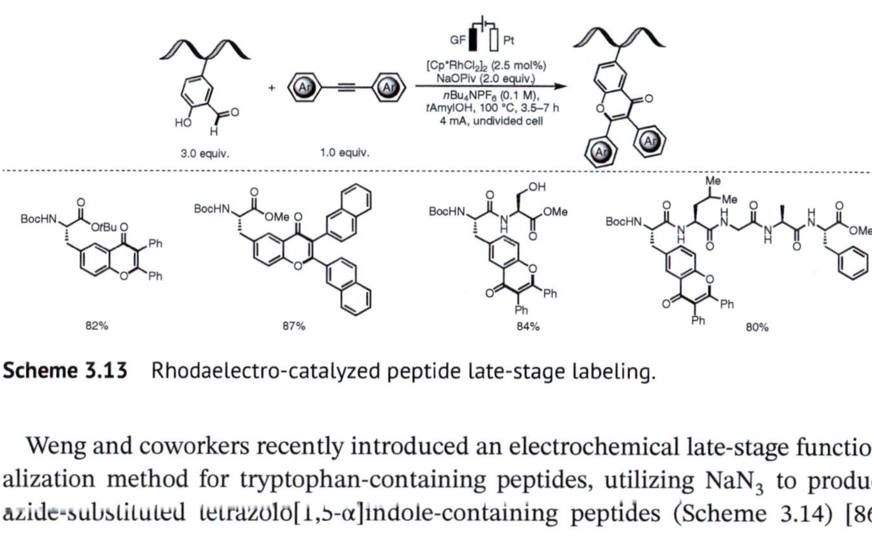

Scheme 3.13 Rhodaelectro-catalyzed peptide late-stage labeling.

Weng and coworkers recently introduced an electrochemical late-stage functionalization method for tryptophan-containing peptides, utilizing NaN_3 to produce azide-substituted tetrazolo[1,5-α]indole-containing peptides (Scheme 3.14) [86].

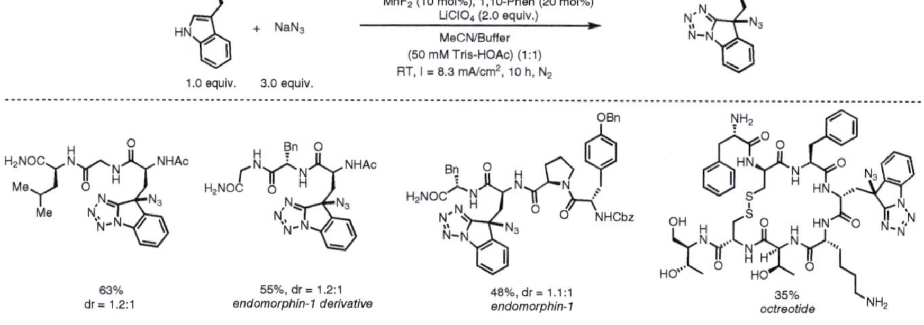

Scheme 3.14 Electrochemical oxidative azidation/heterocyclization of peptides.

This process employed an earth-abundant manganese catalyst under mild, buffered conditions, providing good functional-group tolerance and high site-selectivity. Furthermore, the azide-functionalized peptides could be further transformed into triazole derivatives via a copper-catalyzed "click" reaction. The reaction is proposed to proceed via manganese-mediated diazidation of the indole, followed by dehydrogenation and heterocyclization to yield the final products.

3.2.2 Electrochemical Late-Stage Functionalization of C(sp^3)—H Bonds

Benzylic C(sp^3)—H bonds are common in natural products and drug molecules. The relatively low bond dissociation energy of benzylic C—H bonds allows for high site-selective C—H functionalization in structurally complex molecules [87]. Significant progress has been made in the field of electrochemical benzylic C(sp^3)—H late-stage functionalization. The reaction mechanism typically involves (Scheme 3.15) anodic oxidation of the hydrocarbon substrate to produce an arene-centered radical cation, rapid proton transfer, a second electron transfer oxidation to form a benzylic cation, and subsequently nucleophile trapping. Meanwhile, hydrogen gas is generated at the cathode.

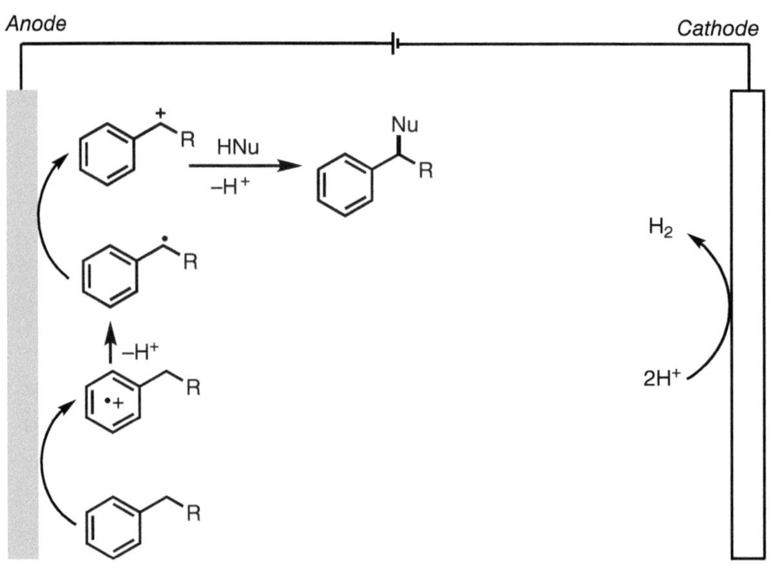

Scheme 3.15 Possible mechanism for electrochemical benzylic C—H functionalization.

The formyl group is a synthetically versatile functional group that can be converted to a wide range of functionalities. The oxygenation of methylarenes to benzaldehyde derivatives is of considerable practical approach for late-stage functionalization. However, the control of chemoselective oxidation with highly functionalized methylarenes remains a significant challenge due to the product overoxidation and selectivity issues for substrates featuring multiple oxidizable C—H bonds [88]. Recently, the Xu group developed an electrochemical method for site-selective oxidation of methyl benzoheterocycles to aromatic acetals (Scheme 3.16) [89]. These acetals can be readily hydrolyzed to the corresponding aldehydes either in a one-pot process or as a separate step. This electrooxidation method is compatible with a variety of functionalized benzoheterocycles and medicinally relevant molecules. Its utility was further demonstrated through the efficient synthesis of the antihypertensive drug telmisartan, where the key dimethyl acetal intermediate was prepared on a 14.2-g scale via a site-selective electrooxidation reaction.

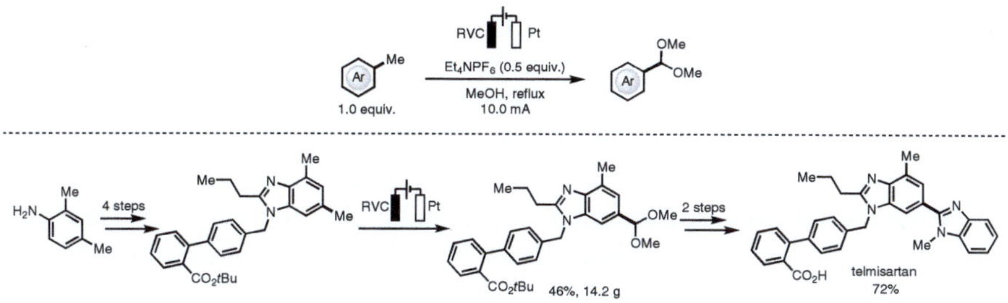

Scheme 3.16 Electrochemical late-stage oxidation of methylarenes.

The oxidation of methylarenes is generally ineffective for electron-neutral and electron-deficient arenes due to their elevated redox potentials, which result in poor selectivity or competitive solvent oxidation. To address these challenges, Stahl and coworkers developed an *N*-hydroxyphthalimide (NHPI)-mediated electrosynthetic method (Scheme 3.17) [90]. In this approach, proton-coupled electrochemical oxidation of NHPI generates the phthalimide-*N*-oxyl (PINO) radical, which acts as a hydrogen atom transfer (HAT) mediator and a metastable persistent radical to capture *in situ*-generated benzylic radicals. The PINOylation reaction operates at electrode potentials ~0.5–1.5 V lower than those required for direct electrolysis, thereby enabling the mediated electrolysis method to tolerate a diverse range of methylarenes with varying electronic properties and functional groups.

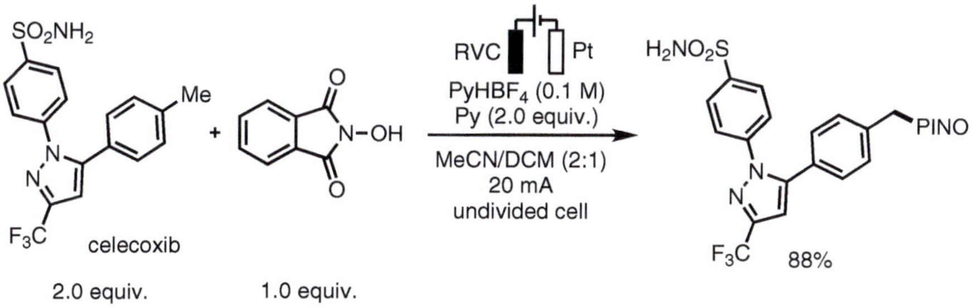

Scheme 3.17 Late-stage benzylic oxidation of methylarenes.

In 2021, Xu and coworkers reported on a site-selective electrochemical benzylic C—H amination utilizing Celebrex as the aminating source on a gram scale [91]. Meanwhile, the Ackermann group disclosed an effective method for electrochemical C—H aminations of 1,3-diarylpropenes via direct oxidative C(sp^3)—H functionalizations with various substituted amides, including the chiral auxiliary (−)-10,2-camphorsultam, as well as the sulfonamide drugs like Celebrex and Topiramate [92].

Furthermore, the Wang group achieved a comparable benzylic C—H amination of bioactive molecule substrates with an array of pyrazoles [93]. To enhance the electrolysis efficiency, 2,3-dichloro-5,6-dicyano-1,4-benzoquinone (DDQ) was utilized as a redox mediator. Subsequently, Ruan and coworkers demonstrated that azoles were effective amination reagents for electrochemical C—H/N—H cross-coupling reactions in an undivided cell [94]. The azolation was achieved with high efficiency and selectivity at primary, secondary, and even challenging tertiary benzylic positions in a range of structurally complex drug molecules.

Direct C(sp^3)—H isothiocyanation represents a straightforward strategy for introducing the versatile isothiocyanate functional group [95]. Recently, Guo and Wen disclosed an electrochemical late-stage benzylic C(sp^3)—H isothiocyanation using TMSNCS (Scheme 3.18) [96]. A wide range of drug and bioactive molecules underwent the isothiocyanation smoothly under mild conditions with high chemo- and regioselectivity. The high chemoselectivity was attributed to the facile isomerization of *in situ*-generated thiocyanates to isothiocyanates under electrolysis conditions.

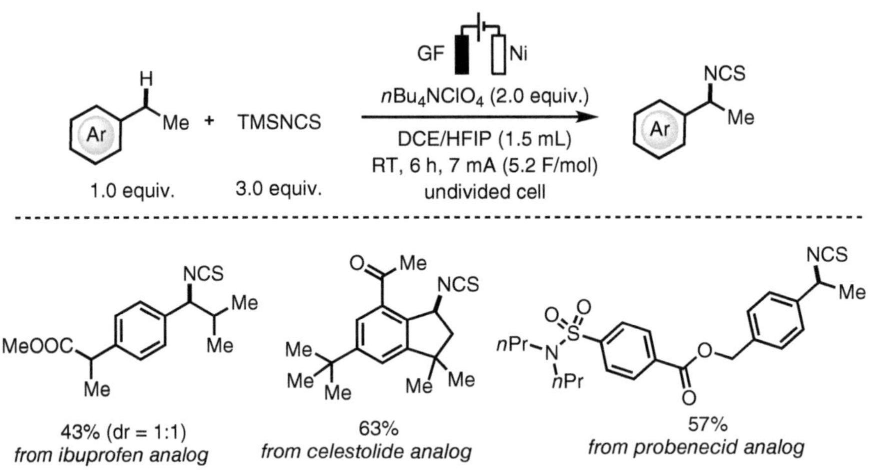

Scheme 3.18 Electrochemical late-stage benzylic C—H isothiocyanation.

The straightforward and efficient introduction of fluorine is of great importance in medicinal chemistry due to the unique properties of the C—F bond [97, 98]. In recent years, there has been a significant increase in interest in the electrochemical fluorination of C(sp^3)—H bonds using nucleophilic fluoride sources [99]. In this context, the Ackermann group developed a selective electrochemical C(sp^3)—H fluorination method employing readily available NEt_3·3HF instead of expensive electrophilic fluorine reagents (Scheme 3.19) [100]. This approach demonstrated broad functional group tolerance, paving the way for late-stage fluorination of bioactive drugs. The practical utility of the method was further validated by the successful fluorination of ibuprofen in a large-scale (2.46 g) synthesis.

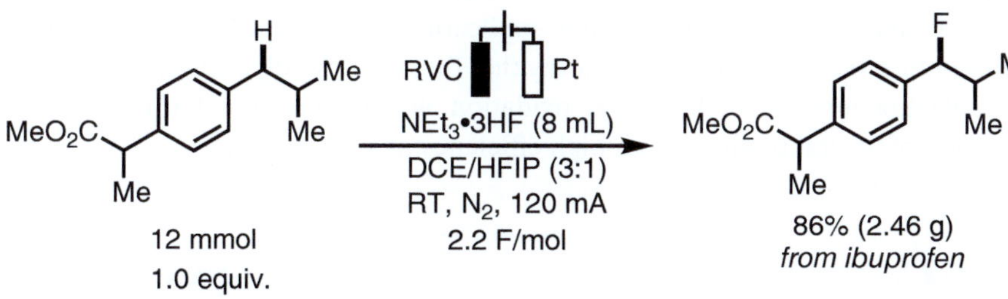

Scheme 3.19 Electrochemical late-stage benzylic C–H fluorination.

Ketones are versatile functional groups and are ubiquitous in natural products and biologically active compounds [101]. Recently, Liu and coworkers developed a sustainable protocol for the direct benzylic C—H bond oxidation of alkylarenes to synthesize ketones using *tert*-butyl hydroperoxide as both radical and oxygen source (Scheme 3.20) [102]. This approach was successfully applied to the late-stage functionalization of bioactive molecules, including celestolide, ibuprofen methyl ester, and papaverine, thereby providing synthetically useful yields without affecting other functional groups.

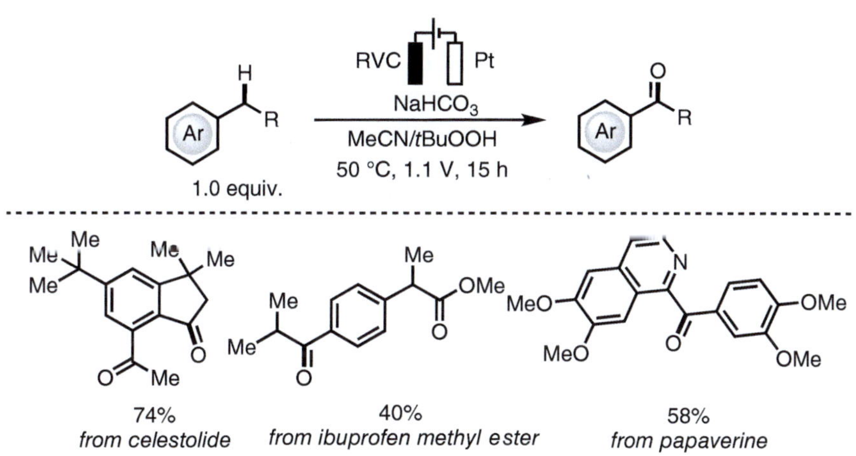

Scheme 3.20 Electrochemical late-stage benzylic C–H bonds oxidation.

Over the past decades, late-stage allylic C(sp^3)—H functionalization has attracted considerable interest among organic chemists, primarily due to the prevalence of carbon–carbon double bonds in natural products and drug molecules. The relatively low bond dissociation energy of the allylic C(sp^3)—H bond enables high site-selectivity in structurally complex molecules. Significant progress has been achieved in electrochemical late-stage allylic C(sp^3)—H functionalization [103–105]. In 2016, Baran and coworkers introduced an elegant electrochemical strategy for allylic C(sp^3)—H oxidation utilizing 20 mol% Cl_4NHPI as a redox mediator (Scheme 3.21) [106]. This method permitted the oxidation of a range of representative terpenes, producing monoterpenes, sesquiterpenes, diterpenes, triterpenes, and steroids with potential applications in the food, fragrance, and pharmaceutical industries. It is noteworthy that the user-friendly and robust nature of this electrochemical allylic C—H oxidation was demonstrated through the preparation of several products with high efficiency at a 100-g scale.

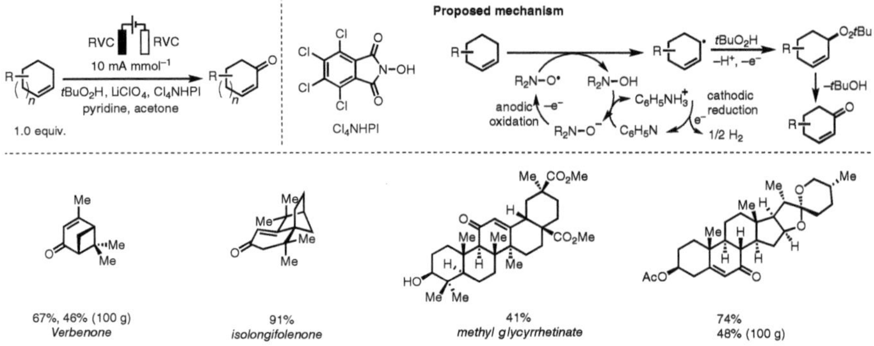

Scheme 3.21 Electrochemical late-stage allylic C(sp^3)−H oxidation.

Allylic amines are indispensable components in the synthesis of molecular structures and are frequently present in a multitude of biologically active compounds [107–109]. In 2021, Wickens and coworkers devised a straightforward electrochemical approach for synthesizing aliphatic allylic amines (Scheme 3.22) [110]. This reaction involves the electrochemical generation of a dicationic alkene-bisthianthrene adduct, formed between thianthrene (TT) and an alkene substrate. The investigation of the mechanism revealed that the crucial reactive intermediates are vinylthianthrenium salts.

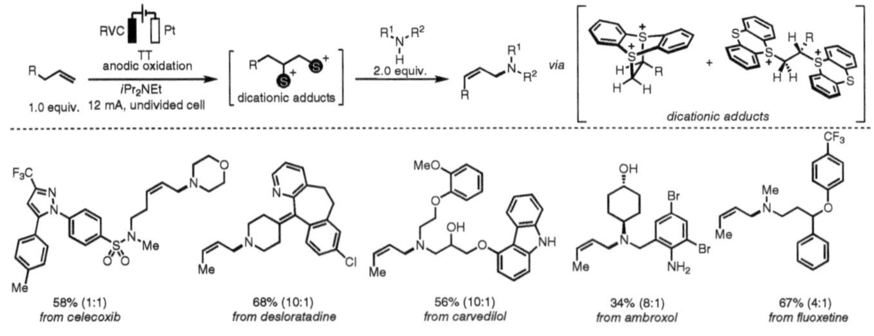

Scheme 3.22 Electrochemical late-stage allylic C(sp^3)−H amination.

The diversification of C(sp^3)—H bonds adjacent to carbonyl groups represents a fundamental and highly valuable transformation in the field of molecular chemistry. Notable examples include the Claisen condensation, aldol reactions, and the Mannich reaction. Recent advances have notably accelerated developments in this area, particularly in electrochemical late-stage functionalization and the synthesis of pharmaceutical derivatives. In 2020, Li and Song introduced a practical electrooxidative dehydrogenative cross-coupling method for ketones and xanthenes (Scheme 3.23) [111]. Mechanistic investigations revealed that the reaction begins with the anodic oxidation of xanthene, generating a stabilized carbocation. The intermediate subsequently reacts with a nucleophilic enol, forming the cross-coupling product.

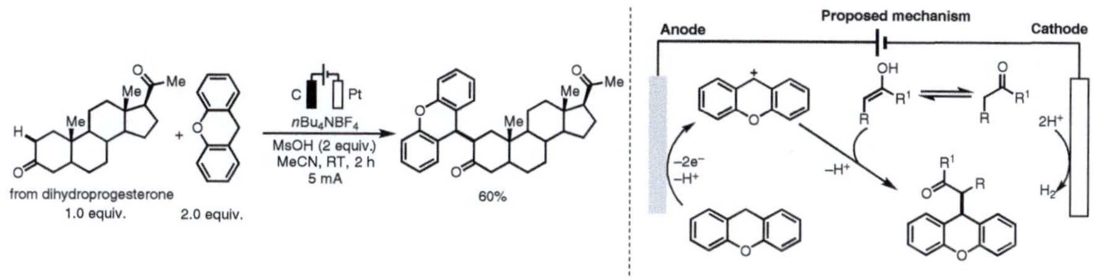

Scheme 3.23 Electrochemical dehydrogenative cross-coupling of ketones with xanthenes.

The desaturation of carbonyl compounds to enones represents a fundamental oxidation reaction that is widely utilized in organic synthesis [112]. Traditional methods for this transformation typically rely on transition metals, such as copper or palladium, or stoichiometric oxidative reagents [113–116]. In 2021, Baran and coworkers reported an operationally straightforward electrochemical method to achieve this transformation using enol phosphates or silanes, which are readily derived from carbonyl compounds (Scheme 3.24) [117]. This electrochemically driven desaturation (EDD) featured a broad substrate scope, accommodating a wide range of ketones and lactams. Notably, the method enabled late-stage, site-selective desaturation of structurally complex molecules, yielding enones in synthetically

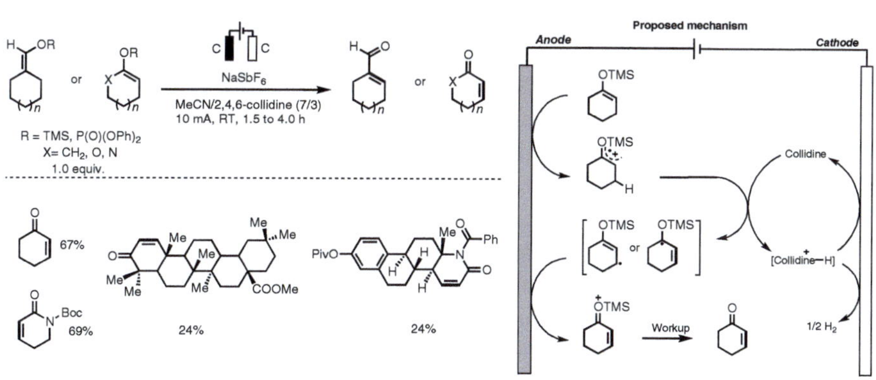

Scheme 3.24 Electrochemically driven late-stage desaturation of carbonyl compounds.

useful quantities. Mechanistic studies suggested a radical-based pathway involving two consecutive single-electron oxidations of the enol silane to form an oxonium ion, which undergoes hydrolysis to release the desired enone.

In 2019, the Meggers group combined organic electrosynthesis with asymmetric catalysis to develop a versatile, electricity-driven, chiral Lewis acid-catalyzed method for the asymmetric coupling of 2-acyl imidazoles with silyl enol ethers. This approach allows the synthesis of 1,4-dicarbonyl compounds, including those containing all-carbon quaternary stereocenters (Scheme 3.25) [118].

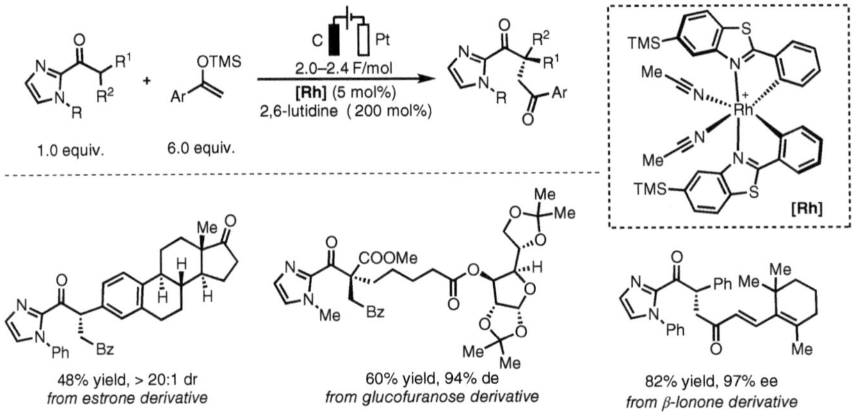

Scheme 3.25 Electrochemical late-stage α-C(sp^3)—H functionalization of carbonyls.

Thereafter, Meggers introduced a conceptually related approach for enantioselective α-C(sp^3)—H alkenylation of ketones using potassium alkenyl trifluoroborates [119]. Catalytic amounts of ferrocene were employed as the redox mediator, facilitating a key chiral rhodium-catalyzed single-electron transfer (SET) reaction that occurred homogeneously in solution rather than at the electrode surface, thereby enabling mild electrochemical conditions. The electrochemical late-stage functionalization alkenylation of complex molecules, including those derived from Oxepinac, abietic acid, and lithocholic acid, was achieved in excellent yields.

The electrochemical functionalization of C(sp^3)—H bonds adjacent to nitrogen atoms, exemplified by the well-established Shono oxidation, has found widespread applications in organic synthesis [120, 121]. However, it was traditionally limited to the functionalization of relatively simple compounds. Building on the classic Shono oxidation, Lin and Terrett recently reported a modular and practical strategy for electrochemical late-stage functionalization α-methylation of complex amine derivatives (Scheme 3.26) [122]. The electrooxidation generates *N*,*O*-acetals, which readily react with organozinc reagents, facilitating the efficient installation of methyl groups and other important functional groups. This improved electrochemical protocol is characterized by operational simplicity and high compatibility with various functional groups. It enables the site-selective late-stage methylation of a range of bioactive compounds with high efficiency. Notably, a TRPA1 inhibitor, which has been studied for its "magic methyl" effect, demonstrating a >10-fold

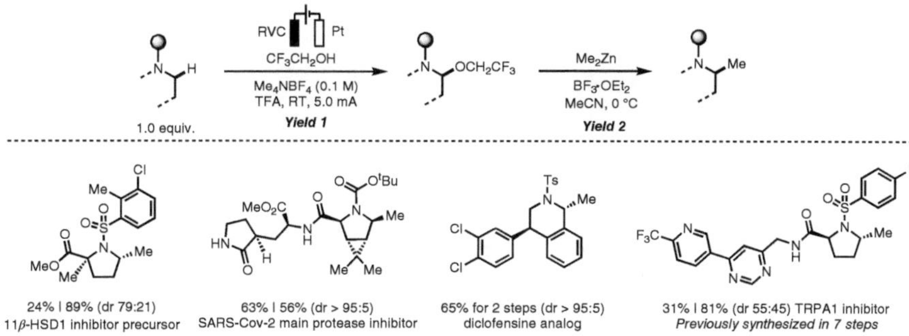

Scheme 3.26 Electrochemical late-stage α-methylation of amines.

increase in potency, was previously synthesized through a *de novo* route in seven steps [123]. In contrast, this compound can be efficiently prepared from its parent inhibitor using the electrochemical approach, highlighting the advantages of this method.

Electrochemical dehydrogenation via α-C(sp^3)—H functionalization of amines represents a significant organic transformation for the synthesis of ubiquitous unsaturated compounds. In 2018, the Lei group introduced a TEMPO-mediated dehydrogenation of *N*-heterocycles, enabling the formation of various five- and six-membered nitrogen heteroarenes without the need for sacrificial hydrogen acceptors [124]. More recently, Qiu and coworkers reported a simple and robust electrochemical approach for the desaturative β-C(sp^3)—H functionalization of cyclic amines [125]. This transformation occurs through multiple single-electron oxidation steps, with catalytic amounts of ferrocene serving as a redox mediator.

The precise and selective activation of oxidation-sensitive sulfur-containing compounds presents a significant challenge due to their inherent reactivity and complex valence states [126]. In 2021, Lei and coworkers reported an electrochemical protocol for constructing α-acyloxy sulfides (Scheme 3.27) [127]. Mechanistic studies suggested a synergistic effect in the self-assembly-driven C(sp^3)—H/O—H coupling pathway. Specifically, the formation of an adduct between sulfide, acetic acid, and methanol is facilitated by hydrogen bonding between acetic acid and the sulfur atom, which enables a SET oxidation of the sulfide. The methanol selectively captures the proton, forming an intermediate with high regioselectivity. Subsequently, a thionium ion is generated through the loss of a proton and electron, and the desired product is formed following a nucleophilic attack by acetic acid on the thionium ion.

The electrolysis of unactivated C(sp^3)—H bonds typically exhibits a high redox potential exceeding 3.0 V *vs.* SCE, which presents a significant challenge in direct electrolysis [128]. The oxidation of other functionalities or solvents often occurs before the desired C—H oxidation of simple alkanes can take place. Despite these challenges, significant progress has been made in electrochemical late-stage functionalization of unactivated C(sp^3)—H bonds through the use of redox mediators or transition metal catalysts [129, 130].

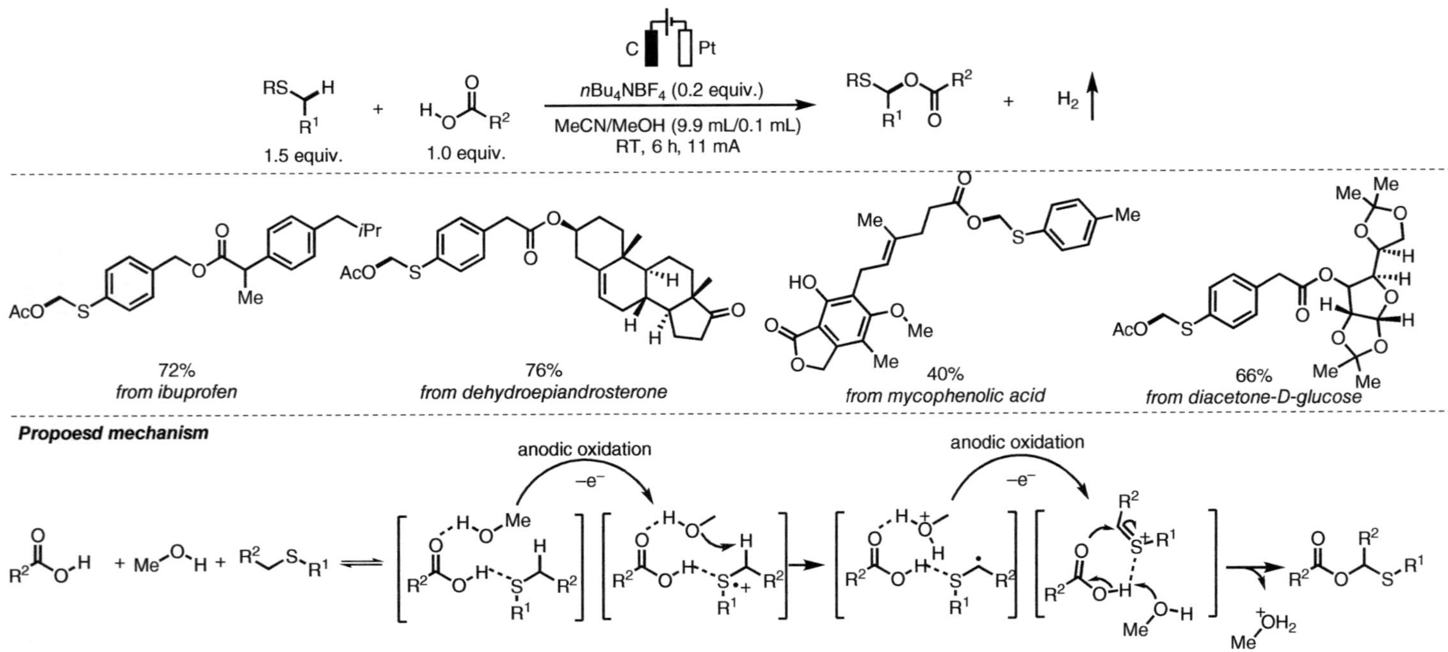

Scheme 3.27 Electrochemical late-stage α-C(sp^3)—H acyloxylation of sulfides.

Baran introduced a practical electrochemical oxidation method for unactivated C—H bonds (Scheme 3.28) [129]. The identification of an appropriate redox mediator was crucial for achieving high yields and chemoselectivity. Mechanistically, *in situ* anodic oxidation generates a quinuclidine radical cation, which serves as a HAT reagent. Meanwhile, oxygen plays a role in the subsequent aerobic oxidation step.

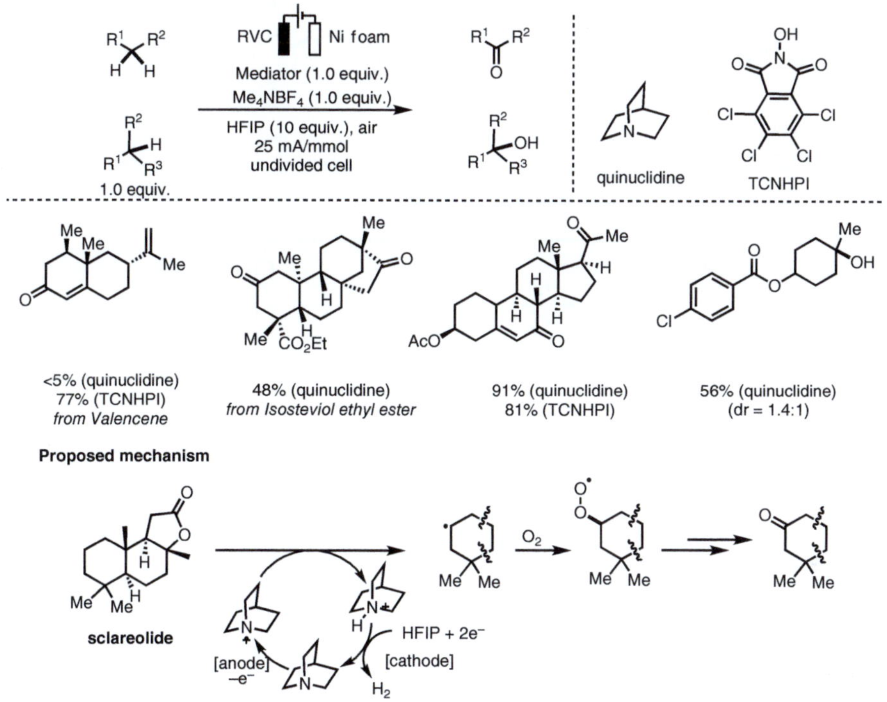

Scheme 3.28 Electrochemical late-stage oxidation of unactivated $C(sp^3)$—H bonds.

Organic azides are important intermediates in various fields, including medicinal chemistry, peptide chemistry, and molecular biology [131–134]. In this context, the Ackermann group introduced a manganese-electrocatalyzed C—H azidation of unactivated $C(sp^3)$—H bonds, using NaN_3 as a user-friendly nitrogen source and traceless electrons as the sole redox reagent (Scheme 3.29) [130]. Detailed mechanistic studies revealed a unique manganese(III/IV) catalytic cycle, which prevents overoxidation to a carbocation, thereby minimizing undesired side reactions leading to oxygenated products.

3.2.3 Electrochemical Late-stage of C(sp)—H bonds

Oxidative carbonylation of alkynes is a significant transformation in molecular synthesis, typically using O_2 as the oxidant. Nevertheless, the high explosiveness of CO/O_2 gas mixtures represents a significant limitation to the scalable

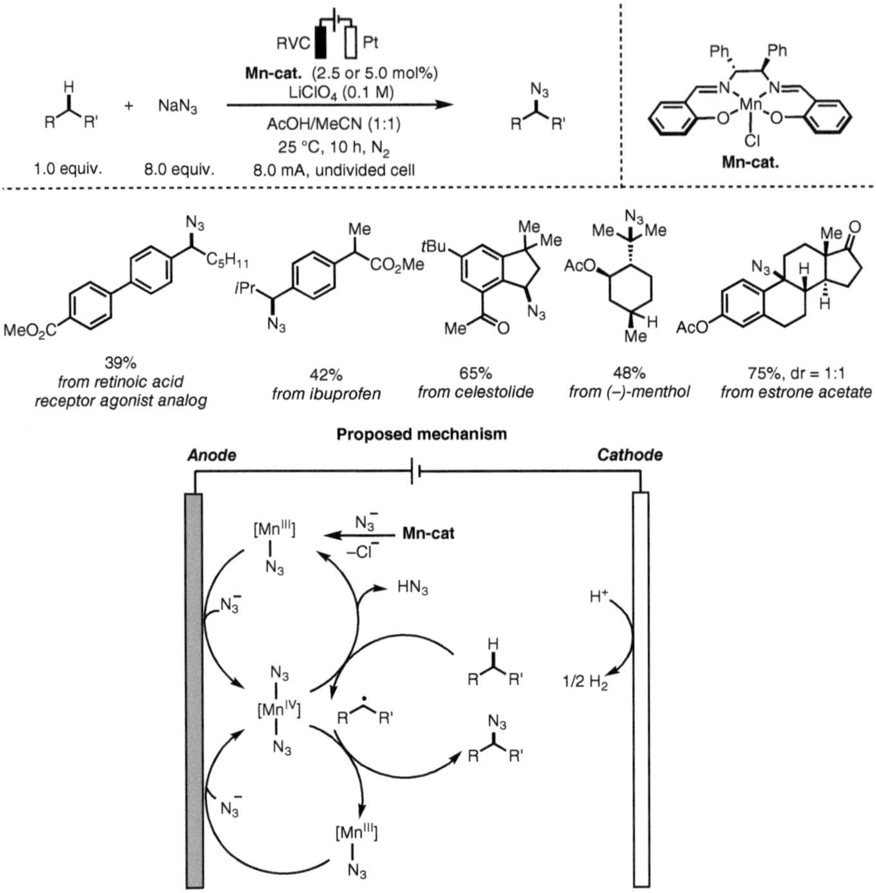

Scheme 3.29 Electrochemical late-stage C(sp^3)–H azidation.

application of this process. In 2020, the Lei group introduced an electrooxidative palladium-catalyzed carbonylation of alkynes to 2-ynamides under copper- and O_2-free conditions (Scheme 3.30) [135]. Electrochemical late-stage functionalization of propyzamide with CO (1 atm) and ammonium nitrate yielded the corresponding propiolamide in 63% yield. Additionally, drug molecules such

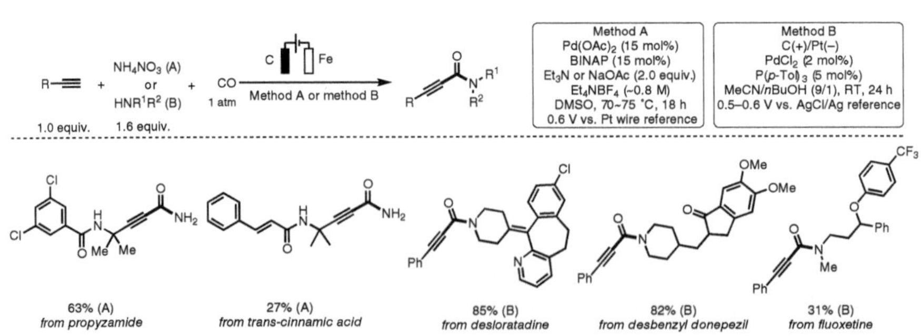

Scheme 3.30 Electrochemical palladium-catalyzed aminocarbonylation of terminal alkynes.

as desloratadine, fluoxetine, and desbenzyl donepezil underwent electrochemical late-stage functionalization, yielding the desired 2-ynamides in satisfactory yields.

In 2021, Lei and coworkers extended their work by employing arylhydrazines instead of amines, achieving the electrochemical palladium-catalyzed oxidative carbonylation of alkynes to synthesize ynones [136]. This method offers an alternative to the carbonylative Sonogashira–Hagihara reaction. The electrochemical late-stage functionalization of bioactive molecules, including propyzamide, estrone, naproxen, ibuprofen, and levulinic acid, yielded the corresponding ynones in excellent yields.

More recently, Xie and coworkers reported an electrochemical gold-catalyzed C(sp)—C(sp^2) coupling reaction between complex alkynes and arylhydrazines (Scheme 3.31) [137]. Mechanistic studies indicated that anodic oxidation of the aromatic hydrazine generates an aryl radical, which recombines with gold(I) and undergoes further anodic oxidation to form the Ar—Au(III) species. This species then undergoes subsequent σ-activation of the alkyne.

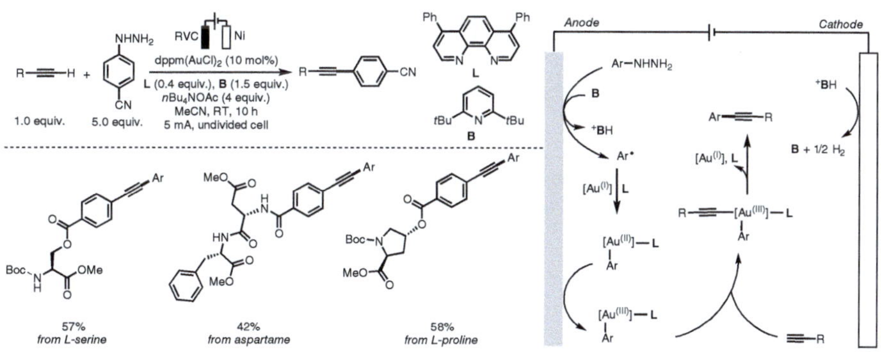

Scheme 3.31 Electrochemical gold-catalyzed oxidative C(sp)—C(sp^2) coupling.

3.3 Electrochemical Late-Stage Functionalization Through Electrochemical Substitution of Functional Groups

While electrochemical late-stage functionalization reactions were generally defined as the functionalization of those unreactive C—H bonds on structurally complex molecules, there are recently developed methodologies, such as functional group addition and substitution, which could be considered as a branch of electrochemical late-stage functionalization, complementary to C—H functionalizations [138]. Electrochemical interconversion of common organic functionalities offers unique potential for late-stage derivatization of complex organic molecules. In the succeeding section, the robustness and versatility of electrochemical late-stage functional group modification strategies will be discussed.

3.3.1 Electrochemical Late-Stage Functionalization of Alkenes and Alkynes

Olefins are common reactive structural motifs, of which the reactivity have often been garnered for incorporating new functional groups in biologically relevant molecules and natural products [139–148]. Lin reported on an electrooxidative heterodifunctionalization of olefins enabled by anodic oxidation of CF_3SO_2Na (Scheme 3.32) [149]. The anodically generated trifluoromethyl radical attacked the terminal olefin to form a secondary alkyl radical intermediate, which then trapped a chloride radical to form the heterodifunctionalized product. The use of catalytic $Mn(OAc)_2$ enhanced the electrochemical process through the formation of an alleged Mn(III) radical chlorinating agent, which assisted the chloride radical recombination step. This electrocatalytic process was employed for the late-stage functionalization of several natural product analogs. Similarly, electrochemical late-stage hydroetherification via cobalt catalysis and hydrofluorination via palladium catalysis were uncovered subsequently by Kim and Shin [150, 151].

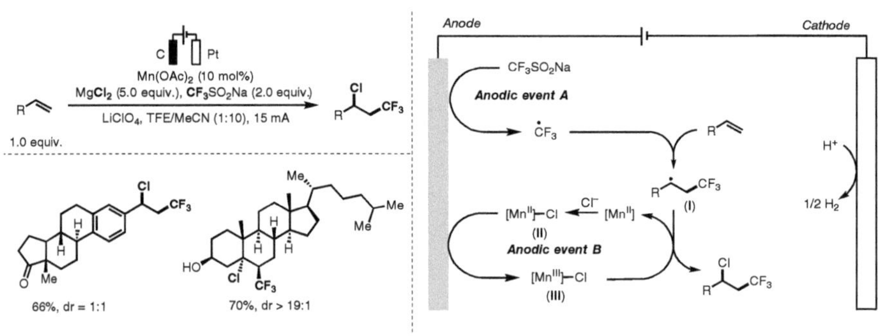

Scheme 3.32 Electrooxidative chlorotrifluoromethylation of olefins.

In addition to anodic oxidative late-stage functionalization, Baran and coworkers have developed an electroreductive cobalt-catalyzed regioselective olefin isomerization involving a transition metal hydride intermediate. This strategy was applied to the late-stage derivatization of structurally complex organic architectures (Scheme 3.33) [152]. While the merger of electrosynthesis and transition metal catalysis offers significant advantages for late-stage functionalization on olefin-based complex molecules, transition metal-free electrochemical late-stage functionalization of alkenes has also been extensively studied, showcasing excellent compatibility with biologically relevant olefins [153–155].

Alkynes are present in a wide range of natural products and biologically relevant molecules [156], making late-stage alkynylation and diversification of alkynes attractive in the drug-discovery regime. Wang reported an alkynylation of 4-acyl-1,4-dihydropyridines (DHPs) into ynones through anodic oxidation, employing boron-doped diamond (BDD) as the electrode material (Scheme 3.34) [157], thus diversifying several pharmaceutical and biologically relevant molecules.

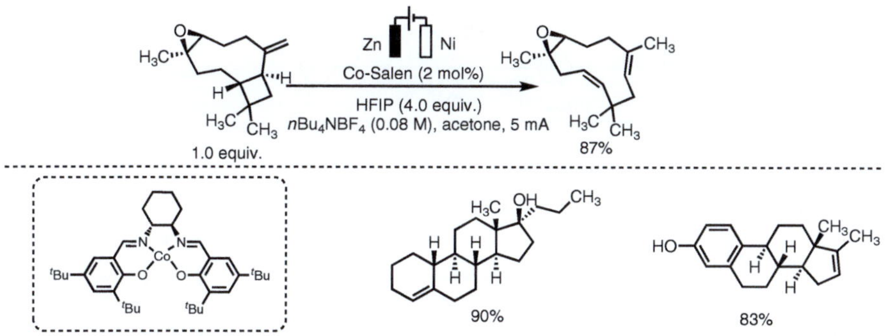

Scheme 3.33 Electroreductive cobalt-catalyzed late-stage functionalization of olefins.

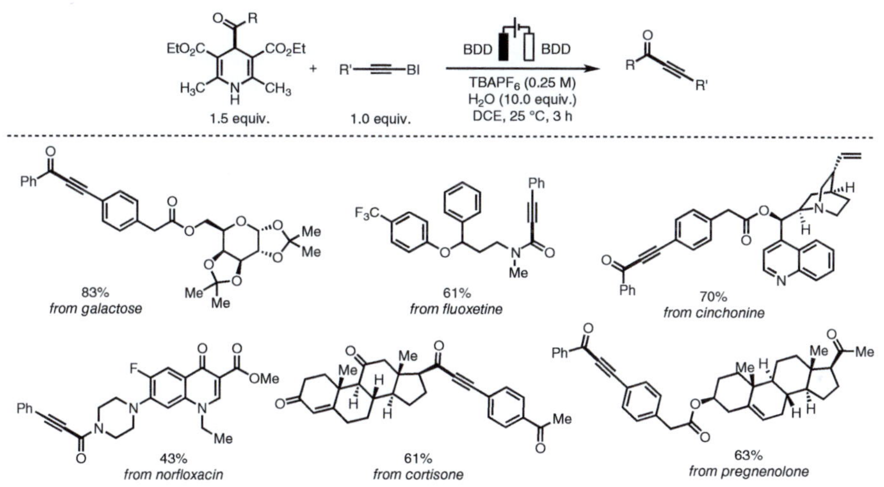

Scheme 3.34 Electrochemical synthesis of ynones.

3.3.2 Electrochemical Late-Stage Functionalization of Organic Halides

For many years, organic halides have served as convenient synthetic handles to access a wide range of functionalities in complex molecules through transition metal catalysis [158–160]. In particular, the exploitation of electrochemical conditions has enabled the transformations of these halides in a more sustainable manner. Baran reported a versatile electrochemical nickel-catalyzed amination of aryl halides (Scheme 3.35) [161], which has been demonstrated viable with the amination of diverse amino acids, peptides, and sugar derivatives. The robustness of the electrochemical nickel-catalyzed strategy was further validated by the amination of anilines, sulfonamides, sulfoximines, carbamates, imines, and even tertiary amines as nucleophiles [162, 163].

The electrochemical nickel-catalyzed functionalization was not limited to organic halides, the corresponding pseudohalides were suitable substrates as well [164–166]. Wang and coworkers reported an electrochemical nickel-catalyzed deoxygenative

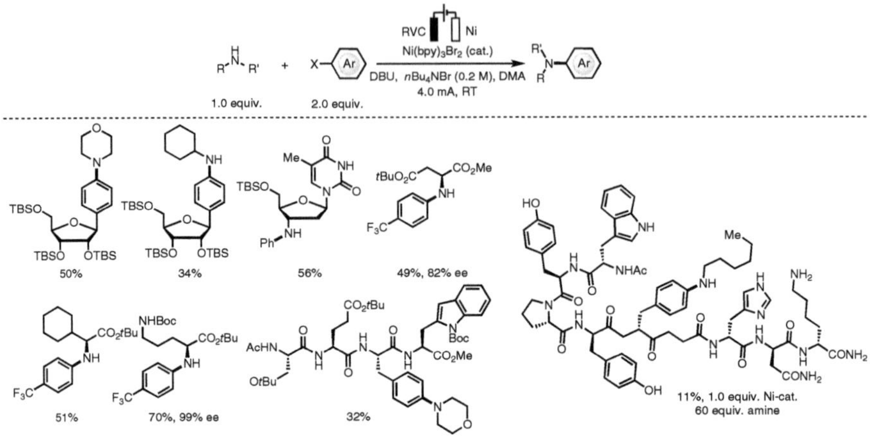

Scheme 3.35 Nickel-catalyzed electrochemical amination of aryl halides.

thiolation of bio-relevant ketones (Scheme 3.36) [167], which showed excellent functional group compatibility and enabled the late-stage diversification of a glucide derivative and various steroid analogs.

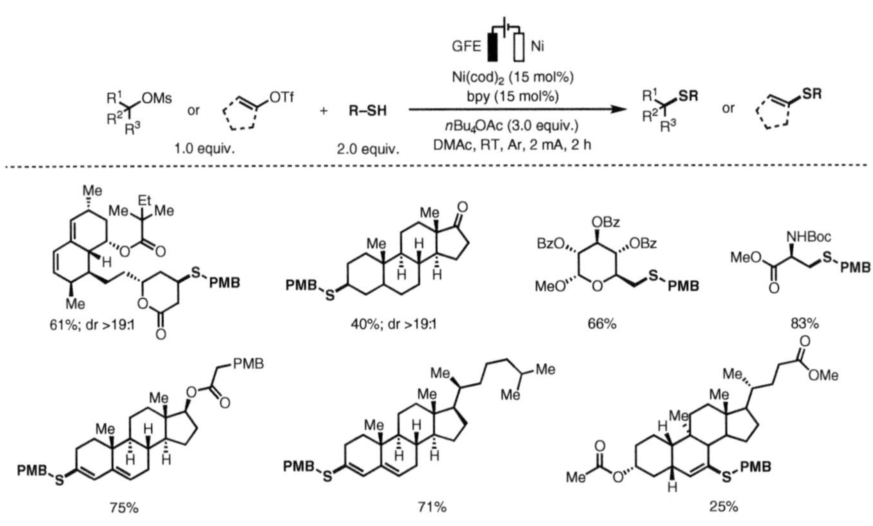

Scheme 3.36 Electrochemical nickel-catalyzed deoxygenative thiolation reactions.

A practical electroreductive borylation of organic halides was developed by Qi and Lu (Scheme 3.37) [168]. The reduction of organic halides generated an alkyl radical, which was then intercepted with the diborane to construct the borylated product. The efficacy of this method was demonstrated through the late-stage borylation of natural products and drug analogs.

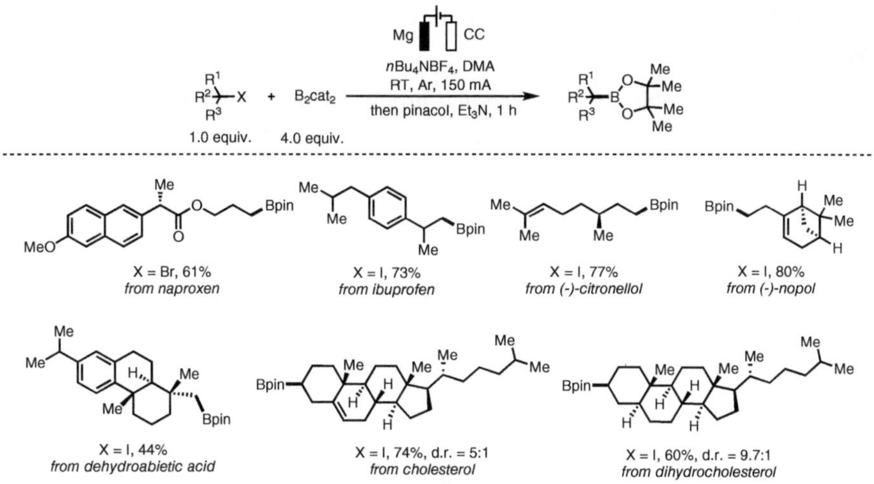

Scheme 3.37 Electrochemical late-stage borylation of alkyl halides.

C(sp^3)-rich organic structural motifs have potentially enhanced the biological activity of drug candidates in clinical trials [169]. In order to construct C(sp^3)—C(sp^3) bonds, Lin and coworkers developed an electroreductive approach using alkyl halides as the alkyl source (Scheme 3.38) [170]. The cathodically generated carbanion from highly substituted alkyl halide exhibited a privileged coupling with less substituted alkyl halide, thus allowing highly chemoselective C(sp^3)—C(sp^3) bond formation. Notably, the mild electrochemical reaction conditions allowed the late-stage modification of complex organic molecules and drug derivatives.

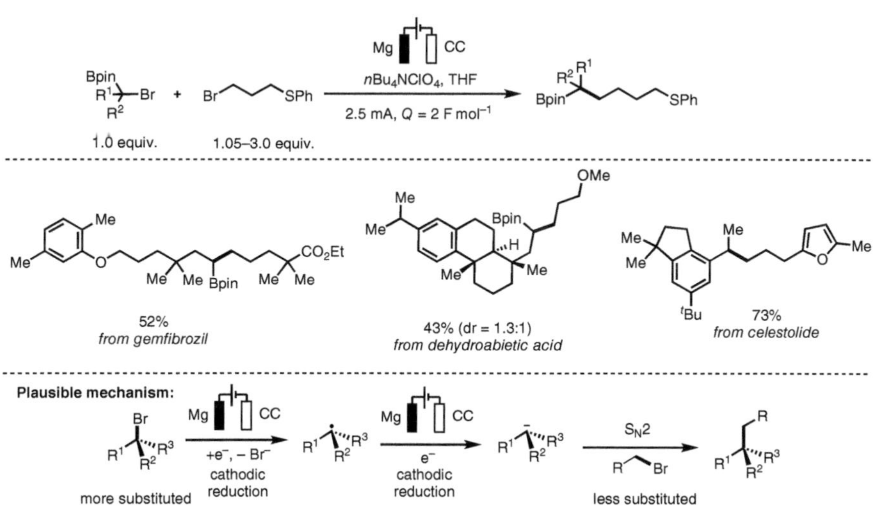

Scheme 3.38 Electrochemically driven late-stage C(sp^3)−C(sp^3) bond formation.

3.3.3 Electrochemical Late-Stage Functionalization of Organic Carboxylic Acids

Carboxylic acids are ubiquitous functional groups in natural products, bioactive molecules, and pharmaceuticals [171]. Therefore, garnering the properties of these carboxylic acid functionalities by electrochemical decarboxylation has been an important strategy for the late-stage functionalization of valuable organic molecules [2, 172]. Baran disclosed an electrochemical decarboxylative assembly of hindered aliphatic dialkylethers (Scheme 3.39) [173], which successfully accomplished late-stage functionalization of a variety of complex and biologically relevant organic molecules. The electrochemical decarboxylation method was further found viable for the synthesis of various bioactive peptides, including biseokeaniamide analogs [174].

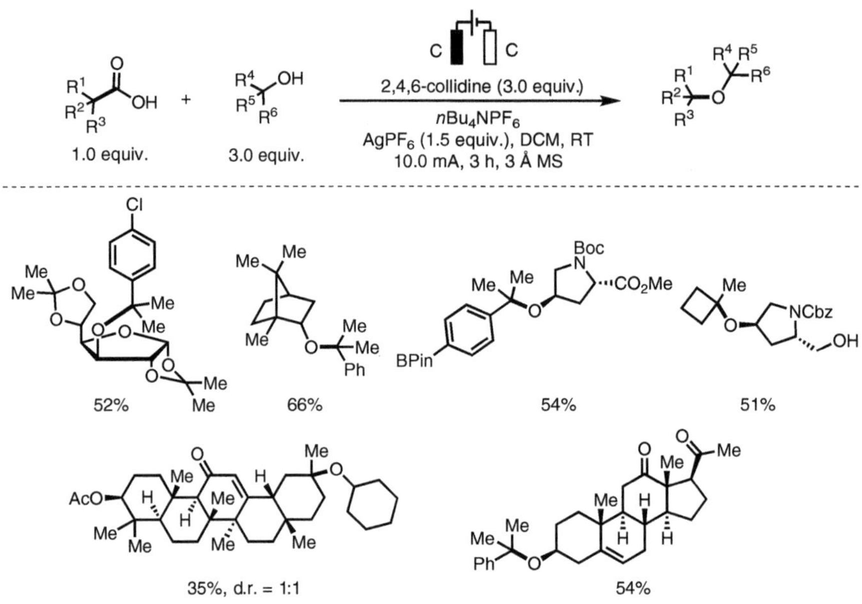

Scheme 3.39 Decarboxylative synthesis of sterically hindered ethers.

3.3.4 Electrochemical Late-Stage Functionalization of Alcohols

Given their prevalence as motifs in a wide variety of natural products, pharmaceuticals, and biologically relevant molecules [175], alcohols and their analogs have offered electrochemical alcohol derivatization great practicability for late-stage functionalizations [176]. An early example by Ohmori and coworkers has demonstrated that anodic oxidation of triphenylphosphine could lead to an efficient C—O bond activation through alkoxy phosphonium intermediate (Scheme 3.40) [177–179]. This versatile electrochemical process has allowed the late-stage deoxygenative functionalization of glycosides with fluoride and chloride nucleophiles [180]. Recently, electrochemical deoxygenative C—N bond formation reaction was also reported, affording the glycosylation of azoles under mild conditions [181].

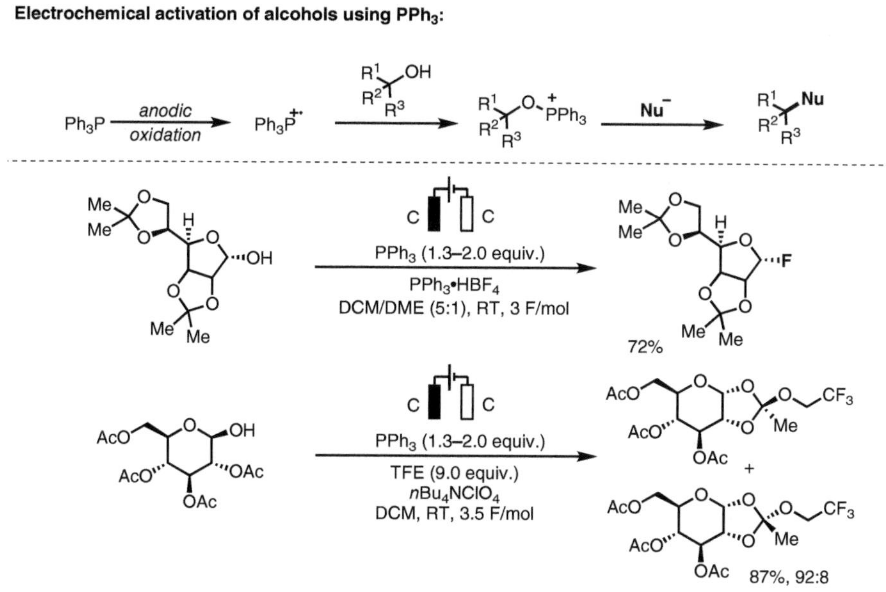

Scheme 3.40 Late-stage dehydroxylative halogenation and alkoxylation.

3.4 Photoelectrochemical Late-Stage Functionalization of Drug Molecules

Photoelectrochemical synthesis represents an enabling approach that integrates the principles of photochemistry and electrocatalysis, thereby facilitating efficient, selective, and environmentally benign organic transformations [182–188]. The method employs a synergistic approach, utilizing light and electrical energy in conjunction with one another to facilitate chemical reactions through the introduction of electrocatalytic elements alongside photocatalysis. Photoelectrochemistry enables redox transformations under milder electrode potentials, reducing the use of hazardous redox reagents and enabling precise regulation of reaction pathways and intermediate generation, thereby enhancing chemical selectivity and efficiency. This approach is particularly well-suited for constructing complex molecules and achieving highly selective transformations, offering broad application prospects in pharmaceutical chemistry, materials science, environmental remediation, and energy conversion [189–198].

3.4.1 Photoelectrochemical Late-Stage Functionalization for C—C Bond Formation

In 2020, Ackermann developed a mild photoelectrochemical C(sp^2)—H trifluoromethylation of arenes with CF_3SO_2Na (Scheme 3.41) [199]. This strategy featured a broad substrate scope and high functional groups tolerance. The late-stage functionalization of natural products, including pentoxifylline, doxofylline, theobromine, methyl estrone, and tryptophan, occurred efficiently. Notably,

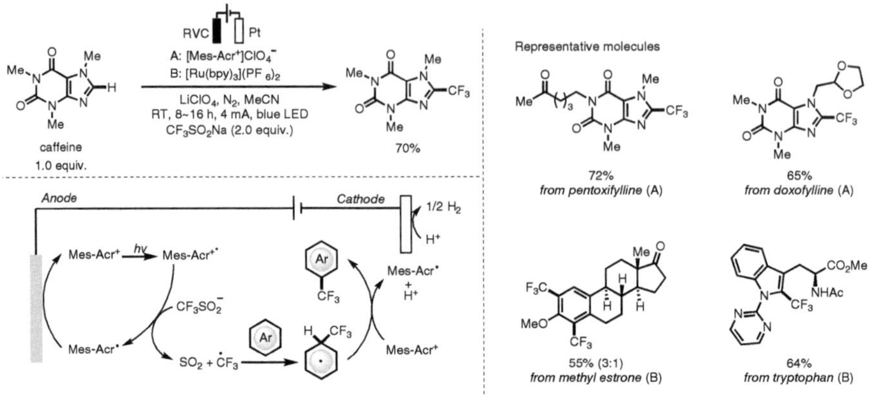

Scheme 3.41 Photoelectrochemical C(sp^2)—H trifluoromethylation.

this photoelectrochemical transformation was also achieved in a flow setup with operationally simple online NMR monitoring. The results of the mechanistic studies indicated that the irradiation of the photocatalyst Mes-Acr$^+$ resulted in the formation of its highly oxidizing excited state, Mes-Acr^{+*}. Subsequently, a SET occurred between CF_3SO_2Na and Mes-Acr^{+*}, resulting in the formation of the acridinyl radical and a controlled sulfinate radical. This radical was rapidly converted to a fluoroalkyl radical through the process of cleavage and the release of SO_2. The stable radical was oxidized at the anode to regenerate Mes-Acr$^+$, while the alkyl radical was trapped by the heteroarene, resulting in the formation of a radical cation. This cation subsequently underwent a proton and electron transfer, leading to the final product.

Xu and coworkers developed a photoelectrochemical approach for the late-stage C(sp^2)—H alkylation of bioactive heteroarenes with stable and readily available organotrifluoroborates (Scheme 3.42) [200]. This approach relied on the generation of alkyl radicals from organotrifluoroborates without the use of an exogenous chemical oxidant, enabling the functionalization of various heteroarenes with

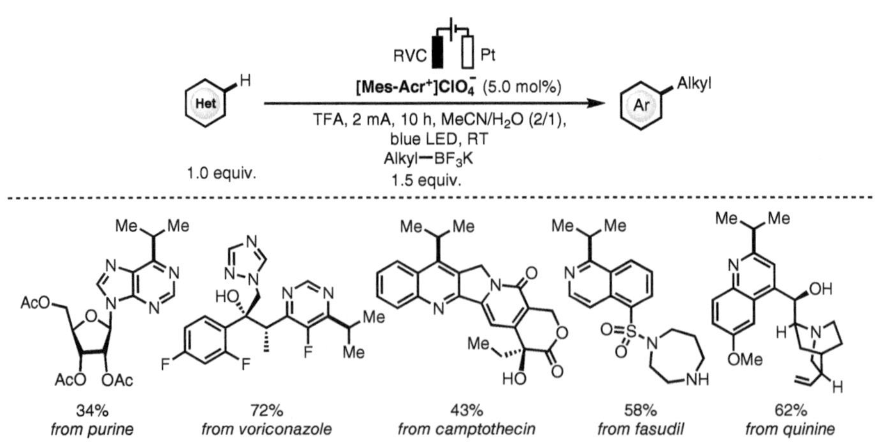

Scheme 3.42 Photoelectrochemical C—H alkylation.

excellent site- and chemoselectivity. Notably, even the challenging α-alkoxyl and tertiary radicals also reacted successfully under mild conditions [201]. The reaction has been effectively applied to the modification of natural products and drug molecules, including purine, voriconazole, camptothecin, fasudil, and quinine. The catalytic cycle is initiated by the irradiation and oxidation of Mes-Acr$^+$, which renders it susceptible to SET with organotrifluoroborates.

In 2020, Xu and coworkers described a photoelectrocatalytic decarboxylative C(sp^2)—H alkylation of heteroarenes catalyzed by $CeCl_3$ (Scheme 3.43) [202]. The carboxylic acids were employed as the alkyl source. These reactions are initiated by a photoinduced ligand-to-metal charge transfer (LMCT), which, upon decarboxylation, generates alkyl radicals. Subsequently, these radicals are trapped with the protonated lepidine, resulting in the production of the desired product. The method is advantageous in that it can be scaled up to produce quantities of up to 10 g and applied to the direct C—H alkylation of drug molecules, including fasudil, quinoxyfen, quinine, and voriconazole.

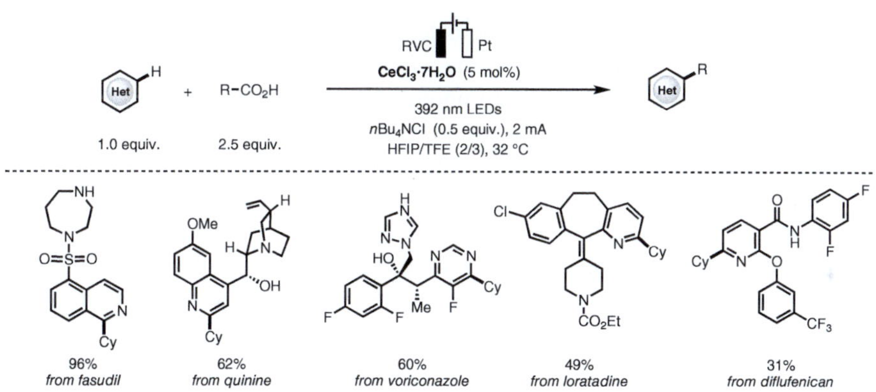

Scheme 3.43 Photoelectrochemical decarboxylative C—H alkylation.

Subsequently, the same research group identified an innovative method for dehydrogenative alkylation of heteroarenes with aliphatic C(sp^3)—H bonds under electrophotocatalytic conditions (Scheme 3.44) [203]. This approach eliminated the need

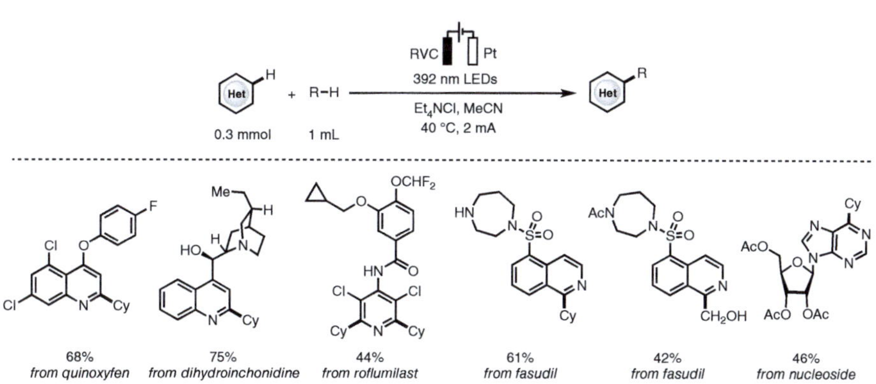

Scheme 3.44 Photoelectrochemical C—H alkylation with alkanes.

for transition metal catalysts or chemical oxidants, enabling efficient coupling of diverse C(sp^3)—H donors with a broad spectrum of heteroaromatic drug molecules, such as quinoxyfen, roflumilast, and fasudil.

The Katritzky salts have demonstrated their versatility as alkylation reagents in various transformations, exhibiting stability when used in a bench setting. Building on this foundation, Chen and coworkers recently developed a photoelectrocatalytic deaminative alkylation strategy (Scheme 3.45) [204]. Investigations into the mechanism revealed that the homogeneous dispersion of the photocatalyst is crucial for the efficient generation of alkyl radicals. Notably, the practicality and robustness of this approach were demonstrated through the late-stage functionalization of a wide array of bioactive molecules, including fasudil, quinoxyfen, and camptothecin, highlighting its broad applicability.

3.4.2 Photoelectrochemical Late-Stage Functionalization for C—X Bond Formation

In 2023, Wang and Hou presented a photoelectrocatalytic C—H silylation method for heteroarenes via dehydrogenative cross-coupling with H_2 evolution, utilizing 9,10-phenanthrenequinone (PQ) as the photoelectrocatalyst (Scheme 3.46) [205]. This approach eliminates the need for external oxidants or HAT reagents, achieving excellent site-selectivity and desirable yields across a wide range of heteroaromatic substrates. It is noteworthy that the late-stage photoelectrocatalytic C—H silylation strategy has been demonstrated to be highly effective in the synthesis of heteroaromatic bioactive molecules, including fasudil, cinchonidine, famciclovir, desloratadine, fenazaquin, and purine. This highlights the broad applicability and synthetic utility of this approach.

In 2021, Lambert reported a versatile method for the photoelectrocatalytic diamination of vicinal C—H bonds using trisaminocyclopropenium (TAC) ion as the catalyst, without requiring external oxidants (Scheme 3.47) [206]. Under photoelectrochemical conditions, anodic oxidation activates the TAC catalyst into a stable radical dication, while the cathodic reaction reduces protons to H_2. Upon illumination, the TAC radical dication forms a highly oxidizing photoexcited intermediate, which undergoes electron transfer with olefins. Subsequently, a Ritter-type C—H bond functionalization is conducted using acetonitrile as both the nitrogen source and the solvent. Notably, the nature of the electrolyte influences the product outcome. Under mild electrochemical conditions with visible light, either 3,4-dihydroimidazoles or 2-oxazolines can be obtained. This innovative approach facilitated the difunctionalization of numerous medicinal compounds with antitumor and antiviral activity, demonstrating its potential for complex molecular synthesis.

Lambert and Ye subsequently devised a photoelectrocatalytic methodology for the selective oxygenation of two or three adjacent C(sp^3)—H bonds, employing TAC ion intermediates as efficacious oxidative catalysts. This approach demonstrated exceptional selectivity, effectively suppressing overoxidative decomposition (Scheme 3.48) [207]. It is noteworthy that the selection of acid determines the extent

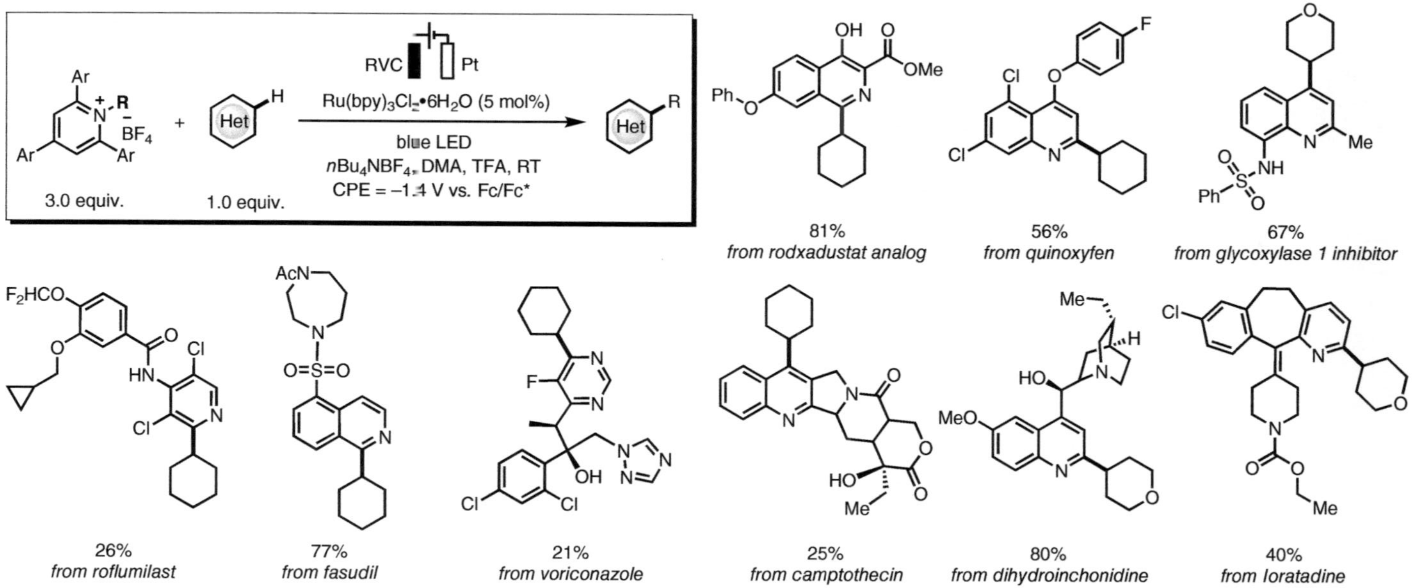

Scheme 3.45 Photoelectrochemical C−H alkylation using bench-stable Katritzky salts.

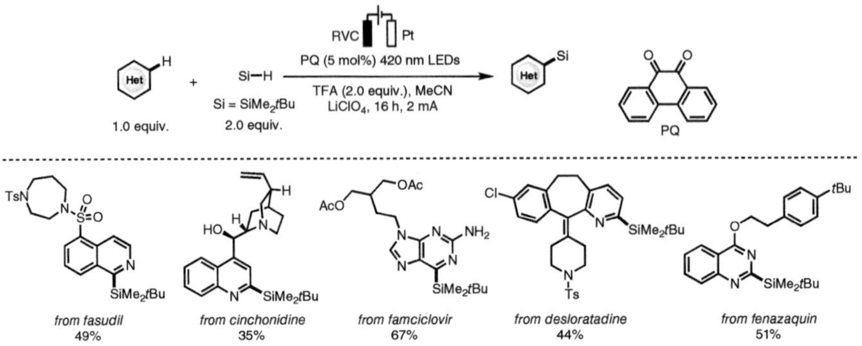

Scheme 3.46 Photoelectrochemical C—H silylation of heteroarenes.

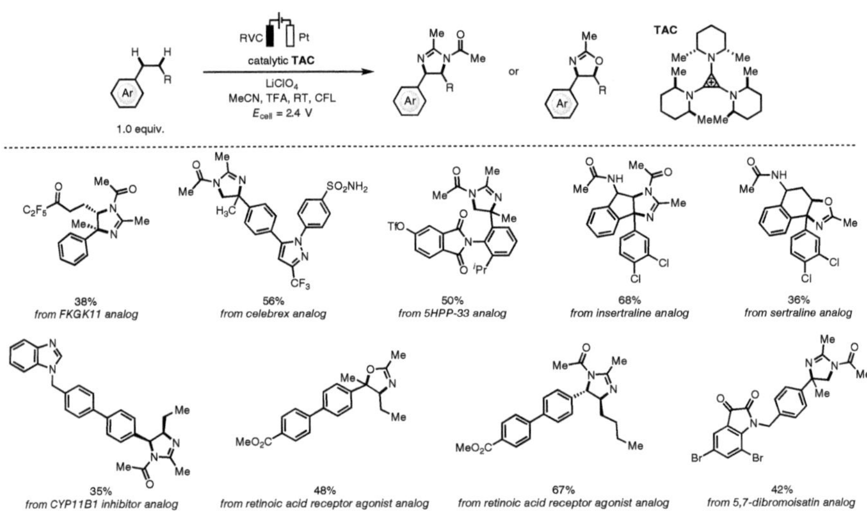

Scheme 3.47 Photoelectrochemical diamination of C—H bond.

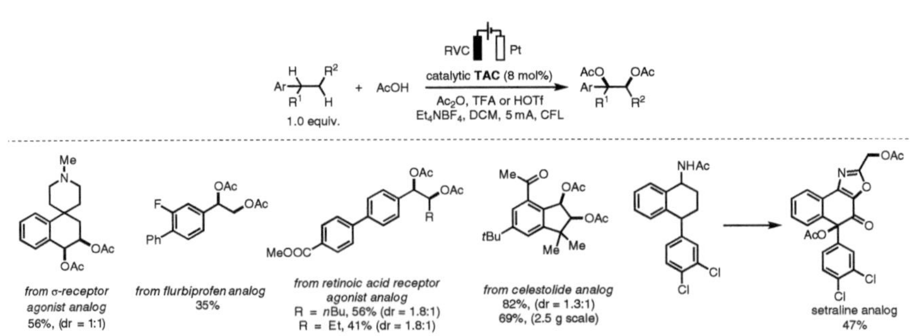

Scheme 3.48 Photoelectrocatalytic C—H bonds oxygenation.

of oxygenation, thereby enabling the selective synthesis of two or three contiguous C—O bonds. The versatility and applicability of this method were demonstrated through the late-stage modification of bioactive molecules, showcasing its utility in complex molecular synthesis.

The azide group is a highly versatile functional group that can be readily reduced to free primary amines, making it invaluable in the context of pharmaceutical discovery and late-stage functionalization. In 2020, Lei and coworkers reported a manganese-catalyzed C(sp^3)—H azidation method that synergistically combines visible-light catalysis with electrochemical oxidation (Scheme 3.49) [208]. This photoelectrocatalytic strategy has the potential to facilitate the efficient late-stage azidation of a variety of valuable drug-like molecules, thereby highlighting its potential for complex molecule modification in medicinal chemistry.

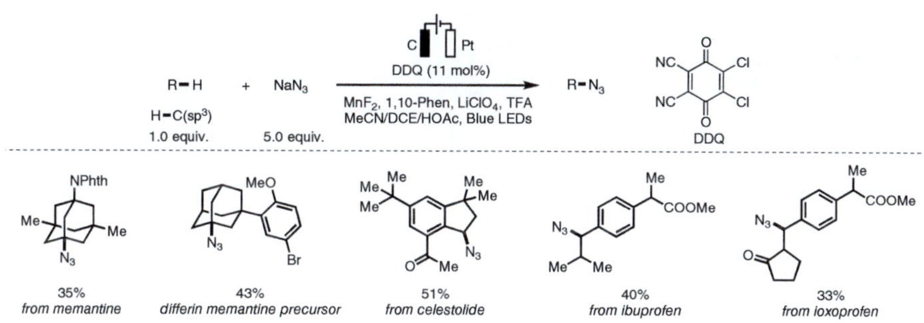

Scheme 3.49 Photoelectrochemical manganese-catalyzed C(sp^3)—H azidation.

3.4.3 Late-Stage Functionalization of Asymmetric Photoelectrocatalysis

Asymmetric C(sp^3)—H functionalization has emerged as a powerful strategy to streamline the late-stage modification of complex molecules [209, 210]. Recently, Xu and Liu independently reported the photoelectrocatalytic asymmetric C—H bond functionalization (Scheme 3.50) [211, 212]. Their groundbreaking asymmetric cyanation reactions utilized a copper catalyst in combination with a photocatalyst, offering a broad substrate scope without requiring chemical oxidants. Mechanistic studies revealed the following pathway: upon irradiation, the photocatalyst AQDS was excited to its electronically active state AQDS*, which facilitated a single-electron oxidation to generate a benzylic radical. This radical then interacted with a chiral copper complex, (L*)Cu(II)$(CN)_2$, leading to the formation of the cyanation product via reductive elimination. Concurrently, the reduced (L*)Cu(I)(CN) and intermediate $(AQDS—H)^{\cdot}$ were anodically oxidized to regenerate (L*)Cu(II)$(CN)_2$ and AQDS, accompanied by H_2 evolution at the cathode. This approach achieved late-stage cyanation of diverse complex structures with high site-selectivity and remarkable enantioselectivity. It can be applied to the direct C—H asymmetric cyanation of drug molecules. This innovative combination of photoelectrocatalysis and asymmetric copper-catalyzed radical cyanation has paved

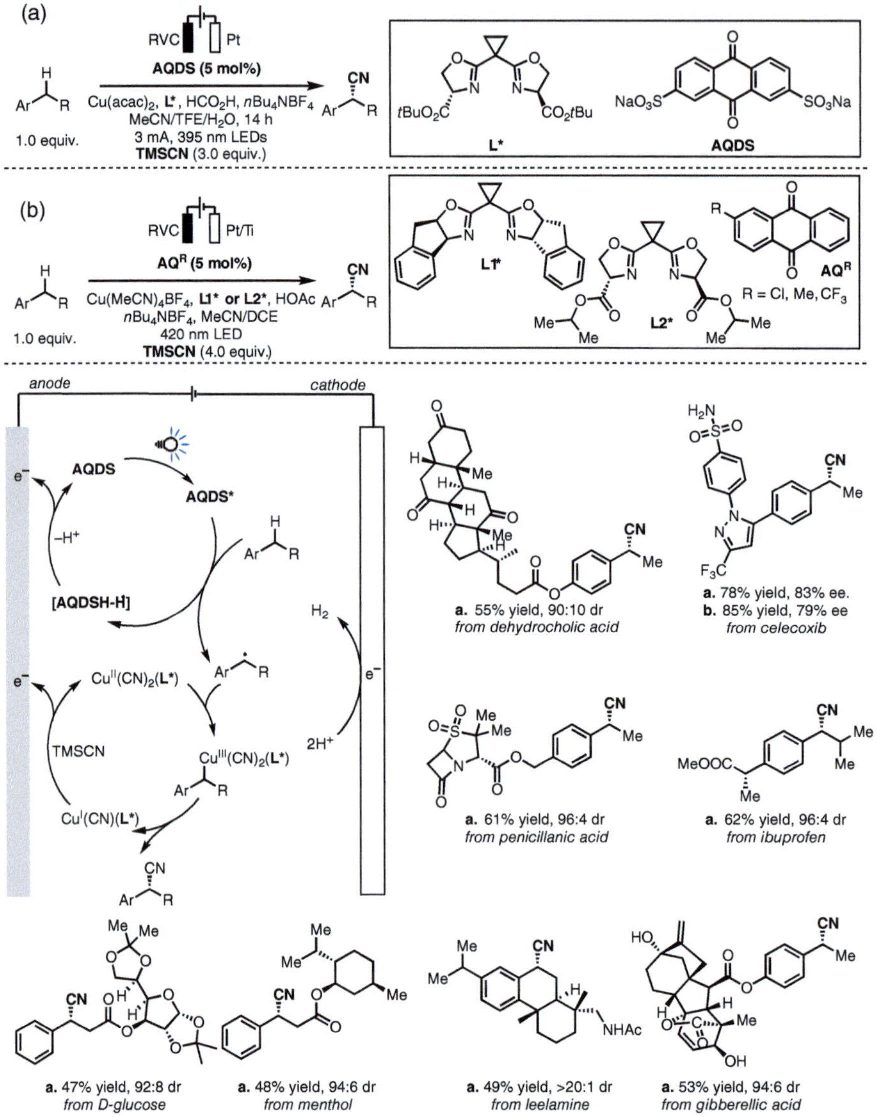

Scheme 3.50 Photoelectrochemical enantioselective cyanation of benzylic C—H bonds. (a) Xu's and (b) Liu's work.

the way for modular control in radical relay catalysis, enabling a diverse range of C—H functionalization reactions.

Building on the success of the asymmetric photoelectrocatalytic strategy, Xu subsequently developed a photoelectrochemical enantioselective decarboxylative cyanation, enabling the direct conversion of racemic carboxylic acids into enantio-enriched nitriles (Scheme 3.51) [213]. This method utilized cerium(III) salts and a chiral Cu(II)(L*) complex in a relay catalytic system. The process relied on cerium salts for photoelectrocatalytic decarboxylation and a chiral copper

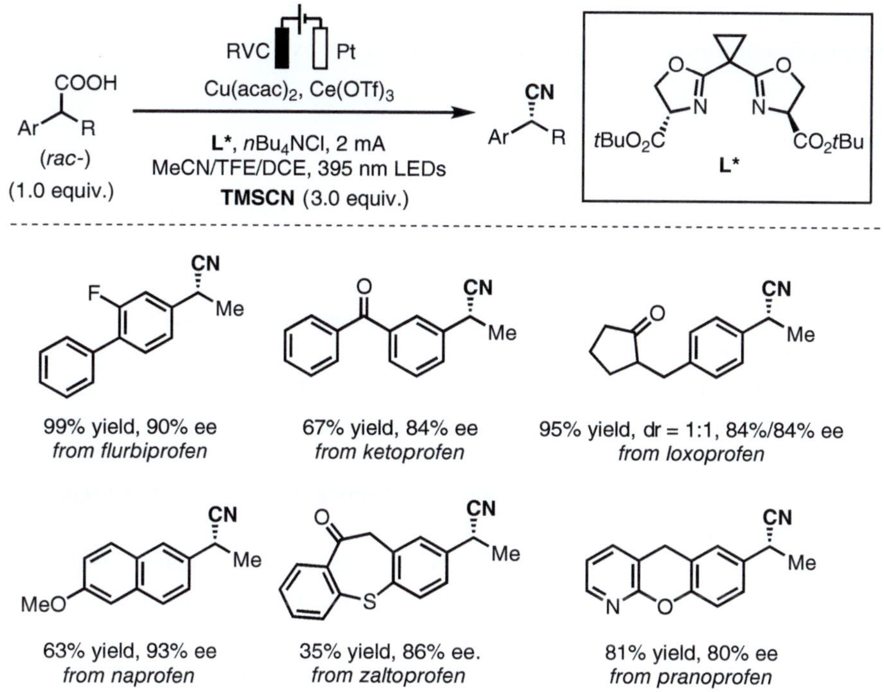

Scheme 3.51 Photoelectrochemical enantioselective decarboxylative cyanation.

complex for enantioselectivity control. A key factor in the success of this approach was the seamless integration of photoelectrocatalysis with the asymmetric copper catalytic cycle, with both catalysts undergoing regeneration via anodic oxidation. This photoelectrochemical approach was effective in transforming a range of commercially available carboxylic acid-based drug molecules into their corresponding enantio-enriched nitriles, including flurbiprofen, ketoprofen, loxoprofen, naproxen, zaltoprofen, and pranoprofen.

3.5 Conclusions

Organic electrosynthesis has witnessed major advances over the past decades, emerging as an increasingly powerful tool in molecular chemistry. The mild and sustainable conditions, precise selectivity control, and exceptional functional group compatibility render this a promising approach for the late-stage functionalization of natural products and drug molecules. Furthermore, the integration of flow chemistry, alternating current, and photochemical strategies with traditional organic electrosynthesis has enabled specific transformations previously unattainable. Despite the indisputable recent progress, challenges and unexplored directions remain in the field of electrochemical late-stage functionalization. These include molecular editing of specific skeletons, the limited advancement of electrochemical asymmetric synthesis, and the opportunities presented by novel electrode materials, such as modified electrodes, metal-organic frameworks (MOFs), and nanomaterials. These represent valuable research directions for future exploration.

References

1 Faraday, M. (1834). Siebente Reihe Von Experimental-Untersuchungen über Elektricität. *Ann. Phys.* 109: 481–520.

2 Kolbe, H. (1847). Beobachtungen über Die Oxydirende Wirkung Des Sauerstoffs, Wenn Derselbe Mit Hülfe Einer Elektrischen Säule Entwickelt Wird. *Journal für Praktische Chemie* 41: 137–139.

3 Simons, J.H. (1949). Production of fluorocarbons: I. The generalized procedure and its use with nitrogen compounds. *Journal of the Electrochemical Society* 95: 47–67.

4 Baizer, M.M. (1964). Electrolytic reductive coupling: I. Acrylonitrile. *Journal of the Electrochemical Society* 111: 215–222.

5 Shono, T., Hamaguchi, H., and Matsumura, Y. (1975). Electroorganic chemistry. XX. Anodic oxidation of carbamates. *Journal of the American Chemical Society* 97: 4264–4268.

6 Yoshida, J.-I., Murata, T., and Isoe, S. (1986). Electrochemical oxidation of organosilicon compounds I. Oxidative cleavage of carbon-silicon bond in allylsilanes and benzylsilanes. *Tetrahedron Letters* 27: 3373–3376.

7 Yoshida, J.-I., Shimizu, A., and Hayashi, R. (2018). Electrogenerated cationic reactive intermediates: the pool method and further advances. *Chemical Reviews* 118: 4702–4730.

8 Yoshida, J.-I., Suga, S., Suzuki, S. et al. (1999). Direct oxidative carbon–carbon bond formation using the "cation pool" method. 1. Generation of iminium cation pools and their reaction with carbon nucleophiles. *Journal of the American Chemical Society* 121: 9546–9549.

9 Steckhan, E. (1987). *Organic Syntheses with Electrochemically Regenerable Redox Systems, Electrochemistry I* (ed. E. Steckhan), 1–69. Berlin, Heidelberg: Springer.

10 Steckhan, E. (1986). Indirect electroorganic syntheses – a modern chapter of organic electrochemistry. *Angewandte Chemie International Edition* 25: 683–701.

11 Schäfer, H.-J. (2005). *Recent Contributions of Kolbe Electrolysis to Organic Synthesis. In Electrochemistry Iv*, 91–151. Springer.

12 Lund, H. (2002). A century of organic electrochemistry. *Journal of the Electrochemical Society* 149: S21–S33.

13 Francke, R. and Little, R.D. (2014). Redox catalysis in organic electrosynthesis: basic principles and recent developments. *Chemical Society Reviews* 43: 2492–2521.

14 Little, R.D. and Schwaebe, M.K. (1997). Reductive cyclizations at the cathode. In: *Electrochemistry Vi Electroorganic Synthesis: Bond Formation at Anode and Cathode* (ed. E. Steckhan), 1–48. Berlin, Heidelberg: Springer.

15 Little, R.D., Fox, D.P., Van Hijfte, L. et al. (1988). Electroreductive cyclization. Ketones and aldehydes tethered to .alpha.,.beta.-unsaturated esters (nitriles). Fundamental investigations. *Journal of Organometallic Chemistry* 53: 2287–2294.

16 Moeller, K.D. (2000). Synthetic applications of anodic electrochemistry. *Tetrahedron* 56: 9527–9554.

17 Jutand, A. (2008). Contribution of electrochemistry to organometallic catalysis. *Chemical Reviews* 108: 2300–2347.

18 Amatore, C., Cammoun, C., and Jutand, A. (2008). $Pd(OAc)_2$/*p*-benzoquinone-catalyzed anaerobic electrooxidative homocoupling of arylboronic acids, arylboronates and aryltrifluoroborates in DMF and/or water. *European Journal of Organic Chemistry* 2008: 4567–4570.

19 Ackermann, L. (2020). Metalla-electrocatalyzed C—H activation by earth-abundant 3d metals and beyond. *Accounts of Chemical Research* 53: 84–104.

20 Rosen, B.R., Werner, E.W., O'Brien, A.G., and Baran, P.S. (2014). Total synthesis of dixiamycin B by electrochemical oxidation. *Journal of the American Chemical Society* 136: 5571–5574.

21 Xiong, P. and Xu, H.-C. (2019). Chemistry with electrochemically generated N-centered radicals. *Accounts of Chemical Research* 52: 3339–3350.

22 Yuan, Y. and Lei, A. (2019). Electrochemical oxidative cross-coupling with hydrogen evolution reactions. *Accounts of Chemical Research* 52: 3309–3324.

23 Huang, H., Steiniger, K.A., and Lambert, T.H. (2022). Electrophotocatalysis: combining light and electricity to catalyze reactions. *Journal of the American Chemical Society* 144: 12567–12583.

24 Siu, J.C., Fu, N., and Lin, S. (2020). Catalyzing electrosynthesis: a homogeneous electrocatalytic approach to reaction discovery. *Accounts of Chemical Research* 53: 547–560.

25 Alvarez-Dorta, D., Thobie-Gautier, C., Croyal, M. et al. (2018). Electrochemically promoted tyrosine-click-chemistry for protein labeling. *Journal of the American Chemical Society* 140: 17120–17126.

26 Barjau, J., Schnakenburg, G., and Waldvogel, S.R. (2011). Diversity-oriented synthesis of polycyclic scaffolds by modification of an anodic product derived from 2,4-dimethylphenol. *Angewandte Chemie International Edition* 50: 1415–1419.

27 Wang, F. and Stahl, S.S. (2020). Electrochemical oxidation of organic molecules at lower overpotential: accessing broader functional group compatibility with electron–proton transfer mediators. *Accounts of Chemical Research* 53: 561–574.

28 Mackay, A.S., Payne, R.J., and Malins, L.R. (2022). Electrochemistry for the chemoselective modification of peptides and proteins. *Journal of the American Chemical Society* 144: 23–41.

29 Jiao, K.-J., Xing, Y.-K., Yang, Q.-L. et al. (2020). Site-selective C—H functionalization via synergistic use of electrochemistry and transition metal catalysis. *Accounts of Chemical Research* 53: 300–310.

30 Ke, J., Liu, W., Zhu, X. et al. (2021). Electrochemical radical silyl-oxygenation of activated alkenes. *Angewandte Chemie International Edition* 60: 8744–8749.

31 Mei, Y., Yan, Z., and Liu, L.L. (2022). Facile synthesis of the dicyanophosphide anion via electrochemical activation of white phosphorus: an avenue to organophosphorus compounds. *Journal of the American Chemical Society* 144: 1517–1522.

32 Chen, M., Zhao, D., Xu, J. et al. (2021). Electrooxidative B—H functionalization of nido-carboranes. *Angewandte Chemie International Edition* 60: 7838–7844.

33 Fried, A.D., Wilson, B.J., Galan, N.J., and Brantley, J.N. (2022). Electroediting of soft polymer backbones. *Journal of the American Chemical Society* 144: 8885–8891.

34 Siddiqi, Z. and Sarlah, D. (2021). Electrochemical dearomatization of commodity polymers. *Journal of the American Chemical Society* 143: 21264–21269.

35 Fagnani, D.E., Kim, D., Camarero, S.I. et al. (2023). Using waste poly(vinyl chloride) to synthesize chloroarenes by plasticizer-mediated electro(de)chlorination. *Nature Chemistry* 15: 222–229.

36 Peterson, B.M., Lin, S., and Fors, B.P. (2018). Electrochemically controlled cationic polymerization of vinyl ethers. *Journal of the American Chemical Society* 140: 2076–2079.

37 Dong, X., Roeckl, J.L., Waldvogel, S.R., and Morandi, B. (2021). Merging shuttle reactions and paired electrolysis for reversible vicinal dihalogenations. *Science* 371: 507–514.

38 Aynetdinova, D., Callens, M.C., Hicks, H.B. et al. (2021). Installing the "magic methyl" – C—H methylation in synthesis. *Chemical Society Reviews* 50: 5517–5563.

39 Ma, C., Zhao, C.-Q., Li, Y.-Q. et al. (2017). Palladium-catalyzed C—H activation/C—C cross-coupling reactions via electrochemistry. *Chemical Communications* 53: 12189–12192.

40 Yang, Q.-L., Li, C.-Z., Zhang, L.-W. et al. (2019). Palladium-catalyzed electrochemical C—H alkylation of arenes. *Organometallics* 38: 1208–1212.

41 Kuciński, K., Simon, H., and Ackermann, L. (2022). Rhoda-electrocatalyzed C—H methylation and paired electrocatalyzed C—H ethylation and propylation. *Chemistry – A European Journal* 28: e202103837.

42 Yang, Q.-L., Liu, Y., Liang, L. et al. (2022). Facilitating Rh-catalyzed C—H alkylation of (hetero)arenes and 6-arylpurine nucleosides (nucleotides) with electrochemistry. *Journal of Organometallic Chemistry* 87: 6161–6178.

43 Müller, K., Faeh, C., and Diederich, F. (2007). Fluorine in pharmaceuticals: looking beyond intuition. *Science* 317: 1881–1886.

44 O'Brien, A.G., Maruyama, A., Inokuma, Y. et al. (2014). Radical C—H functionalization of heteroarenes under electrochemical control. *Angewandte Chemie, International Edition* 53: 11868–11871.

45 Panda, B. (2020). Joy and challenges of alkynylation of arenes and heteroarenes through double C—H functionalizations. *Asian Journal of Organic Chemistry* 9: 492–507.

46 Dudnik, A.S. and Gevorgyan, V. (2010). Formal inverse Sonogashira reaction: direct alkynylation of arenes and heterocycles with alkynyl halides. *Angewandte Chemie International Edition* 49: 2096–2098.

47 Caspers, L.D. and Nachtsheim, B.J. (2018). Directing-group-mediated C—H-alkynylations. *Chemistry – An Asian Journal* 13: 1231–1247.

48 Ye, X., Wang, C., Zhang, S. et al. (2020). Facilitating Ir-catalyzed C—H alkynylation with electrochemistry: anodic oxidation-induced reductive elimination. *ACS Catalysis* 10: 11693–11699.

49 Chakrabarty, S., Wang, Y., Perkins, J.C., and Narayan, A.R.H. (2020). Scalable biocatalytic C—H oxyfunctionalization reactions. *Chemical Society Reviews* 49: 8137–8155.

50 So, Y.-H., Becker, J.Y., and Miller, L.L. (1975). Hydroxylation and amidation of aromatic carbonyl compounds. *Journal of the Chemical Society, Chemical Communications* 262–263.

51 Fujimoto, K., Tokuda, Y., Maekawa, H. et al. (1996). Selective and one-pot formation of phenols by anodic oxidation. *Tetrahedron* 52: 3889–3896.

52 Long, H., Chen, T.-S., Song, J. et al. (2022). Electrochemical aromatic C—H hydroxylation in continuous flow. *Nature Communications* 13: 3945–3951.

53 Hu, X., Zhang, G., Nie, L. et al. (2019). Electrochemical oxidation induced intermolecular aromatic C—H imidation. *Nature Communications* 10: 5467–5476.

54 Hu, X., Zhang, G., Bu, F. et al. (2018). Electrochemical-oxidation-induced site-selective intramolecular C(sp^3)—H amination. *ACS Catalysis* 8: 9370–9375.

55 Zhang, Y., Lin, Z., and Ackermann, L. (2021). Electrochemical C—H amidation of heteroarenes with *N*-alkyl sulfonamides in aqueous medium. *Chemistry – A European Journal* 27: 242–246.

56 Zhang, S.-K., Samanta, R.C., Sauermann, N., and Ackermann, L. (2018). Nickel-catalyzed electrooxidative C—H amination: support for nickel(iv). *Chemistry – A European Journal* 24: 19166–19170.

57 Sauermann, N., Mei, R., and Ackermann, L. (2018). Electrochemical C–H amination by cobalt catalysis in a renewable solvent. *Angewandte Chemie International Edition* 57: 5090–5094.

58 Gao, X., Wang, P., Zeng, L. et al. (2018). Cobalt(II) catalyzed electrooxidative C—H amination of arenes with alkylamines. *Journal of the American Chemical Society* 140: 4195–4199.

59 Yang, Q.-L., Wang, X.-Y., Lu, J.-Y. et al. (2018). Copper-catalyzed electrochemical C—H amination of arenes with secondary amines. *Journal of the American Chemical Society* 140: 11487–11494.

60 Ban, H.; Gavrilyuk, J.; Barbas, C. F., III Tyrosine bioconjugation through aqueous ene-type reactions: a click-like reaction for tyrosine. *Journal of the American Chemical Society* 2010, 132, 1523–1525.

61 Cui, L., Ma, Y., Li, M. et al. (2021). Tyrosine-reactive cross-linker for probing protein three-dimensional structures. *Analytical Chemistry* 93: 4434–4440.

62 Song, C., Liu, K., Wang, Z. et al. (2019). Electrochemical oxidation induced selective tyrosine bioconjugation for the modification of biomolecules. *Chemical Science* 10: 7982–7987.

63 Finkbeiner, P., Hehn, J.P., and Gnamm, C. (2020). Phosphine oxides from a medicinal chemist's perspective: physicochemical and in vitro parameters relevant for drug discovery. *Journal of Medicinal Chemistry* 63: 7081–7107.

64 Yurko, E.O., Gryaznova, T.V., and Budnikova, Y.H. (2019). Electrochemical C—H phosphonation of caffeine. *Phosphorus, Sulfur and Silicon and the Related Elements* 194: 343–344.

65 Wu, Z.-J., Su, F., Lin, W. et al. (2019). Scalable rhodium(III)-catalyzed aryl C—H phosphorylation enabled by anodic oxidation induced reductive elimination. *Angewandte Chemie International Edition* 58: 16770–16774.

66 Long, H., Huang, C., Zheng, Y.-T. et al. (2021). Electrochemical C—H phosphorylation of arenes in continuous flow suitable for late-stage functionalization. *Nature Communications* 12: 6629–6635.

67 Latham, J., Brandenburger, E., Shepherd, S.A. et al. (2018). Development of halogenase enzymes for use in synthesis. *Chemical Reviews* 118: 232–269.

68 Chen, W., Wang, H., Tay, N. et al. (2022). Arene radiofluorination enabled by photoredox-mediated halide interconversion. *Nature Chemistry* 14: 216–223.

69 Hassan, J., Sévignon, M., Gozzi, C. et al. (2002). Aryl–aryl bond formation one century after the discovery of the Ullmann reaction. *Chemical Reviews* 102: 1359–1470.

70 Petrone, D.A., Ye, J., and Lautens, M. (2016). Modern transition-metal-catalyzed carbon–halogen bond formation. *Chemical Reviews* 116: 8003–8104.

71 Kakiuchi, F., Kochi, T., Mutsutani, H. et al. (2009). Palladium-catalyzed aromatic C—H halogenation with hydrogen halides by means of electrochemical oxidation. *Journal of the American Chemical Society* 131: 11310–11311.

72 Wang, Y., Simon, H., Chen, X. et al. (2022). Distal ruthenaelectro-catalyzed *meta*-C—H bromination with aqueous Hbr. *Angewandte Chemie International Edition* 61: e202201595.

73 Yang, Q.-L., Wang, X.-Y., Wang, T.-L. et al. (2019). Palladium-catalyzed electrochemical C—H bromination using NH_4Br as the brominating reagent. *Organic Letters* 21: 2645–2649.

74 Tan, Z., Liu, Y., Helmy, R. et al. (2017). Electrochemical bromination of late stage intermediates and drug molecules. *Tetrahedron Letters* 58: 3014–3018.

75 Liang, Y., Lin, F., Adeli, Y. et al. (2019). Efficient electrocatalysis for the preparation of (hetero)aryl chlorides and vinyl chloride with 1,2-dichloroethane. *Angewandte Chemie, International Edition* 58: 4566–4570.

76 Li, J., Liu, T., Liu, J. et al. (2024). Construction of acenaphthylenes via C–H activation-based tandem penta- and hexaannulation reactions. *Nature Communications* 15: 8319.

77 Gandeepan, P., Müller, T., Zell, D. et al. (2019). 3D transition metals for C—H activation. *Chemical Reviews* 119: 2192–2452.

78 Rej, S., Ano, Y., and Chatani, N. (2020). Bidentate directing groups: an efficient tool in c–h bond functionalization chemistry for the expedient construction of C—C bonds. *Chemical Reviews* 120: 1788–1887.

79 Xu, F., Li, Y.-J., Huang, C., and Xu, H.-C. (2018). Ruthenium-catalyzed electrochemical dehydrogenative alkyne annulation. *ACS Catalysis* 8: 3820–3824.

80 Zeng, L., Li, H., Tang, S. et al. (2018). Cobalt-catalyzed electrochemical oxidative C—H/N—H carbonylation with hydrogen evolution. *ACS Catalysis* 8: 5448–5453.

81 Tian, C., Dhawa, U., Scheremetjew, A., and Ackermann, L. (2019). Cupraelectro-catalyzed alkyne annulation: evidence for distinct C—H alkynylation and decarboxylative C—H/C—C manifolds. *ACS Catalysis* 9: 7690–7696.

82 Mei, R., Sauermann, N., Oliveira, J.C.A., and Ackermann, L. (2018). Electroremovable traceless hydrazides for cobalt-catalyzed electro-oxidative C—H/N—H activation with internal alkynes. *Journal of the American Chemical Society* 140: 7913–7921.

83 Meyer, T.H., Oliveira, J.C.A., Sau, S.C. et al. (2018). Electrooxidative allene annulations by mild cobalt-catalyzed C—H activation. *ACS Catalysis* 8: 9140–9147.

84 Yang, Q.-L., Xing, Y.-K., Wang, X.-Y. et al. (2019). Electrochemistry-enabled Ir-catalyzed vinylic C—H functionalization. *Journal of the American Chemical Society* 141: 18970–18976.

85 Stangier, M., Messinis, A.M., Oliveira, J.C.A. et al. (2021). Rhodaelectro-catalyzed access to chromones via formyl C—H activation towards peptide electro-labeling. *Nature Communications* 12: 4736–4743.

86 Weng, Y., Xu, X., Chen, H. et al. (2022). Tandem electrochemical oxidative azidation/heterocyclization of tryptophan-containing peptides under buffer conditions. *Angewandte Chemie International Edition* 61: e202206308.

87 Blanksby, S.J. and Ellison, G.B. (2003). Bond dissociation energies of organic molecules. *Accounts of Chemical Research* 36: 255–263.

88 Wang, H., Liang, K., Xiong, W. et al. (2020). Electrochemical oxidation-induced etherification via C(sp^3)—H/O—H cross-coupling. *Science Advances* 6: eaaz0590.

89 Xiong, P., Zhao, H.-B., Fan, X.-T. et al. (2020). Site-selective electrooxidation of methylarenes to aromatic acetals. *Nature Communications* 11: 2706–2713.

90 Hoque, M.A., Twilton, J., Zhu, J. et al. (2022). Electrochemical pinoylation of methylarenes: Improving the scope and utility of benzylic oxidation through mediated electrolysis. *Journal of the American Chemical Society* 144: 15295–15302.

91 Hou, Z.-W., Liu, D.-J., Xiong, P. et al. (2021). Site-selective electrochemical benzylic C—H amination. *Angewandte Chemie International Edition* 60: 2943–2947.

92 Wang, Y., Lin, Z., Oliveira, J.C.A., and Ackermann, L. (2021). Electro-oxidative intermolecular allylic C(sp^3)—H aminations. *Journal of Organometallic Chemistry* 86: 15935–15945.

93 Hou, Z.-W., Li, L., and Wang, L. (2021). Organocatalytic electrochemical amination of benzylic C—H bonds. *Organic Chemistry Frontiers* 8: 4700–4705.

94 Ruan, Z., Huang, Z., Xu, Z. et al. (2021). Late-stage azolation of benzylic C—H bonds enabled by electrooxidation. *Science China. Chemistry* 64: 800–807.

95 Mukerjee, A.K. and Ashare, R. (1991). Isothiocyanates in the chemistry of heterocycles. *Chemical Reviews* 91: 1–24.

96 Zhang, S., Li, Y., Wang, T. et al. (2022). Electrochemical benzylic C(sp^3)—H isothiocyanation. *Organic Letters* 24: 1742–1746.

97 O'Hagan, D. (2008). Understanding organofluorine chemistry. An introduction to the C—F bond. *Chemical Society Reviews* 37: 308–319.

98 Wang, J., Sánchez-Roselló, M., Aceña, J.L. et al. (2014). Fluorine in pharmaceutical industry: fluorine-containing drugs introduced to the market in the last decade (2001–2011). *Chemical Reviews* 114: 2432–2506.

99 Tajima, T., Ishii, H., and Fuchigami, T. (2002). Anodic benzylic fluorination of toluene, ethylbenzene, and cumene derivatives. *Electrochemistry Communications* 4: 589–592.

100 Stangier, M., Scheremetjew, A., and Ackermann, L. (2022). Chemo- and site-selective electro-oxidative alkane fluorination by C(sp^3)—H cleavage. *Chemistry – A European Journal* 28: e202201654.

101 Recupero, F. and Punta, C. (2007). Free radical functionalization of organic compounds catalyzed by *N*-hydroxyphthalimide. *Chemical Reviews* 107: 3800–3842.

102 Marko, J.A., Durgham, A., Bretz, S.L., and Liu, W. (2019). Electrochemical benzylic oxidation of C—H bonds. *Chemical Communications* 55: 937–940.

103 Masui, M., Hosomi, K., Tsuchida, K., and Ozaki, S. (1985). Electrochemical oxidation of olefins using *N*-hydroxyphthalimide as a mediator. *Chemical & Pharmaceutical Bulletin* 33: 4798–4802.

104 Masui, M., Hara, S., Ueshima, T. et al. (1983). Anodic oxidation of compounds having benzylic or allylic carbon and α-carbon to hetero atom using *N*-hydroxyphthalimide as a mediator. *Chemical & Pharmaceutical Bulletin* 31: 4209–4211.

105 Shono, T. and Ikeda, A. (1972). Electroorganic chemistry. X. Anodic allylic substitution. *Journal of the American Chemical Society* 94: 7892–7898.

106 Horn, E.J., Rosen, B.R., Chen, Y. et al. (2016). Scalable and sustainable electrochemical allylic C—H oxidation. *Nature* 533: 77–81.

107 Trowbridge, A., Walton, S.M., and Gaunt, M.J. (2020). New strategies for the transition-metal catalyzed synthesis of aliphatic amines. *Chemical Reviews* 120: 2613–2692.

108 Richter, M.F., Drown, B.S., Riley, A.P. et al. (2017). Predictive compound accumulation rules yield a broad-spectrum antibiotic. *Nature* 545: 299–304.

109 Vitaku, E., Smith, D.T., and Njardarson, J.T. (2014). Analysis of the structural diversity, substitution patterns, and frequency of nitrogen heterocycles among U.S. FDA approved pharmaceuticals. *Journal of Medicinal Chemistry* 57: 10257–10274.

110 Wang, D.J., Targos, K., and Wickens, Z.K. (2021). Electrochemical synthesis of allylic amines from terminal alkenes and secondary amines. *Journal of the American Chemical Society* 143: 21503–21510.

111 Yang, Y.-Z., Wu, Y.-C., Song, R.-J., and Li, J.-H. (2020). Electrochemical dehydrogenative cross-coupling of xanthenes with ketones. *Chemical Communications* 56: 7585–7588.

112 Turlik, A., Chen, Y., and Newhouse, T.R. (2016). Dehydrogenation adjacent to carbonyls using palladium–allyl intermediates. *Synlett* 27: 331–336.

113 Diao, T. and Stahl, S.S. (2011). Synthesis of cyclic enones via direct palladium-catalyzed aerobic dehydrogenation of ketones. *Journal of the American Chemical Society* 133: 14566–14569.

114 Chen, M. and Dong, G. (2019). Copper-catalyzed desaturation of lactones, lactams, and ketones under PH-neutral conditions. *Journal of the American Chemical Society* 141: 14889–14897.

115 Huang, D., Szewczyk, S.M., Zhang, P., and Newhouse, T.R. (2019). Allyl-nickel catalysis enables carbonyl dehydrogenation and oxidative cycloalkenylation of ketones. *Journal of the American Chemical Society* 141: 5669–5674.

116 Nicolaou, K.C., Zhong, Y.L., and Baran, P.S. (2000). A new method for the one-step synthesis of α,β-unsaturated carbonyl systems from saturated alcohols and carbonyl compounds. *Journal of the American Chemical Society* 122: 7596–7597.

117 Gnaim, S., Takahira, Y., Wilke, H.R. et al. (2021). Electrochemically driven desaturation of carbonyl compounds. *Nature Chemistry* 13: 367–372.

118 Huang, X., Zhang, Q., Lin, J. et al. (2019). Electricity-driven asymmetric lewis acid catalysis. *Nature Catalysis* 2: 34–40.

119 Xiong, P., Hemming, M., Ivlev, S.I., and Meggers, E. (2022). Electrochemical enantioselective nucleophilic α-C(sp^3)—H alkenylation of 2-acyl imidazoles. *Journal of the American Chemical Society* 144: 6964–6971.

120 Jones, A.M. and Banks, C.E. (2014). The shono-type electroorganic oxidation of unfunctionalised amides. Carbon–carbon bond formation via electrogenerated *N*-acyliminium ions. *Beilstein Journal of Organic Chemistry* 10: 3056–3072.

121 Kabeshov, M.A., Musio, B., Murray, P.R.D. et al. (2014). Expedient preparation of nazlinine and a small library of indole alkaloids using flow electrochemistry as an enabling technology. *Organic Letters* 16: 4618–4621.

122 Novaes, L.F.T., Ho, J.S.K., Mao, K. et al. (2022). Exploring electrochemical C(sp^3)—H oxidation for the late-stage methylation of complex molecules. *Journal of the American Chemical Society* 144: 1187–1197.

123 Chen, H., Volgraf, M., Do, S. et al. (2018). Discovery of a potent (4*R*,5*S*)-4-fluoro-5-methylproline sulfonamide transient receptor potential ankyrin 1 antagonist and its methylene phosphate prodrug guided by molecular modeling. *Journal of Medicinal Chemistry* 61: 3641–3659.

124 Wu, Y., Yi, H., and Lei, A. (2018). Electrochemical acceptorless dehydrogenation of N-heterocycles utilizing tempo as organo-electrocatalyst. *ACS Catalysis* 8: 1192–1196.

125 Feng, T., Wang, S., Liu, Y. et al. (2022). Electrochemical desaturative β-acylation of cyclic *N*-aryl amines. *Angewandte Chemie International Edition* 61: e202115178.

126 Kaiser, D., Klose, I., Oost, R. et al. (2019). Bond-forming and -breaking reactions at sulfur(IV): sulfoxides, sulfonium salts, sulfur ylides, and sulfinate salts. *Chemical Reviews* 119: 8701–8780.

127 Wang, H., He, M., Li, Y. et al. (2021). Electrochemical oxidation enables regioselective and scalable α-C(sp^3)—H acyloxylation of sulfides. *Journal of the American Chemical Society* 143: 3628–3637.

128 Eberson, L. and Nyberg, K. (1976). Synthetic uses of anodic substitution reactions. *Tetrahedron* 32: 2185–2206.

129 Kawamata, Y., Yan, M., Liu, Z. et al. (2017). Scalable, electrochemical oxidation of unactivated C—H bonds. *Journal of the American Chemical Society* 139: 7448–7451.

130 Meyer, T.H., Samanta, R.C., Del Vecchio, A., and Ackermann, L. (2021). Mangana(III/IV)electro-catalyzed C(sp^3)—H azidation. *Chemical Science* 12: 2890–2897.

131 Tron, G.C., Pirali, T., Billington, R.A. et al. (2008). Click chemistry reactions in medicinal chemistry: applications of the 1,3-dipolar cycloaddition between azides and alkynes. *Medicinal Research Reviews* 28: 278–308.

132 Smith, D.B., Kalayanov, G., Sund, C. et al. (2009). The design, synthesis, and antiviral activity of monofluoro and difluoro analogues of 4′-azidocytidine against hepatitis C virus replication: the discovery of 4′-azido-2′-deoxy-2′-fluorocytidine and 4′-azido-2′-dideoxy-2′,2′-difluorocytidine. *Journal of Medicinal Chemistry* 52: 2971–2978.

133 Geurink, P.P., van der Linden, W.A., Mirabella, A.C. et al. (2013). Incorporation of non-natural amino acids improves cell permeability and potency of specific inhibitors of proteasome trypsin-like sites. *Journal of Medicinal Chemistry* 56: 1262–1275.

134 Bräse, S., Gil, C., Knepper, K., and Zimmermann, V. (2005). Organic azides: an exploding diversity of a unique class of compounds. *Angewandte Chemie International Edition* 44: 5188–5240.

135 Zeng, L., Li, H., Hu, J. et al. (2020). Electrochemical oxidative aminocarbonylation of terminal alkynes. *Nature Catalysis* 3: 438–445.

136 Wu, Y., Zeng, L., Li, H. et al. (2021). Electrochemical palladium-catalyzed oxidative Sonogashira carbonylation of arylhydrazines and alkynes to ynones. *Journal of the American Chemical Society* 143: 12460–12466.

137 Liang, H., Julaiti, Y., Zhao, C.-G., and Xie, J. (2023). Electrochemical gold-catalysed biocompatible C(sp^2)—C(sp) coupling. *Nature Synthesis* 2: 338–347.

138 Castellino, N.J., Montgomery, A.P., Danon, J.J., and Kassiou, M. (2023). Late-stage functionalization for improving drug-like molecular properties. *Chemical Reviews* 123: 8127–8153.

139 Liu, J., Liu, X., Wu, J., and Li, C.-C. (2020). Total synthesis of natural products containing a bridgehead double bond. *Chem* 6: 579–615.

140 Ertl, P. and Schuhmann, T. (2019). A systematic cheminformatics analysis of functional groups occurring in natural products. *Journal of Natural Products* 82: 1258–1263.

141 Zhang, W. and Lin, S. (2020). Electroreductive carbofunctionalization of alkenes with alkyl bromides via a radical-polar crossover mechanism. *Journal of the American Chemical Society* 142: 20661–20670.

142 Lu, L., Siu, J.C., Lai, Y., and Lin, S. (2020). An electroreductive approach to radical silylation via the activation of strong Si—Cl Bond. *Journal of the American Chemical Society* 142: 21272–21278.

143 Fu, N., Song, L., Liu, J. et al. (2019). New bisoxazoline ligands enable enantioselective electrocatalytic cyanofunctionalization of vinylarenes. *Journal of the American Chemical Society* 141: 14480–14485.

144 Fu, N., Sauer, G.S., Saha, A. et al. (2017). Metal-catalyzed electrochemical diazidation of alkenes. *Science* 357: 575–579.

145 Fu, N., Sauer, G.S., and Lin, S. (2017). Electrocatalytic radical dichlorination of alkenes with nucleophilic chlorine sources. *Journal of the American Chemical Society* 139: 15548–15553.

146 Sauer, G.S. and Lin, S. (2018). An electrocatalytic approach to the radical difunctionalization of alkenes. *ACS Catalysis* 8: 5175–5187.

147 Siu, J.C., Sauer, G.S., Saha, A. et al. (2018). Electrochemical azidooxygenation of alkenes mediated by a TEMPO–N_3 charge-transfer complex. *Journal of the American Chemical Society* 140: 12511–12520.

148 Siu, J.C., Parry, J.B., and Lin, S. (2019). Aminoxyl-catalyzed electrochemical diazidation of alkenes mediated by a metastable charge-transfer complex. *Journal of the American Chemical Society* 141: 2825–2831.

149 Ye, K.-Y., Pombar, G., Fu, N. et al. (2018). Anodically coupled electrolysis for the heterodifunctionalization of alkenes. *Journal of the American Chemical Society* 140: 2438–2441.

150 Park, S.H., Jang, J., Shin, K., and Kim, H. (2022). Electrocatalytic radical-polar crossover hydroetherification of alkenes with phenols. *ACS Catalysis* 12: 10572–10580.

151 Mandal, A., Jang, J., Yang, B. et al. (2023). Palladium-catalyzed electrooxidative hydrofluorination of aryl-substituted alkenes with a nucleophilic fluorine source. *Organic Letters* 25: 195–199.

152 Gnaim, S., Bauer, A., Zhang, H.-J. et al. (2022). Cobalt-electrocatalytic HAT for functionalization of unsaturated C—C bonds. *Nature* 605: 687–695.

153 Sun, X., Ma, H.-X., Mei, T.-S. et al. (2019). Electrochemical radical formyloxylation–bromination, –chlorination, and –trifluoromethylation of alkenes. *Organic Letters* 21: 3167–3171.

154 Wei, B.-Y., Xie, D.-T., Lai, S.-Q. et al. (2021). Electrochemically tuned oxidative [4+2] annulation and dioxygenation of olefins with hydroxamic acids. *Angewandte Chemie International Edition* 60: 3182–3188.

155 Kim, S., Hirose, K., Uematsu, J. et al. (2012). Electrochemically active cross-linking reaction for fluorescent labeling of aliphatic alkenes. *Chemistry – A European Journal* 18: 6284–6288.

156 Chai, Q.-Y., Yang, Z., Lin, H.-W., and Han, B.-N. (2016). Alkynyl-containing peptides of marine origin: a review. *Marine Drugs* 14: 216–233.

157 Luo, X. and Wang, P. (2021). Ynonylation of acyl radicals by electroinduced homolysis of 4-acyl-1,4-dihydropyridines. *Organic Letters* 23: 4960–4965.

158 Alonso, F., Beletskaya, I.P., and Yus, M. (2002). Metal-mediated reductive hydrodehalogenation of organic halides. *Chemical Reviews* 102: 4009–4092.

159 Alberico, D., Scott, M.E., and Lautens, M. (2007). Aryl–aryl bond formation by transition-metal-catalyzed direct arylation. *Chemical Reviews* 107: 174–238.

160 Ackermann, L. (2009). *Modern Arylation Methods.* Wiley Online Library.

161 Kawamata, Y., Vantourout, J.C., Hickey, D.P. et al. (2019). Electrochemically driven, ni-catalyzed aryl amination: scope, mechanism, and applications. *Journal of the American Chemical Society* 141: 6392–6402.

162 Zhu, C., Kale, A.P., Yue, H., and Rueping, M. (2021). Redox-neutral cross-coupling amination with weak *N*-nucleophiles: arylation of anilines, sulfonamides, sulfoximines, carbamates, and imines via nickelaelectrocatalysis. *JACS Au* 1: 1057–1065.

163 Ma, Y., Hong, J., Yao, X. et al. (2021). Aminomethylation of aryl bromides by nickel-catalyzed electrochemical redox neutral cross coupling. *Organic Letters* 23: 9387–9392.

164 Korch, K.M. and Watson, D.A. (2019). Cross-coupling of heteroatomic electrophiles. *Chemical Reviews* 119: 8192–8228.

165 Reeves, E.K., Entz, E.D., and Neufeldt, S.R. (2021). Chemodivergence between electrophiles in cross-coupling reactions. *Chemistry – A European Journal* 27: 6161–6177.

166 Ackermann, L., Althammer, A., and Born, R. (2006). Catalytic Arylation Reactions by C–H Bond Activation with Aryl Tosylates. *Angewandte Chemie International Edition* 45: 2619–2622.

167 Zhang, F., Wang, Y., Wang, Y., and Pan, Y. (2021). Electrochemical deoxygenative thiolation of preactivated alcohols and ketones. *Organic Letters* 23: 7524–7528.

168 Wang, B., Peng, P., Ma, W. et al. (2021). Electrochemical borylation of alkyl halides: fast, scalable access to alkyl boronic esters. *Journal of the American Chemical Society* 143: 12985–12991.

169 Lovering, F., Bikker, J., and Humblet, C. (2009). Escape from flatland: increasing saturation as an approach to improving clinical success. *Journal of Medicinal Chemistry* 52: 6752–6756.

170 Zhang, W., Lu, L., Zhang, W. et al. (2022). Electrochemically driven cross-electrophile coupling of alkyl halides. *Nature* 604: 292–297.

171 Gooßen, L.J., Rodríguez, N., and Gooßen, K. (2008). Carboxylic acids as substrates in homogeneous catalysis. *Angewandte Chemie International Edition* 47: 3100–3120.

172 Leech, M.C. and Lam, K. (2020). Electrosynthesis using carboxylic acid derivatives: new tricks for old reactions. *Accounts of Chemical Research* 53: 121–134.

173 Xiang, J., Shang, M., Kawamata, Y. et al. (2019). Hindered dialkyl ether synthesis with electrogenerated carbocations. *Nature* 573: 398–402.

174 Lin, Y. and Malins, L.R. (2020). Total synthesis of biseokeaniamides A–C and late-stage electrochemically-enabled peptide analogue synthesis. *Chemical Science* 11: 10752–10758.

175 Cramer, J., Sager, C.P., and Ernst, B. (2019). Hydroxyl groups in synthetic and natural-product-derived therapeutics: a perspective on a common functional group. *Journal of Medicinal Chemistry* 62: 8915–8930.

176 Villo, P., Shatskiy, A., Kärkäs, M.D., and Lundberg, H. (2023). Electrosynthetic C—O bond activation in alcohols and alcohol derivatives. *Angewandte Chemie International Edition* 62: e202211952.

177 Ohmori, H., Nakai, S., Sekiguchi, M., and Masui, M. (1980). Anodic oxidation of organophosphorus compounds. III. Anodic alkoxylation and thioalkoxylation of triphenylphosphine. *Chemical & Pharmaceutical Bulletin* 28: 910–915.

178 Ohmori, H., Nakai, S., Miyasaka, H., and Masui, M. (1982). An improved electrochemical preparation of alkoxytriphenylphosphonium salts. *Chemical & Pharmaceutical Bulletin* 30: 4192–4194.

179 Yoshida, H., Usami, N., Ohishi, Y. et al. (1995). Synthesis and pharmacological effects in mice of halogenated cannabinol derivatives. *Chemical & Pharmaceutical Bulletin* 43: 335–337.

180 Maeda, H., Matsumoto, S., Koide, T., and Ohmori, H. (1998). Dehydroxy substitution reactions of the anomeric hydroxy groups in some protected sugars initiated by anodic oxidation of triphenylphosphine. *Chemical & Pharmaceutical Bulletin* 46: 939–943.

181 Xu, Z., Zheng, Y., Wang, Z. et al. (2019). Triphenylphosphine-Assisted Dehydroxylative Csp^3—N bond formation via electrochemical oxidation. *Chemical Communications* 55: 15089–15092.

182 Buglioni, L., Raymenants, F., Slattery, A. et al. (2022). Technological innovations in photochemistry for organic synthesis: flow chemistry, high-throughput experimentation, scale-up, and photoelectrochemistry. *Chemical Reviews* 122: 2752–2906.

183 Wu, S., Kaur, J., Karl, T.A. et al. (2022). Synthetic molecular photoelectrochemistry: new frontiers in synthetic applications, mechanistic insights and scalability. *Angewandte Chemie International Edition* 61: e202107811.

184 Capaldo, L., Quadri, L.L., and Ravelli, D. (2019). Merging photocatalysis with electrochemistry: the dawn of a new alliance in organic synthesis. *Angewandte Chemie International Edition* 58: 17508–17510.

185 Yang, G., Wang, Y., and Qiu, Y. (2021). Advances in organic photoelectrochemical synergistic catalysis. *Chinese Journal of Organic Chemistry* 41: 3935–3947.

186 Li, P., Zhang, T., Mushtaq, M.A. et al. (2021). Research progress in organic synthesis by means of photoelectrocatalysis. *Chemical Record* 21: 841–857.

187 Barham, J.P. and König, B. (2020). Synthetic photoelectrochemistry. *Angewandte Chemie International Edition* 59: 11732–11747.

188 Yu, Y., Guo, P., Zhong, J.-S. et al. (2020). Merging photochemistry with electrochemistry in organic synthesis. *Organic Chemistry Frontiers* 7: 131–135.

189 Zhang, L., Liardet, L., Luo, J. et al. (2019). Photoelectrocatalytic arene C—H amination. *Nature Catalysis* 2: 366–373.

190 Chen, Y.-J., Deng, W.-H., Guo, J.-D. et al. (2022). Transition-metal-free, site-selective C—F arylation of polyfluoroarenes via electrophotocatalysis. *Journal of the American Chemical Society* 144: 17261–17268.

191 Chen, Y.-J., Lei, T., Hu, H.-L. et al. (2021). Tandem photoelectrochemical and photoredox catalysis for efficient and selective aryl halides functionalization by solar energy. *Matter* 4: 2354–2366.

192 Tan, Z., He, X., Xu, K., and Zeng, C. (2022). Electrophotocatalytic C—H functionalization of N-heteroarenes with unactivated alkanes under external oxidant-free conditions. *ChemSusChem* 15: e202102360.

193 Jiang, Y., Xu, K., and Zeng, C. (2022). Electrophotocatalytic Si—H activation governed by polarity-matching effects. *CCS Chemistry* 4: 1796–1805.

194 Yan, H., Song, J., Zhu, S., and Xu, H.-C. (2021). Synthesis of acridinium photocatalysts via site-selective C—H alkylation. *CCS Chemistry* 3: 317–325.

195 Huang, H., Strater, Z.M., and Lambert, T.H. (2020). Electrophotocatalytic C—H functionalization of ethers with high regioselectivity. *Journal of the American Chemical Society* 142: 1698–1703.

196 Zhang, W., Carpenter, K.L., and Lin, S. (2020). Electrochemistry broadens the scope of flavin photocatalysis: photoelectrocatalytic oxidation of unactivated alcohols. *Angewandte Chemie, International Edition* 59: 409–417.

197 Wang, F. and Stahl, S.S. (2019). Merging photochemistry with electrochemistry: functional-group tolerant electrochemical amination of C(sp^3)—H bonds. *Angewandte Chemie International Edition* 58: 6385–6390.

198 Huang, H., Strater, Z.M., Rauch, M. et al. (2019). Electrophotocatalysis with a trisaminocyclopropenium radical dication. *Angewandte Chemie International Edition* 58: 13318–13322.

199 Qiu, Y., Scheremetjew, A., Finger, L.H., and Ackermann, L. (2020). Electrophotocatalytic undirected C—H trifluoromethylations of (het)arenes. *Chemistry – A European Journal* 26: 3241–3246.

200 Yan, H., Hou, Z.-W., and Xu, H.-C. (2019). Photoelectrochemical C—H alkylation of heteroarenes with organotrifluoroborates. *Angewandte Chemie International Edition* 58: 4592–4595.

201 Suzuki, J.; Tanigawa, M.; Inagi, S.; Fuchigami, T. Electrochemical oxidation of organotrifluoroborate compounds *ChemElectroChem* 2016, 3, 2078–2083.

202 Lai, X.-L., Shu, X.-M., Song, J., and Xu, H.-C. (2020). Electrophotocatalytic decarboxylative C—H functionalization of heteroarenes. *Angewandte Chemie International Edition* 59: 10626–10632.

203 Xu, P., Chen, P.-Y., and Xu, H.-C. (2020). Scalable photoelectrochemical dehydrogenative cross-coupling of heteroarenes with aliphatic C—H bonds. *Angewandte Chemie International Edition* 59: 14275–14280.

204 Wang, K., Liu, X., Yang, S. et al. (2022). In situ alkyl radical recycling-driven decoupled electrophotochemical deamination. *Organic Letters* 24: 3471–3476.

205 Wan, Q., Hou, Z.-W., Zhao, X.-R. et al. (2023). Organoelectrophotocatalytic C—H silylation of heteroarenes. *Organic Letters* 25: 1008–1013.

206 Shen, T. and Lambert, T.H. (2021). Electrophotocatalytic diamination of vicinal C—H bonds. *Science* 371: 620–626.

207 Shen, T., Li, Y.-L., Ye, K.-Y., and Lambert, T.H. (2023). Electrophotocatalytic oxygenation of multiple adjacent C—H bonds. *Nature* 614: 275–280.

208 Niu, L., Jiang, C., Liang, Y. et al. (2020). Manganese-catalyzed oxidative azidation of C(sp^3)—H bonds under electrophotocatalytic conditions. *Journal of the American Chemical Society* 142: 17693–17702.

209 Davies, H.M.L. and Beckwith, R.E.J. (2003). Catalytic enantioselective C—H activation by means of metal–carbenoid-induced C—H insertion. *Chemical Reviews* 103: 2861–2904.

210 Woźniak, Ł.; Tan, J.-F.; Nguyen, Q.-H.; Madron du Vigné, A.; Smal, V.; Cao, Y.-X.; Cramer, N. Catalytic enantioselective functionalizations of C—H bonds by chiral iridium complexes. *Chemical Reviews* 2020, 120, 10516–10543.

211 Cai, C.-Y., Lai, X.-L., Wang, Y. et al. (2022). Photoelectrochemical asymmetric catalysis enables site- and enantioselective cyanation of benzylic C—H bonds. *Nature Catalysis* 5: 943–951.

212 Fan, W., Zhao, X., Deng, Y. et al. (2022). Electrophotocatalytic decoupled radical relay enables highly efficient and enantioselective benzylic C—H functionalization. *Journal of the American Chemical Society* 144: 21674–21682.

213 Lai, X.-L., Chen, M., Wang, Y. et al. (2022). Photoelectrochemical asymmetric catalysis enables direct and enantioselective decarboxylative cyanation. *Journal of the American Chemical Society* 144: 20201–20206.

4

Flow Chemistry Applied for Late-Stage Functionalization and Other Functional Group-Tolerant Manipulation

Philipp Natho, Marco Colella*, and Renzo Luisi*

University of Bari "A. Moro", FLAME-Lab, Flow Chemistry and Microreactor Technology Laboratory, Department of Pharmacy-Drug Sciences, Via E. Orabona 4, Bari, 70125, Italy

4.1 Introduction

Flow technology has, in particular over the last two decades, significantly widened the synthetic opportunities available in the organic chemists' toolbox. Not only has flow technology enabled the discovery and development of novel reactivity and methodologies which were inefficient, or even impossible, under classical batch conditions, but also can conventional batch processes be significantly improved with a view to enhanced mass and heat transfer, scalability, efficiency, safety, and reproducibility. These benefits were, after a slow adoption period, recognized by synthetic chemists across disciplines and as such combined with other enabling technologies, e.g., photochemistry and electrochemistry. An inestimable benefit of flow technology is that important reaction parameters, such as temperature, pressure, and residence time, can be precisely and reproducibly controlled, thus enabling reaction success and, often equally importantly, fine-tuning of reaction selectivity. In particular, its propensity toward controlling reaction selectivity makes flow technology suited for the late-stage functionalization of complex and functional group-diverse drug molecules. The need for structurally divergent transformations as late as possible in multi-step syntheses is important in drug discovery processes to prevent laborious and lengthy *de novo* syntheses for lead compound identification and optimization and to increase structural diversity in compound libraries.

Late-stage functionalization has been defined as "a desired chemoselective transformation on a complex molecule to provide analogs without the necessity for installation of a functional group that exclusively serves the purpose of enabling this transformation" [1, 2]. As such, high chemoselectivity and functional-group tolerance are a requirement, whereas pre-installation of any group that engages directly or indirectly in a given transformation precludes this reaction from being late-stage

* These authors contributed equally to this work.

Late-Stage Functionalization and Diversification in Organic Synthesis: Methods and Applications, First Edition. Edited by Tatiana Besset.

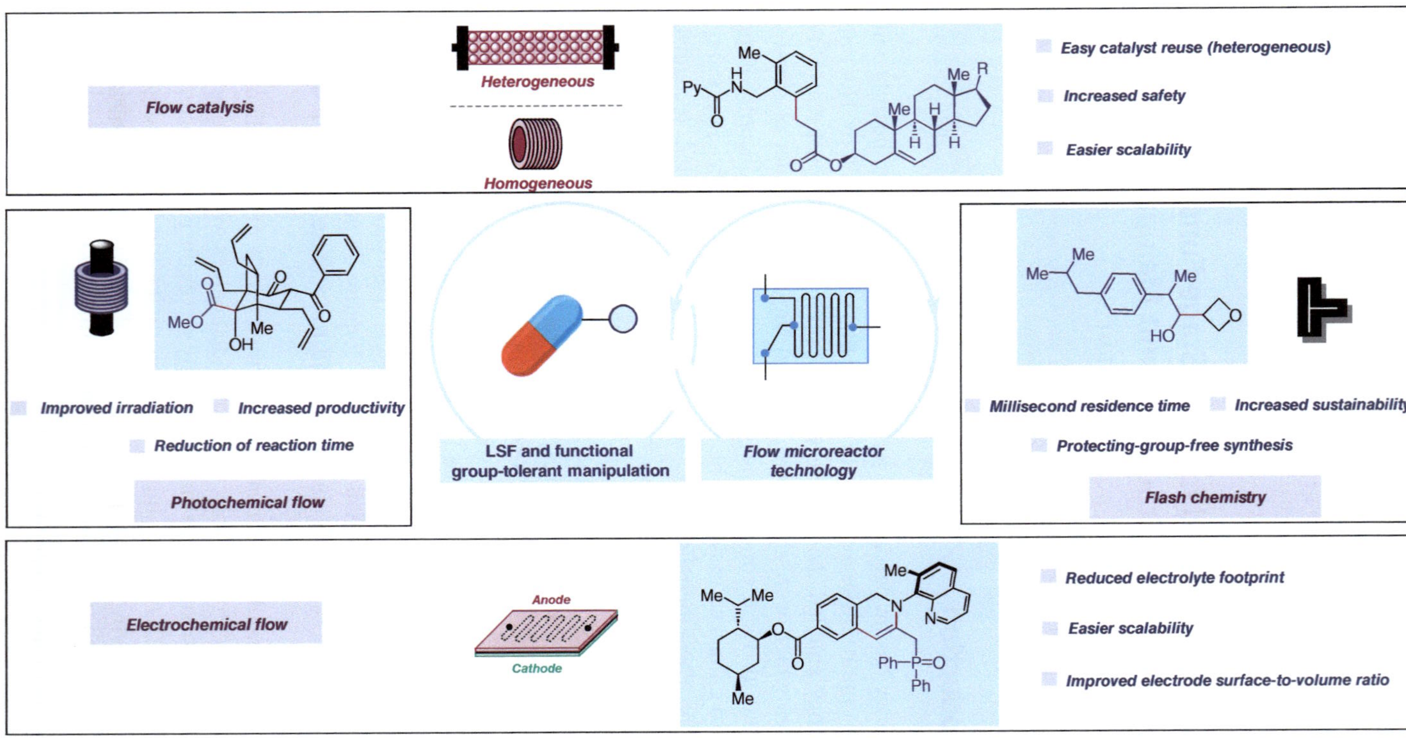

Figure 4.1 Late-stage functionalization leveraging flow technology.

functionalization. In particular, direct C—H functionalization protocols suitable for molecules featuring high structural complexity fit this definition. For the purpose of this chapter, we aim to remain as close as possible to the above-introduced definition. However, in some cases, to showcase the power of flow technology as an enabling technology to perform reactions which are impossible, or significantly less efficient, under classic batch conditions, the definition is slightly loosened to "functional-group tolerant and chemoselective protocols for the functionalization of complex molecules" particularly in subchapters concerning flash chemistry and heterogeneous catalysis under flow conditions (Figure 4.1). This chapter is divided into the following sections:

- Photochemical flow
- Electrochemical flow
- Flash chemistry
- Homogeneous and heterogeneous metal catalysis in flow

Given the scope of this chapter, key advances from the last five years are showcased. For a more exhaustive view, the readers are referred to timely reviews [3–7].

4.2 Photochemical Flow

Over the last ten years, the field of photochemistry, particularly photocatalysis, has revolutionized synthetic organic chemistry, unlocking reaction pathways unattainable to conventional ground-state reactivity [8]. Indeed, through the absorption of light, organic molecules and photocatalysts can reach an excited state with chemical reactivity which is different from that in the ground state, enabling the formation of new chemical bonds. Notably, many of these innovative methodologies operate under mild reaction conditions, providing excellent tolerance for various functional groups, which is essential for developing late-stage functionalization processes [9]. Photoinduced chemical transformations are likely those that have benefited the most from the transition from traditional batch chemistry to continuous flow chemistry. This is due to improved irradiation of the reaction mixture, easier scalability, enhanced reaction selectivity, and increased reproducibility [10–13]. Given these premises, it is unsurprising that several continuous flow approaches for late-stage functionalization and functional group-tolerant manipulation of complex molecules under light irradiation have been developed, and selected examples for further reading are referenced here [14–19].

Christmann and coworkers showcased the importance of flow chemistry in facilitating photocatalytic late-stage transformations in natural product preparation (Scheme 4.1) [20]. The photochemical step involves the oxidation of a methoxyallene to the corresponding ester using singlet oxygen, which is generated with tetraphenylporphyrin (TPP) as the photocatalyst. The limitations of the batch protocol, including long reaction times (26 hours) and scale-up issues beyond 10 mg, were addressed by transitioning to a flow setup.

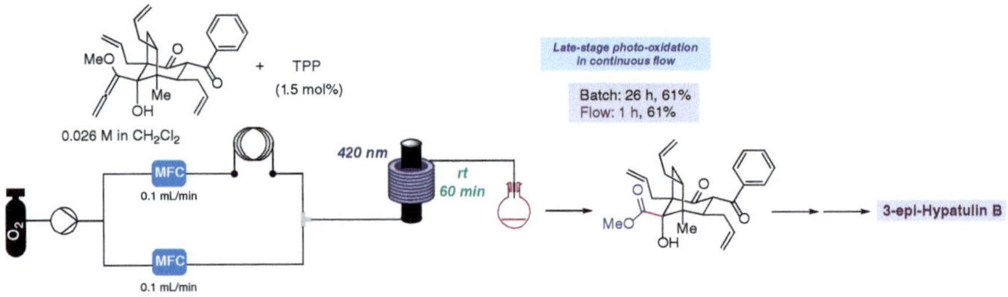

Scheme 4.1 Christmann's late-stage photo-oxidation in continuous flow for the synthesis of 3-*epi*-Hypatulin B.

The value of flow chemistry in enabling late-stage transformations for the synthesis of natural products was further emphasized by Nay and coworkers (Scheme 4.2) [21]. They presented a photochemical ring expansion method for producing radulanin A, a weed killer, from a 2,2-dimethylchromene precursor. Despite the synthesis of the 2,5-dihydro-1-benzoxepine core inevitably being accompanied by the formation of the 5-membered ring isomer, this methodology provides a shorter route compared to previous strategies for obtaining radulanin A through late-stage molecular editing. The continuous flow approach overcomes the reproducibility issues of batch processes and allows the production of the compound on a larger scale (300 mg, 127 mg/24 h), which is useful for biological investigations.

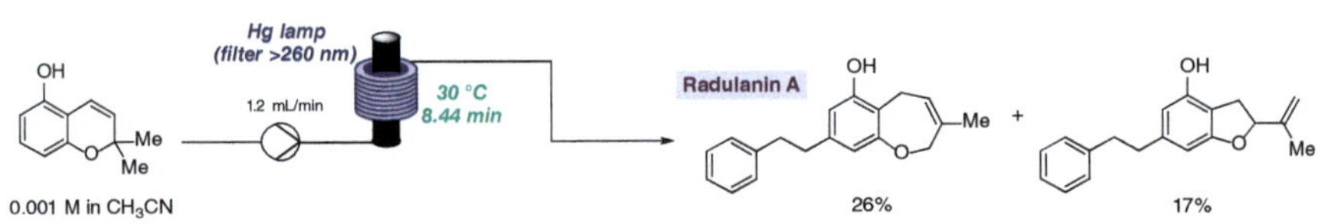

Scheme 4.2 Nay's photochemical ring expansion for the total synthesis of radulanin A.

Singh designed a micro-flow reactor for the ruthenium-catalyzed late-stage functionalization of pyrazolopyridines through reaction with alkynes under visible-light irradiation, without employing any oxidant or photocatalyst (Scheme 4.3) [22]. The fabricated reactor was integrated with a one-flow workup protocol for the selective removal of waste solvent. These optimized conditions were applied to prepare isoquinoline products in up to 89% yield from 12 analogs of sildenafil (a PDE-5 inhibitor), demonstrating the application of the methodology for late-stage functionalization and early-stage diversification purposes.

The mild conditions of photocatalytic protocols have also enabled the diversification of biomolecules containing sensitive functional groups. For instance, Noël and colleagues recently designed a photocatalytic hydroarylation protocol that couples radicals derived from arylthianthrenium salts with dehydroalanine (Dha), which has an electrophilic α,β-unsaturated moiety (Scheme 4.4) [23]. Interestingly, radical addition to Dha using arylthianthrenium salts derived from biologically active molecules has been accomplished, yielding a range of novel amino

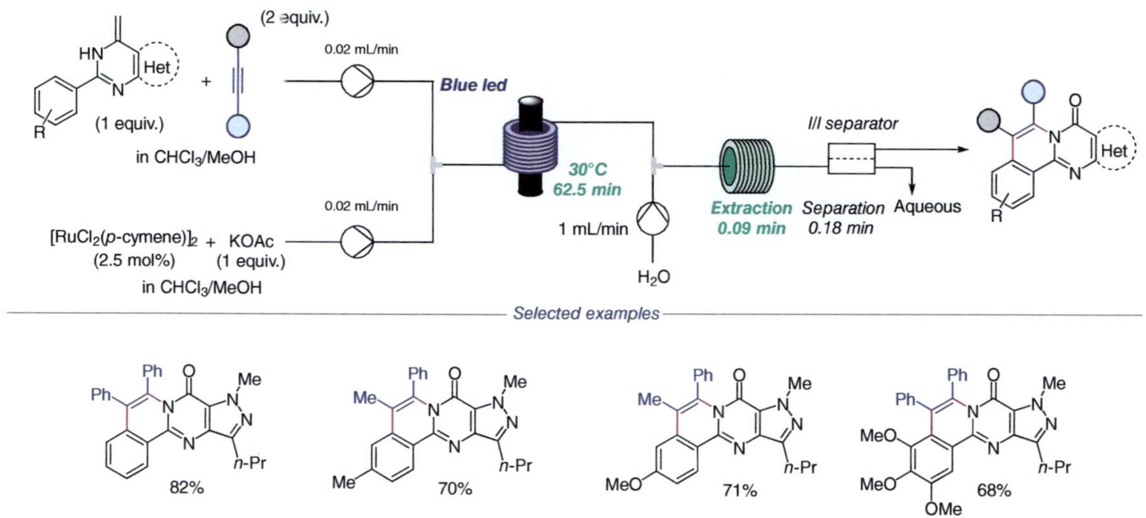

Scheme 4.3 Ruthenium-catalyzed late-stage functionalization of PDE-5 inhibitor intermediates.

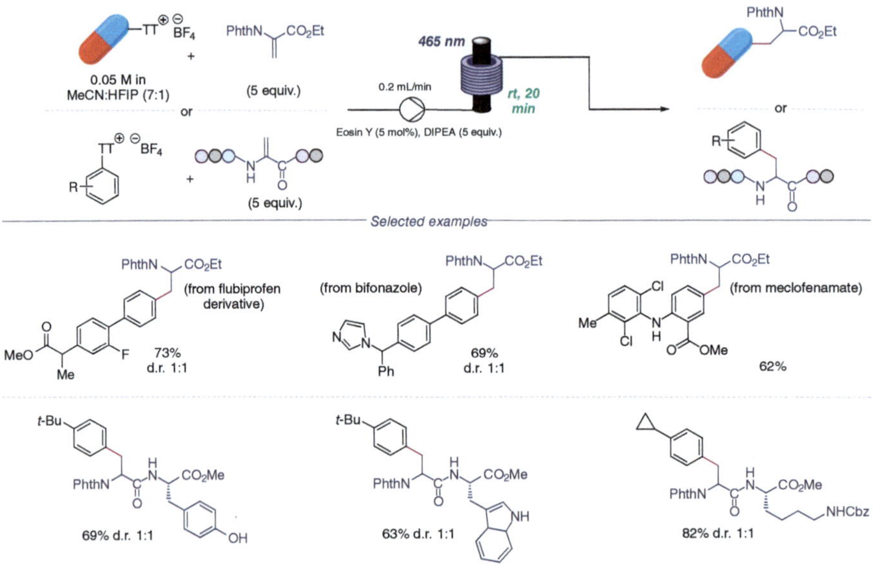

Scheme 4.4 Photocatalytic hydroarylation between arylthianthrenium salts and dehydroalanine.

acid/drug conjugates. Furthermore, the methodology was successfully extended to Dha-containing peptides with residues that have sensitive functionalities such as phenol and indole groups. The continuous flow approach ensures scalability and reduces the reaction time to 20 minutes compared to 4 hours necessary in batch processing.

Photocatalytic hydrogen atom transfer (HAT), which relies on the unique capacity of the excited state of a photocatalyst to cleave $C(sp^3)$—H bonds, delivering carbon-centered radicals, is a powerful approach for the late-stage functionalization of complex architectures [24]. Noël and coworkers have made significant contributions in this field, not only by enabling new synthetic methodologies but also by facilitating scalability through the use of flow microreactor technology (Scheme 4.5) [16, 25–27]. For example, they designed a methodology to forge $C(sp^3)$—N bonds by merging HAT photocatalysis with an oxidative radical-polar crossover (RPC) approach [28]. Specifically, carbon-centered radicals generated via tetrabutylammonium decatungstate (TBADT)-mediated HAT undergo an oxidation event promoted by *tert*-butyl hydroperoxide (TBHP) to form carbocations, which can then be trapped with *N*-heteroaryl nucleophiles. Continuous flow technology enables the scaling up of this strategy, which can be extended to the late-stage functionalization of biologically relevant molecules.

As mentioned earlier, the enhanced irradiation of the reaction mixture provided by microflow reactors often leads to a significant reduction in reaction time. Given that fluorine-18 (^{18}F) has a half-life of 109.771 minutes, this acceleration of the reaction rate could be particularly advantageous in light-mediated fluorine-18 chemistry, as demonstrated by Luxen, Genicot, and colleagues (Scheme 4.6) [29]. They reported

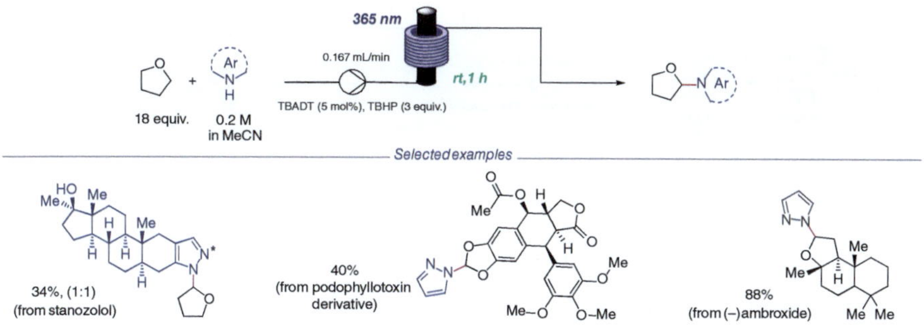

Scheme 4.5 HAT photocatalysis merged with an oxidative radical-polar crossover (RPC) approach.

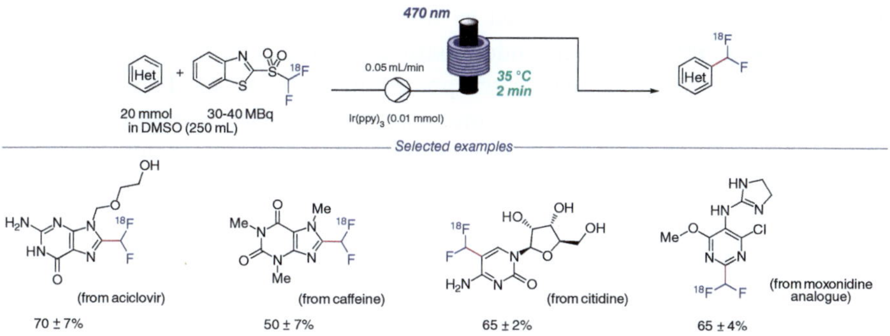

Scheme 4.6 Direct photoredox ^{18}F-difluoromethylation of N-heteroaromatics.

a direct photoredox ^{18}F-difluoromethylation of N-heteroaromatics proceeding through C—H activation using 2-((difluoromethyl)sulfonyl)benzo[*d*]thiazole (Hu's reagent) as the difluoromethyl (·CHF^{18}F) radical precursor. The use of a continuous flow setup enables the labeling of various drug-like compounds with high molar activity in less than 2 minutes, without the need for pre-functionalization.

Staying within the realm of visible-light-mediated fluoroalkylation protocols, Colella, Luisi, and coworkers recently reported a scalable and mild strategy for the monofluoroalkylation of a broad range of Giese acceptors in continuous flow (Scheme 4.7) [30]. This approach leverages readily available α-monofluorocarboxylic acids as versatile precursors for generating diverse secondary and tertiary monofluoroalkyl radicals via visible light-mediated oxidative CO_2 extrusion. While flow technology significantly enhances productivity and scalability, the mild reaction conditions – enabled by the transition-metal-free organic photocatalyst 4CzIPN – ensure broad functional group tolerance. These mild conditions make the strategy particularly well-suited for the late-stage functionalization of biologically relevant molecules, including menthol, amantadine, bepotastine, and estrone derivatives.

In 2020, Murakami designed a light-mediated dehydrogenative C(sp^2)—C(sp^3) strategy to couple alkylarenes and aldehydes using an iridium photocatalyst

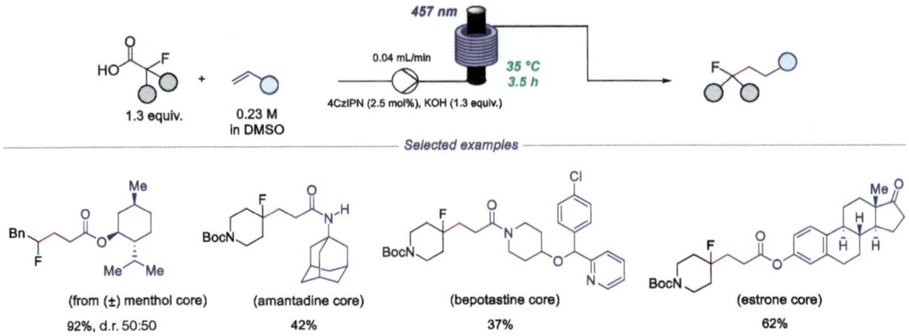

Scheme 4.7 Continuous flow decarboxylative monofluoroalkylation enabled by photoredox catalysis.

and a nickel co-catalyst [31]. Inspired by this methodology, Ley and coworkers reported a process intensification under continuous flow conditions, overcoming scalability challenges that hamper batch processing (Scheme 4.8) [32]. This was achieved by reducing reaction time, improving yields, and increasing reaction concentrations. The optimized continuous flow procedure was applied to the late-stage acylation of a pool of APIs containing methyl-substituted aromatic rings using 7-hydroxycitronellal as the coupling partner. Notably, the corresponding ketones were produced in good to excellent yields in only 120 minutes compared to the 20 hours required under batch conditions. By adopting a higher power flow photoreactor, the reaction can be conducted on a multigram scale while also recycling the precious iridium photocatalyst.

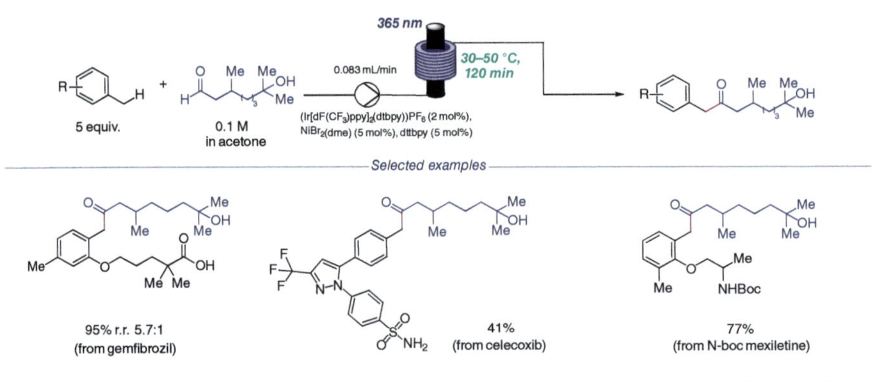

Scheme 4.8 Process intensification of light-mediated dehydrogenative $C(sp^2)-C(sp^3)$ coupling between alkylarenes and aldehydes.

4.3 Electrochemical Flow

In contrast to photocatalysis in flow, flow electrocatalysis, particularly flow metalloelectrocatalysis, is comparatively at the beginning of its evolution, as there are significant challenges associated with its development. First, often harsh reaction conditions are required; second, effective mass and electron transfer needs to be ensured; and finally, if a (metal-) catalyst is present, undesired cathodic reduction needs to be minimized. On the other hand, it offers vast prospects of improved electrode surface-to-volume ratio, easier scale-up, reduced electrolyte footprint, as well as more efficient heat and mass transfer, thus improving catalytic efficiency, so that its development efforts are imperative [33, 34].

Several fused polycyclic systems have been prepared by inter- and intramolecular C—H functionalization under electrochemical flow protocols. From aziridines over isoquinolines to more unusual 1,4-benzoxathiins, resourceful methodologies have been developed and applied in synthesis. One early example includes the flow-rhodaelectro-catalyzed annulation between internal alkynes and imidates to produce isoquinolines (Scheme 4.9) [35]. A purpose-designed flow cell consisting of a meshed thin PTFE plate as a turbulence promoter between the flexible GF anode and the Ni plate cathode enables the desired transformation when run at a constant potential of 1.5 V, a flow rate of 400 μL/min under 1 atm oxygen at room temperature, and $[Cp^*RhCl_2]_2$ (2.5 mol%) as the catalyst. The optimized protocol is amenable for both inter- and intramolecular annulation of (hetero) romatic imidates, independent of its steric or electronic properties, although strongly electron-deficient imidates undergo the desired transformation in reduced yields. Notably non-aromatic cyclohexene imidate is also a suitable precursor for this transformation.

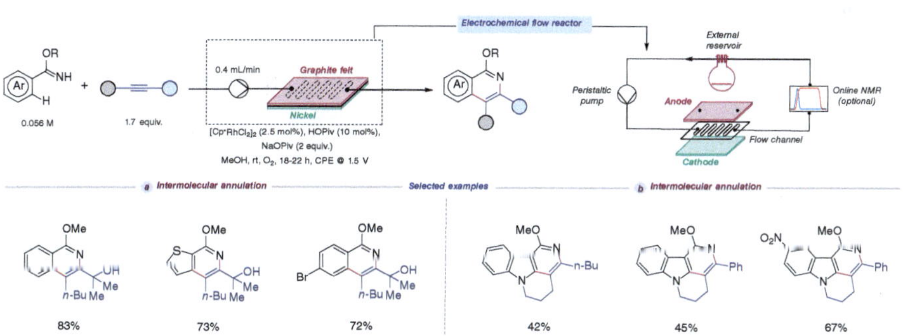

Scheme 4.9 Rhodaelectro-catalyzed annulation between internal alkynes and imidates for the synthesis of isoquinolines.

In a similar vein, Ackermann reported a flow-cobaltelectro-catalyzed C—H annulation between aromatic-substituted amides and allenes for the atroposelective construction C—N axially chiral 1-isoquinolones (Scheme 4.10) [36]. This is the first example of enantioselective electrochemical organometallic C—H activation in continuous flow. Specifically, benzamides, substituted on the nitrogen with 7-methylquinoline to provide a rotational barrier, reacted with allenes in an undivided flow cell at room temperature, constant current of 2 mA in the presence of chiral ligand **L** to provide the desired substituted 1-isoquinolones in generally

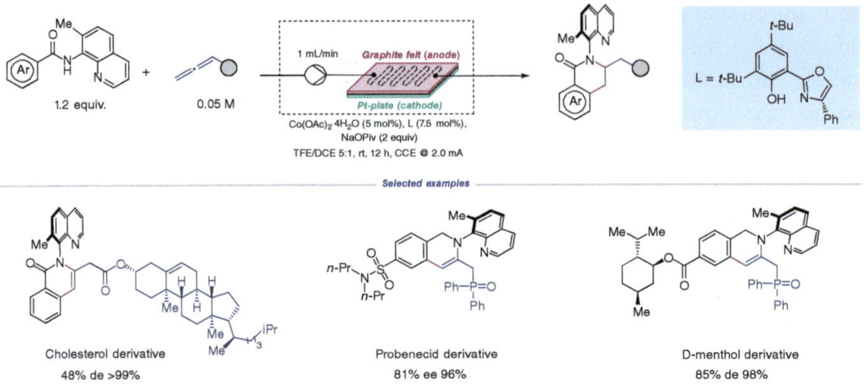

Scheme 4.10 Cobaltelectro-catalyzed C–H annulation between aromatic-substituted amides and allenes for the atroposelective construction C–N axially chiral 1-isoquinolones.

very good yield and excellent enantiomeric excesses (>96% ee). The suitability of this protocol for late-stage functionalization purposes was demonstrated by the transformation of biologically and pharmaceutically relevant motifs, including probenecid and cholesterol.

The synthesis of fused polycycles is, however, not limited to intermolecular annulations. Xu reported that appropriately substituted phenoxyethanethioamides or phenylaminoethanethioamides undergo intramolecular dehydrogenative C—H activation/C—S cross-coupling reactions to afford 1,4-benzoxathiins or 1,4-benzothiazines, respectively, under continuous flow electrosynthesis (Scheme 4.11) [37]. Electrolysis of the respective thioamides was conducted using a flow electrolytic cell consisting of a Pt-cathode and a carbon-filled polyvinylidene fluoride anode at constant current of 35 mA in the presence of scandium triflate in a mildly acidic medium (MeCN:TFA 9:1). Mechanistically, anodic oxidation of the amide affords the thioamidyl radical cation, which is in equilibrium with the neutral thioamidyl radical. The more *S*-nucleophilic radical cation can undergo fast cyclization with the phenyl ring affording the desired fused cycle after aromatization. The presence of Trifluoroacetic acid (TFA) is important to increase the concentration of the radical cation, and hence the rate of cyclization, preventing

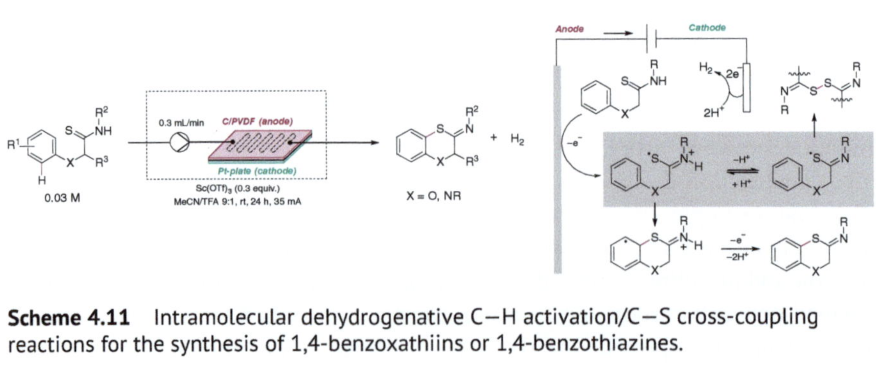

Scheme 4.11 Intramolecular dehydrogenative C–H activation/C–S cross-coupling reactions for the synthesis of 1,4-benzoxathiins or 1,4-benzothiazines.

undesired dimerization and hydrolysis by-products from the neutral thioamidyl radical. The optimized reaction conditions were compatible with aromatic rings of varying electronic properties, differing amide protecting groups, and variations on the α-position of the thioamide, which was shown to tolerate no substitution, mono-substitution, and bis-substitution, without notable variations in yield. Given its wide functional group tolerance, this protocol is thus amenable for late-stage functionalization. Notably, the expensive platinum cathode could be replaced with a more cost-effective sequential titanium- and platinum-plated stainless-steel plate.

Phenols are common motifs in a wide range of natural products and pharmaceuticals. Given its reactivity toward a range of electrophiles, and activation of the aromatic ring, laborious protection and deprotection steps are usually required to mask this functional group to prevent undesired side reactions. The directing-group free hydroxylation of arene C—H bonds is thus highly desirable in order to introduce the phenol-motif selectively as late as possible in a multi-step synthesis. However, undesired overoxidation of the desired phenol products and electrode fouling in batch processes are common hurdles for efficient processes. Given the low residence time, improved mass, heat transfer, and high surface-to-volume ratio, electrosynthesis in continuous flow is suitable to solve the aforementioned issues. As such, Oseka developed an electrochemical hydroxylation of arenes in continuous flow using trifluoroacetic acid as the oxygen source with a graphite anode/stainless steel cathode system (1.8–5.0 mA cm^{-2}) and a residence time of only 1.25 minute (Scheme 4.12a) [38]. Specifically, single-electron transfer anodic oxidation of the arene initiates the reaction. The resulting radical cation is subsequently trapped with trifluoroacetic acid, and further oxidation/aromatization affords the TFA-ester, which is hydrolyzed to the phenol during basic aqueous workup. As the reaction is dependent on the oxidation potential of the arene, hydroxylation proceeds exclusively on electron-rich arenes, and electron-deficient examples or benzene itself fail. Whereas the hydroxylation is tolerant to potentially sensitive functional groups (e.g., benzyl or propargyl ethers, double bonds, or esters), it was shown to be limited in regioselectivity, as commonly hydroxylation occurred to some extent on all activated positions (e.g., *ortho*- and *para*-hydroxylation of anisole).

The necessity for electron-rich arenes was subsequently alleviated in a protocol developed by Xu and coworkers (Scheme 4.12b) [39]. Operationally and mechanistically similar to the previous protocol, TFA is used as the oxygen source; however, the use of a platinum cathode, 2,6-lutidine as the amine base, and *t*BuOMe/MeOH as the solvent system proved to be significantly more tolerant to the electronic properties of the arene providing the hydroxylated arene within 22 seconds residence time. Although, in contrast to the previous described protocol, this proved to be suitable also to electron-deficient arenes, the lack of site-selectivity often afforded a mixture of hydroxylation products on various activated sites. Nonetheless, late-stage hydroxylation of pharmaceutically relevant compounds, including sulbactam, aniracetam, or fenofibrate, afforded the desired phenols site-selectively in up to 87% yield. In addition, the scalability aspect of continuous flow protocols is impressively showcased by 20 parallel reactors delivering over 200 g of product within 24 hours.

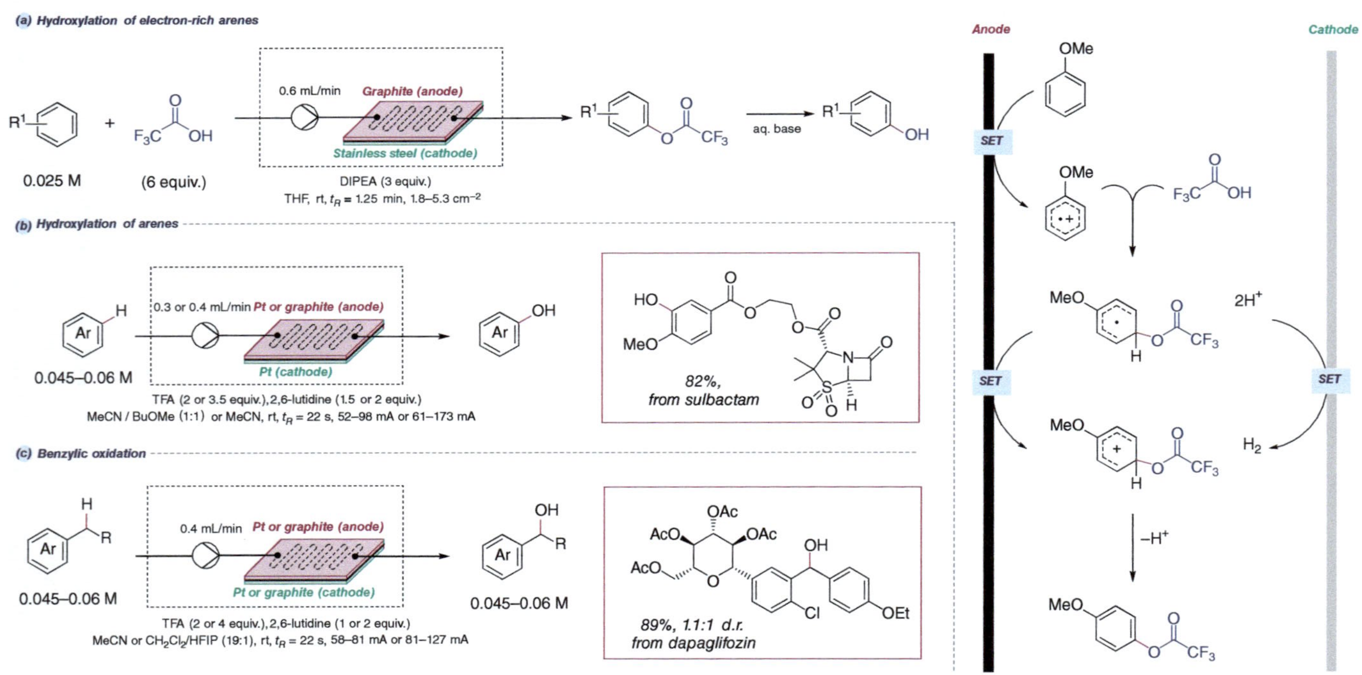

Scheme 4.12 (a) Hydroxylation of electron-rich arenes; (b) hydroxylation of electronically unactivated arenes; and (c) benzylic oxidation.

A closely related protocol allows for the selective benzylic oxidation of alkylarenes to benzylic alcohols, which is often notoriously difficult to achieve due to overoxidation (Scheme 4.12c) [40]. Under these conditions, dapagliflozin and methyl dehydroabietate were site-selectively oxidized to the corresponding benzyl alcohols.

Oxidation by C—H functionalization under electrochemical continuous flow is, however, not limited to arene oxidation, but also benzylic alcohols can be oxidized to aldehydes or ketones. In this regard, Lei developed a direct continuous flow electrochemical oxidation of benzylic primary and secondary alcohols under mild conditions with hydrogen as the sole by-product (Scheme 4.13) [41]. Using a carbon/nickel anode/cathode system, a constant current of 800 mA, and a residence time of only 10 minutes, the corresponding aldehydes and ketones were obtained in generally excellent yields under mild neutral conditions at room temperature. High chemoselectivity was observed as other benzylic positions remained untouched. Equally, benzylic alcohols can be selectively oxidized in the presence of distal aliphatic alcohols. The applicability to late-stage functionalization was demonstrated by the oxidation of a Rosuvastatin precursor.

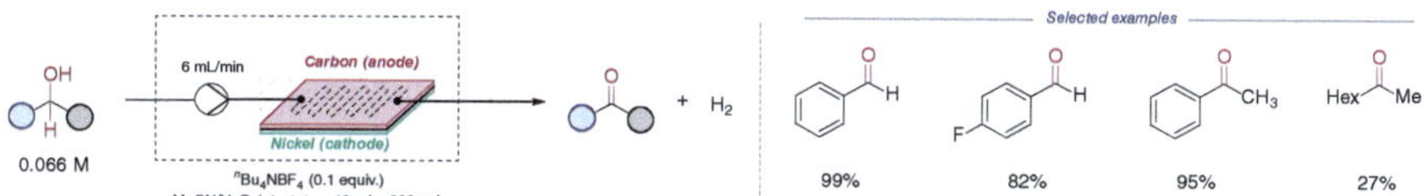

Scheme 4.13 Continuous flow electrochemical oxidation of benzylic primary and secondary alcohols.

In addition to arene oxidation, the selective installation of other functional groups by C—H activation is equally desirable. As such, significant efforts have been devoted to enabling a variety of such protocols in electrochemical continuous flow. For example, cyanated N-heterocycles are common motifs in pharmaceuticals and agrochemicals, including tanaproget or pruvanserin. Despite their propensity, installation often requires transition-metal catalysts or the presence of directing groups. Electrochemically, this transformation can be achieved with a platinum anode/cathode system, a constant current of 30–50 mA, and a residence time of only three to five minutes, using $n\text{Bu}_4\text{NCN}$ as the cyano-source (Scheme 4.14) [42]. Indoles, pyrroles, benzimidazole, and related *N*-heterocycles can be selectively cyanated, tolerating esters, amides, and alkenes.

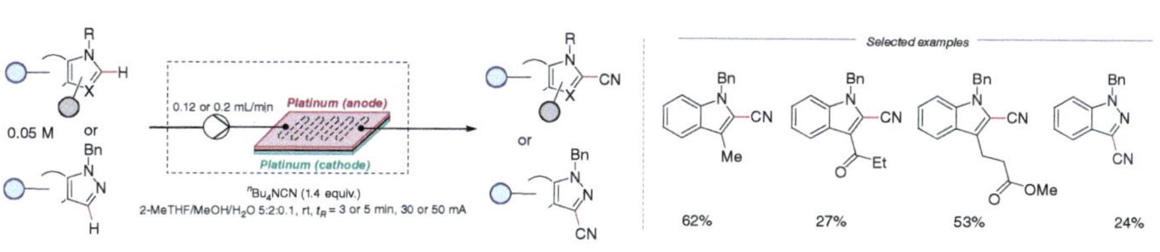

Scheme 4.14 Cyanation of N-heterocycles under electrochemical continuous flow.

In a similar vein, C—H phosphorylation of arenes is highly desirable, yet previous methodologies also suffer from limited scope or the requirement for transition metals or strong chemical oxidants. Direct phosphorylation in electrochemical continuous flow addresses these issues (Scheme 4.15) [43]. Anodic oxidation of a trialkylphosphite to a nucleophilic radical cation enables addition to the arene. Further oxidation and aromatization yields the desired phosphorylated products. Given the mild conditions, drug molecules, including indoprofen or benzafibrate, containing potentially sensitive functional groups (e.g., amides) can be phosphorylated site-selectively.

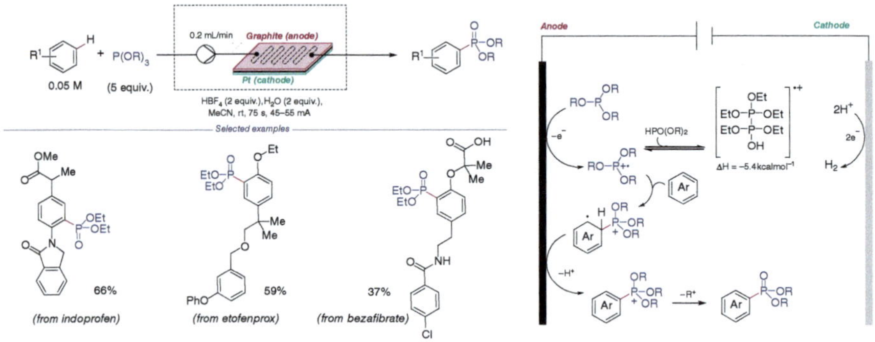

Scheme 4.15 Phosphorylation of arenes under electrochemical continuous flow.

Also other heteroatom-based functional groups can be installed on the aromatic ring. For example, Waldvogel and coworkers demonstrated a sulfonamidation protocol by C-H functionalization of activated arenes using secondary amines and free SO_2 (Scheme 4.16) [44]. This offers a conceptually different approach to the electrochemical oxidative coupling of amines and thiols under continuous flow conditions reported by Noel [45]. Waldvogel's process hinges on the formation of an amine–SO_2 adduct, which adds to the cationic radical intermediate generated from oxidation of the aromatic ring in single pass flow. Subsequent oxidation and aromatization afford the sulfonamidated aromatic motif.

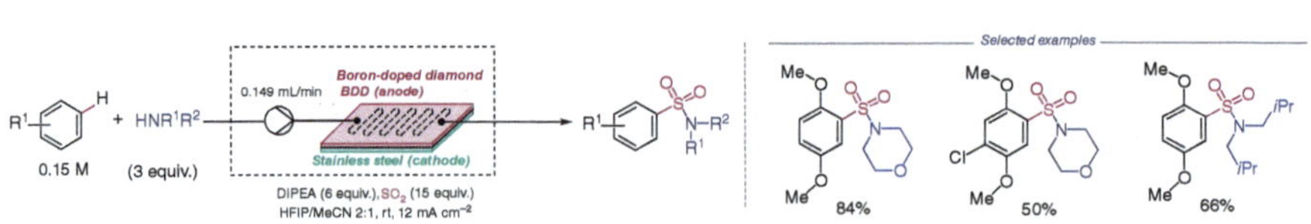

Scheme 4.16 Sulfonamidation of arenes under electrochemical continuous flow.

4.4 Flash Chemistry

Flash chemistry is a discipline within the field of continuous flow research, in which extremely fast reactions involving highly unstable intermediates are conducted in a highly controlled manner to produce the desired products with high selectivity. In many cases, the instability or reactivity of the (most commonly metalated) intermediates involved precludes such reactions from classical batch reactions as their lifetime is often within the (sub-)second timeframe. Continuous flow setups provide a solution for taming such short-lived intermediates as the reaction duration can be precisely controlled by the residence time within the flow apparatus, which is a function of reactor volume and flow rate, enabling residence times of few milliseconds (Scheme 4.17). In addition, microreactors enable rapid mixing, contributing to the success of extremely fast reactions in which the reaction rate exceeds the rate of diffusion. Although not fully coherent with the above-mentioned definition of late-stage functionalization, flash chemistry enables highly functional-group tolerant metalation/functionalization reactions, which are unfeasible under classical batch conditions and thus warrant inclusion in this chapter [46–49].

It is generally accepted that lithiation in the presence of electrophilic functional groups (e.g., isothiocyanates and nitriles) is unfeasible due to their susceptibility toward reactions with the lithiating agent (e.g., *n*-BuLi) or homo-coupling side products, decreasing selectivity and yields. This problem can be alleviated using flash chemistry. Isothiocyanate-substituted aryl iodides can be efficiently lithiated with phenyl lithium within 0.01 seconds at −40 °C and reacted further with a variety of electrophiles, including tributyltin chloride, methyl triflate, or N-fluorobenzenesulfonimide (NFSI), within 2.2 seconds, leaving the isothiocyanate group intact (Scheme 4.18a) [50]. When aryl bromides are used instead, to reduce the rate of lithium halogen exchange, the desired lithiation/quench by electrophile can be conducted at room temperature with a total reaction time of less than three seconds (Scheme 4.18b) [51]. Lowering the temperature to 0 °C or −28 °C, nitrile and nitro-groups are also tolerated, respectively, in the same sequence.

The presence of other reactive functional groups can be determined not only by the instability of metalated intermediates but also by the potential intramolecular decomposition pathways or rearrangements. For example, directed *ortho*-lithiation, followed by transmetallation to the aryl zinc of polyhalo-substituted (hetero)aryl tosylates, is challenging given (1) the presence of another exchangeable halogen on the aromatic ring and (2) the potential competitive elimination to form (hetero)arynes. Using *n*-butyllithium and a residence time of 2.35 seconds, *ortho*-selective lithium–halogen exchange was affected, followed by transmetallation with zinc chloride within less than nine seconds at −40 °C, and collected in batch for a subsequent Negishi coupling (Scheme 4.19) [52]. The synthetic utility of the obtained *ortho*-arylated motifs was showcased by the synthesis of Perampanel.

The presence of a directing group can facilitate selective lithiation by deprotonation or lithium–halogen exchange over other sites prone to lithiation. In the absence of such directing group, selective lithium–halogen exchange of one halogen over another by relying on different rates of exchange, while possible in batch, often requires energy-intense cryogenic conditions and produces a mixture of products which are tedious to separate. Lebel and coworkers demonstrated that such challenges can be overcome in continuous flow (Scheme 4.20) [53]. Thus, bromoiodoarenes can be chemoselectively metalated using Turbo Grignard reagents,

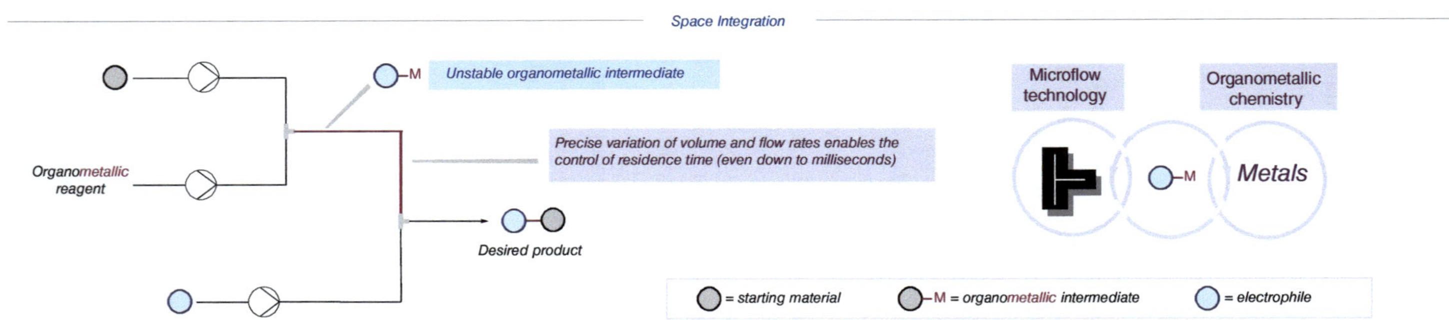

Scheme 4.17 Generation and taming of unstable organometallic intermediates in flow microreactors.

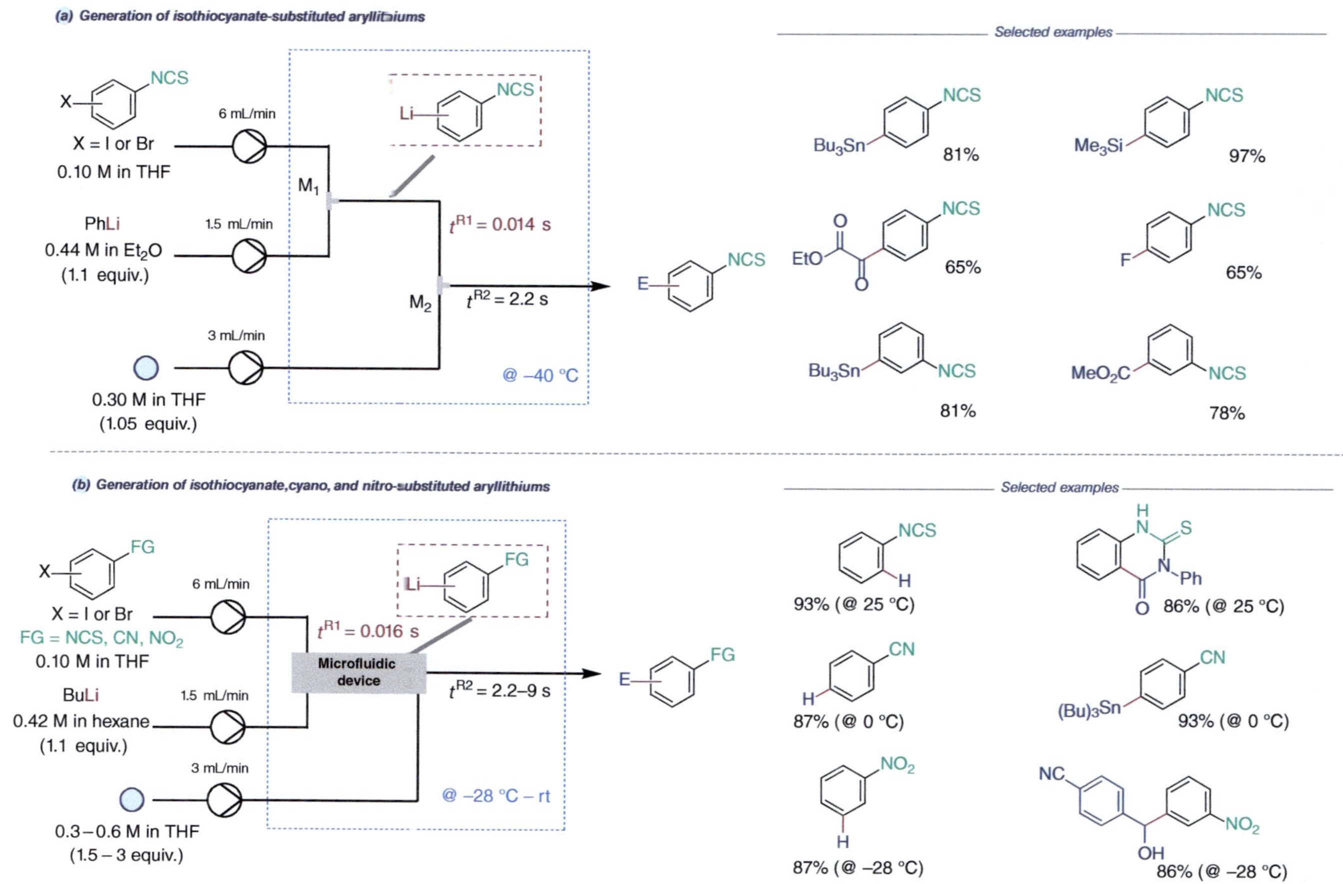

Scheme 4.18 (a) Generation of isothiocyanate-s‿bstituted aryllithiums and (b) generation of aryllithiums in the presence of reactive functional groups.

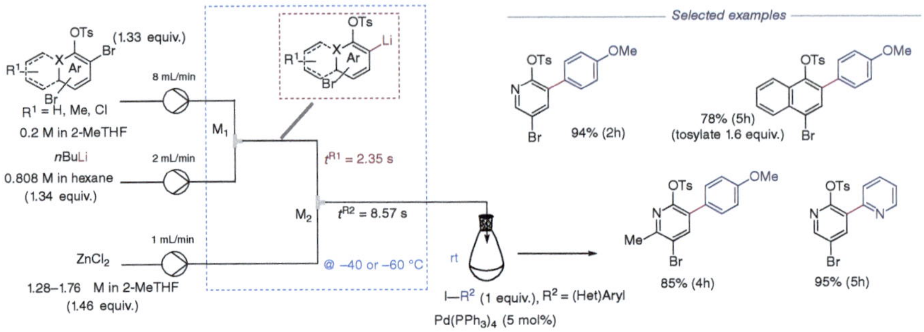

Scheme 4.19 *Ortho*-selective lithiation, transmetallation with zinc, and Negishi coupling of (hetero)aryl tosylates.

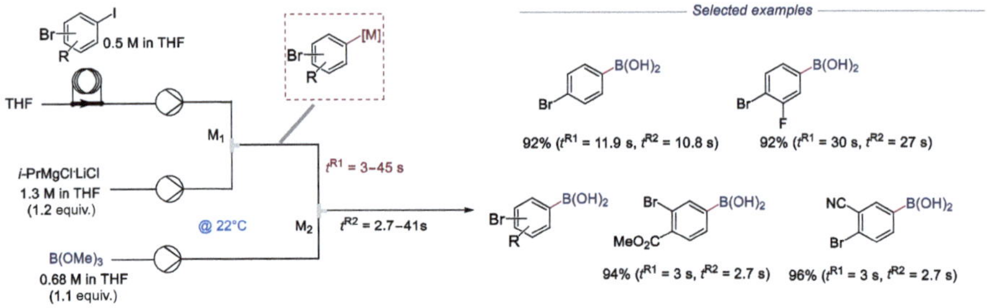

Scheme 4.20 Chemoselective metalation and functionalization of bromoiodoarenes.

and subsequently functionalized, at room temperature, even in the presence of electrophilic groups, including esters, nitriles, or ketones.

Lithiation and functionalization of strained cyclic ethers is a key synthetic challenge given the instability of the required anion. For example, direct C—H metalation of THF generates an α-anion, which is prone to rapid cleavage of the THF core by a [3+2]-cycloreversion in batch conditions. This unimolecular decomposition pathway can be suppressed to produce and functionalize the α-metalated intermediate under flash conditions using Schlosser's base (sec-BuLi/*t*BuOK) (Scheme 4.21) [54]. Functionalization with suitable electrophiles provided the THF-addition products, including alcohol-protected 7-ketolithocholic acid, a precursor to bioactive ursodeoxycholic acid.

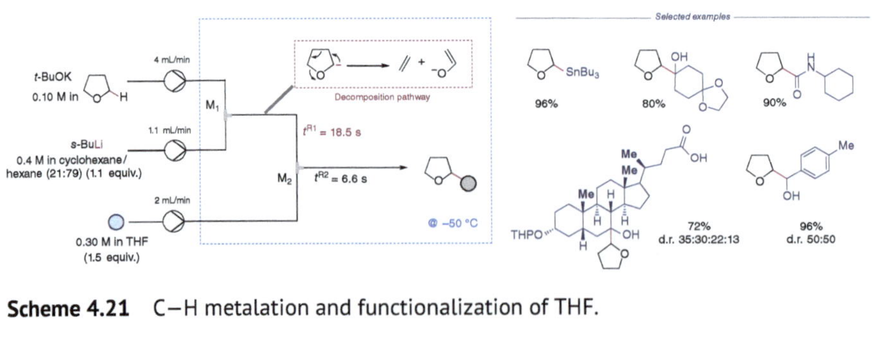

Scheme 4.21 C—H metalation and functionalization of THF.

In a similar vein, 3-oxetanyllithium is prone to undergo unimolecular decomposition to allyl alcohol upon formation. For this reason, 3-iodooxetane has frequently been used as an electrophile, but its use as a nucleophile remained elusive, despite the importance of the oxetane moiety as a bioisostere for dimethyl and carbonyl groups in drug development. Leveraging flash technology, Luisi and coworkers reported that 3-oxetanyllithium can be formed within 52 milliseconds, and subsequently added to electrophiles, which were introduced to the flow system through a second T-mixer (Scheme 4.22) [55]. Late-stage functionalization of ibuprofen and naproxen cores demonstrate the feasibility of the methodology for pharmaceutical development of complex substrates. Flash technology has also proven advantageous for the improved synthesis of other strained heterocycles [56–58].

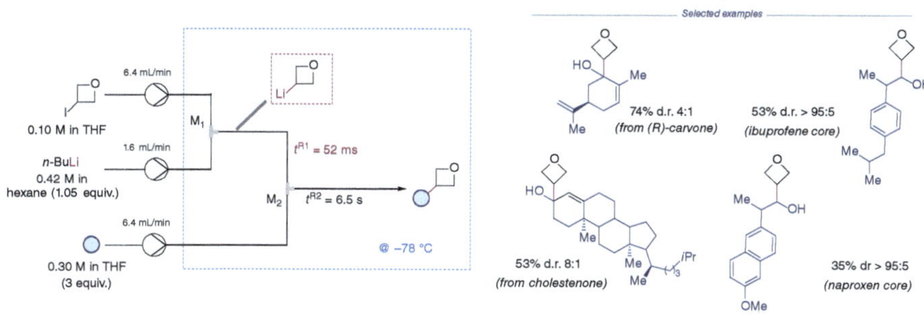

Scheme 4.22 Generation and functionalization of 3-oxetanyllithium from 3-iodooxetane.

In the realm of bioisosterism, fluoroalkanes also play an important role. One promising avenue for their introduction is via fluorinated organolithiums, bearing Li and F on the same carbon. Whereas seemingly straightforward in theory, the chemical and thermal instability of the required intermediate severely limits its synthetic utility. Nagaki, Luisi, and coworkers showed that fluoroiodomethane is a suitable and versatile precursor for such fluoroalkylation protocol (Scheme 4.23) [59]. For example, deprotonation with LDA generates fluoroiodomethyllithium with an estimated lifetime of only 82 milliseconds at −40 °C. Addition to ketones generates highly sensitive fluoroepoxides. In contrast, lithium–halogen exchange with methyllithium within even shorter residence time of 13 milliseconds at −60 °C generates and tames fluoromethyllithium, which in a second reactor reacts efficiently with Weinreb amides providing α-fluoroketones [60]. Fluorinated organolithiums bearing an additional aliphatic substituent, hence prone to carbene formation, can also be generated by lithium–halogen exchange from the corresponding iodide with *n*-butyllithium within 206 milliseconds at −78 °C, which reacted smoothly with ketones, aldehydes, imines, or sulfinimidate esters. Other dihalomethane precursors, including dibromomethane and iodochloromethane, can be exploited as monohaloalkylating agents in a similar vein using methyllithium or Turbo Grignard reagents, respectively [61, 62].

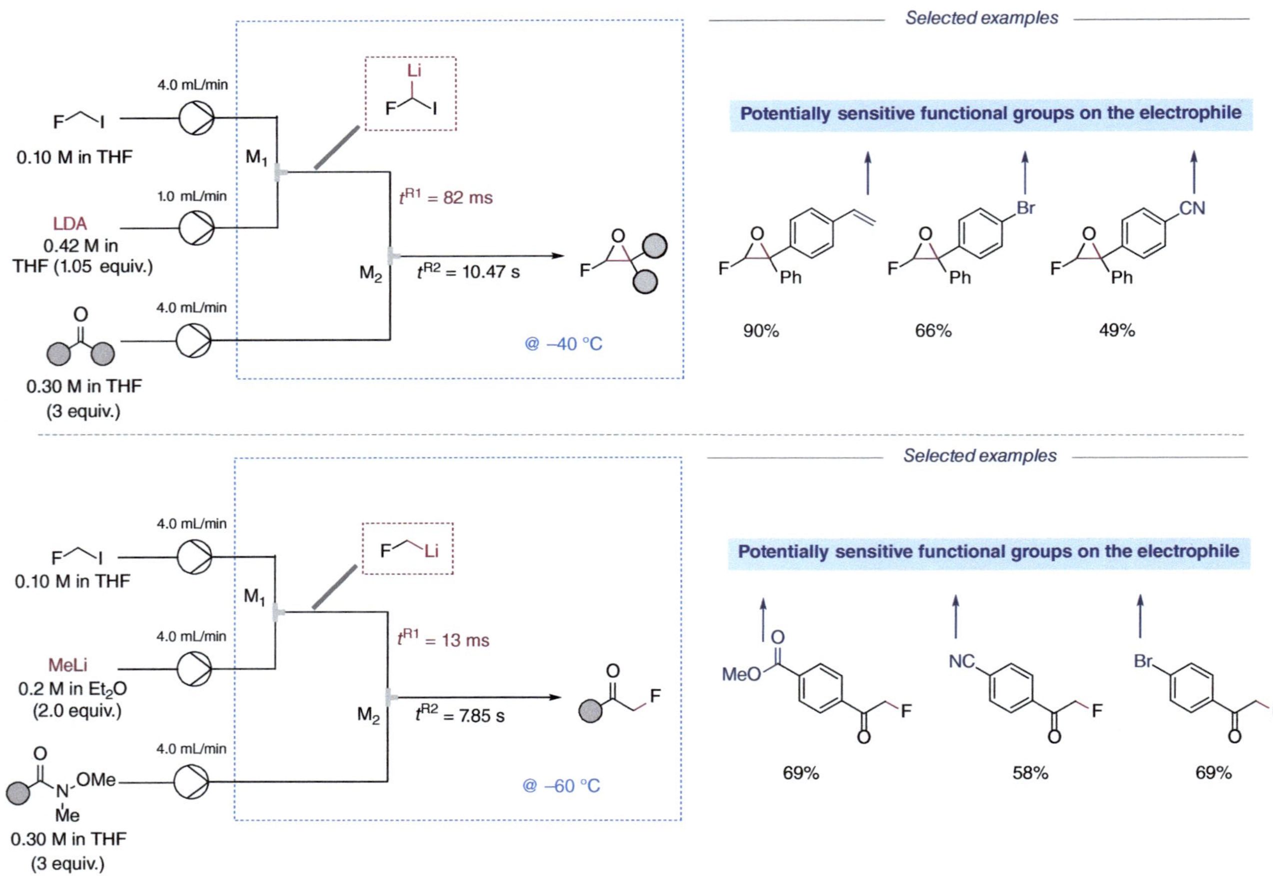

Scheme 4.23 Deprotonation/functionalization or lithium–halogen exchange/functionalization of fluoroiodomethane.

4.5 Homogeneous and Heterogeneous Metal Catalysis in Flow

With the move of the synthetic community toward greener protocols, the development of catalytic reactions is sought after, as they are generally more energy efficient and reduce quantities of waste and other unwanted by-products. Combination of such reactions with flow technology can further improve green metrics and reaction efficiency [63]. Generally, catalysis under continuous flow can be classified into two types [64]. First, homogeneous catalysis, in which the reagents, as well as the metal catalyst, are solubilized in the reaction solvent, requiring separation/purification of the product due to catalyst contamination. Second, heterogeneous catalysis, in which a reaction is promoted by a solid catalyst as a stationary phase, so that the catalyst is less likely to leach into the product solution, hence obviating the requirement for catalyst separation or laborious workup procedures and thus making it suitable for late-stage functionalization. Different reactor setups and types of catalyst immobilization (packed bed, wall coated, or monolith) can be used to enable flow reactions under heterogeneous catalysis categories (Figure 4.2).

With respect to homogeneous catalysis for late-stage functionalization in flow, in particular rhodium and palladium have played a key role for advancements. For example, benzylamines can be selectively *ortho*-alkylated with electron-deficient alkenes by rhodium(I)-catalyzed C—H activation (Scheme 4.24) [65]. *Ortho*-selectivity is induced by an amine-tethered picolinamide directing group, as the rhodium catalyst can be complexed between the benzylamine- and pyridine-nitrogen atoms in spatial vicinity to the *ortho*-position, making it particularly susceptible to C—H insertion of the transiently formed metal carbene. Optimized flow conditions, which were found to require a residence time of 196 minutes and elevated temperatures of 170 °C, produced the alkylated benzylamines in superior yields compared to the batch protocol, including a biologically active steroid derivative.

Dong and coworkers, on the other hand, showed that a rhodium(III)/silver(I) co-catalyst system efficiently enabled [3+2]-annulation between cyclic *N*-sulfonyl ketimines and internal alkenes to produce spirocyclic sultams by C(sp^2)—H activation (Scheme 4.25) [66]. This process proceeded efficiently and tolerated various substitutions on the alkyne, including aromatic and aliphatic substituents,

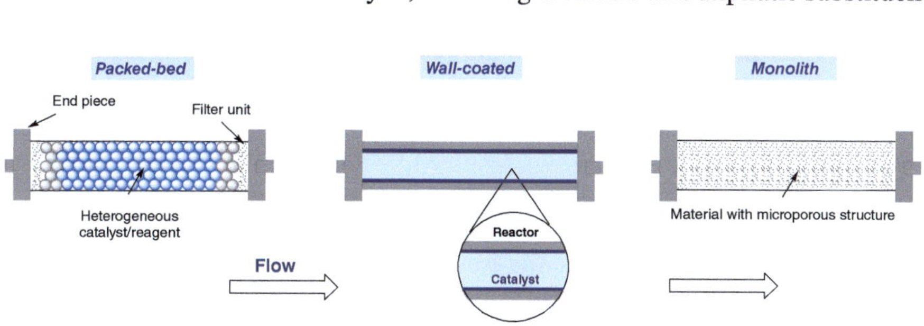

Figure 4.2 Different types of reactors for heterogeneous catalysis in flow.

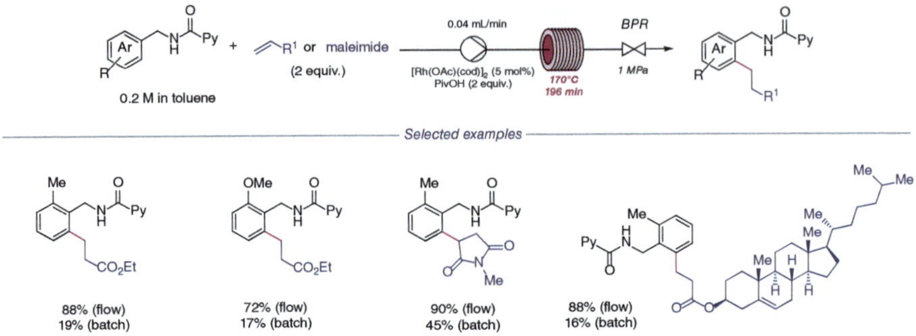

Scheme 4.24 *Ortho*-alkylation of electron-deficient alkenes with benzylamines by rhodium(I)-catalyzed C—H activation.

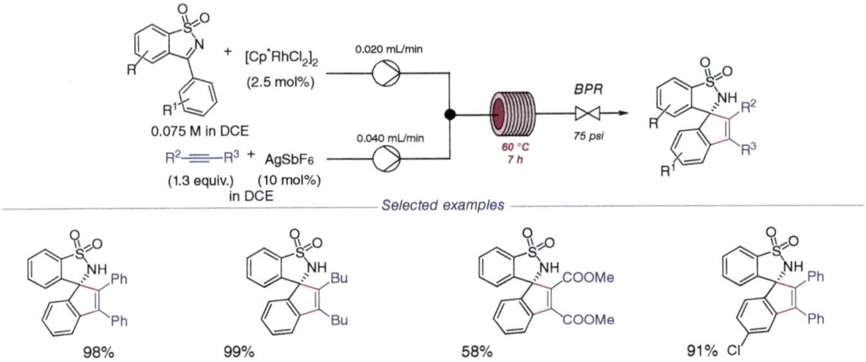

Scheme 4.25 Rhodium(III)/silver(I) co-catalyzed [3+2]-annulation between cyclic *N*-sulfonyl ketimines and internal alkenes.

as well as esters and aldehydes. A notable drawback is that unsymmetrical alkynes generally produced the desired spirocycles with poor regioselectivity.

The increase of molecular complexity by homogeneous catalysis is, however, not limited to the late-stage functionalization of aromatic C—H bonds, but also aliphatic C—H bonds can selectively be activated and transformed. For example, Charette and coworkers reported that bromoalkene-tethered cyclopropylamides can be chemoselectively transformed into hexahydroazepinones through a palladium-catalyzed tandem C(sp^3)—H alkenylation/ring-opening process under continuous flow conditions using 10 mol% Pd(II)-loading and a residence time of 40 minutes at 150 °C (Scheme 4.26) [67].

With a view on heterogeneous catalysis under continuous flow conditions, a conceptually similar protocol involving a C(sp^3)—H activation of unactivated methyl groups/cyclization of *N*-alkyl-2-bromoanisolines furnishes indolines (Scheme 4.27) [68]. Although such transformation is feasible under homogeneous batch conditions, the use of a polymer-supported *N*-heterocyclic carbene Pd(II)-pre-catalyst in a packed-bed-type reactor enables an efficient recovery and reuse of the catalyst. Interestingly, it is postulated that a "release-and-catch" mechanism is operative,

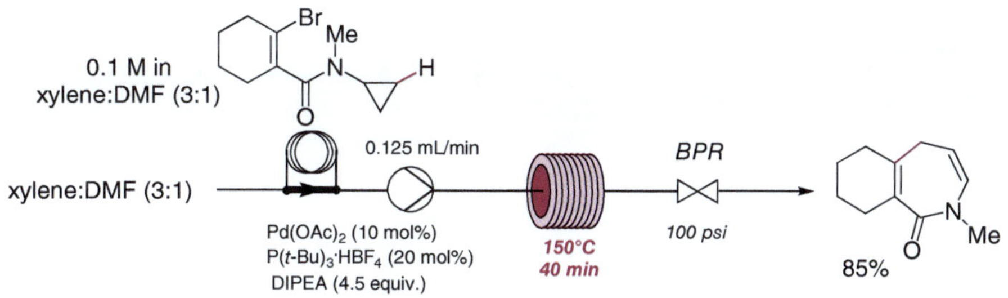

Scheme 4.26 Palladium-catalyzed tandem C(sp^3)—H alkenylation/ring-opening of bromoalkene-tethered cyclopropylamides.

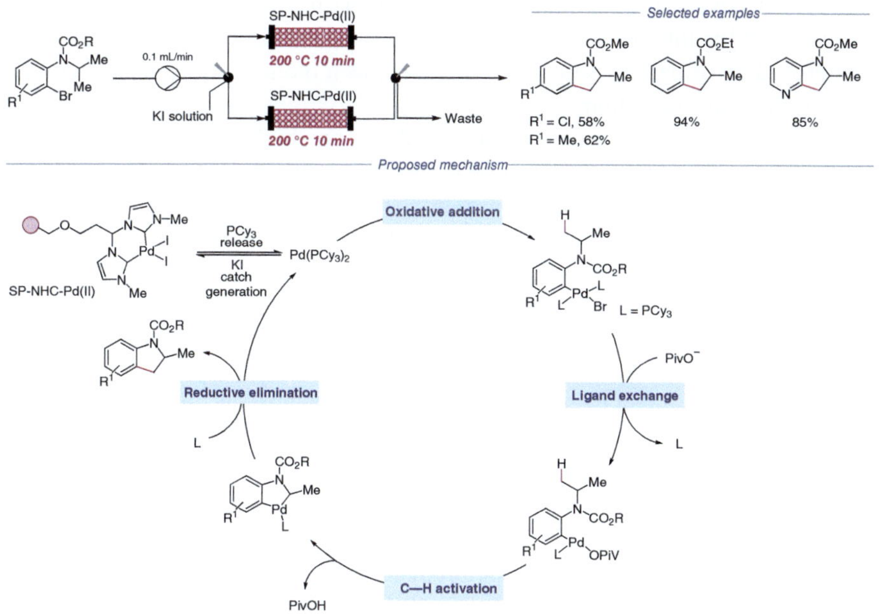

Scheme 4.27 C(sp^3)—H activation of unactivated methyl groups/cyclization of *N*-alkyl-2-bromoanisolines for the synthesis of indolines.

in which the catalytically active homogeneous Pd(0)-complex is released into the solution to catalyze the desired oxidative addition/C—H activation/reductive elimination sequence before being trapped into the polymer-supported bis-NHC ligand again. This is confirmed by minimal catalyst leaching into the product solution (max. 2 ppm) even after an extended steady-state operation of 20 hours, in which the obtained yield was also stable. In contrast, in the batch approach, ca. 17 ppm of palladium were leached into the product solution.

Ackermann and Vaccaro reported the C—H alkenylation of arenes by means of the Fujiwara–Moritani cross-coupling using commercially available Pd/C catalyst in a packed-bed reactor and benzoquinone as the stoichiometric oxidant (Scheme 4.28) [69]. Reaction of acetanilide with a series of acrylates affords the *ortho*-alkenylated

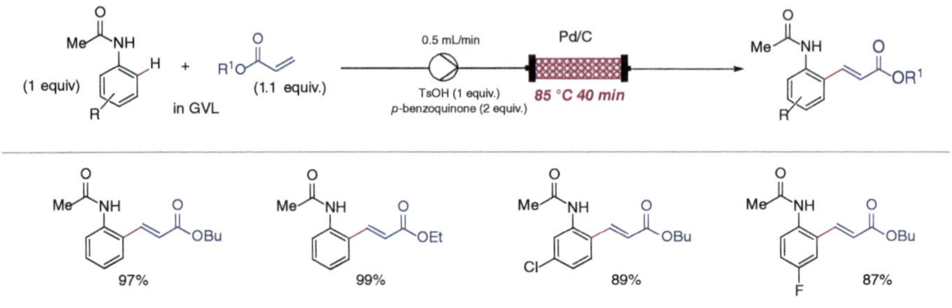

Scheme 4.28 C—H alkenylation of arenes by a Fujiwara–Moritani cross-coupling.

motifs site-selectively and in high productivity (>4 g h^{-1}), notably with minimal palladium-leaching into the solution (ca. 4 ppm). In addition, green metrics are further improved by utilizing γ-valerolactone (GVL) as a biomass-derived alternative solvent for this transformation.

Enantioselective syntheses of complex products are highly desirable in pharmaceutical drug development to reduce waste of potentially biologically inactive enantiomers, so that the research into new ligands, metal complexes, and transformations is highly sought after. Nonetheless, the frequent use and often-times high catalyst loadings of precious and expensive transition metals necessitates the increase of efficiency or facilitated recovery/reuse strategies for the catalysts involved. Heterogeneous continuous flow protocols, in which a chiral catalyst is immobilized in a packed bed, address the shortcomings. Kobayashi showed that substituted *ortho*-vinyl benzaldehydes are enantioselectively acylated with an immobilized rhodium(I)-catalyst and a BINAP-catalyst (Rh-BINAP/STA/SBA-15) (Scheme 4.29) [70]. This particular catalyst/support combination produced the desired indanones in generally high yields and *ee*s over an extended period of six hours (TON >300, compared to batch TON ≤50) before loss of catalyst activity is observed. Notably, Rh-leaching into the product solution was below the detection limit, and various aliphatic and alkyl substitutents are tolerated on the styrene.

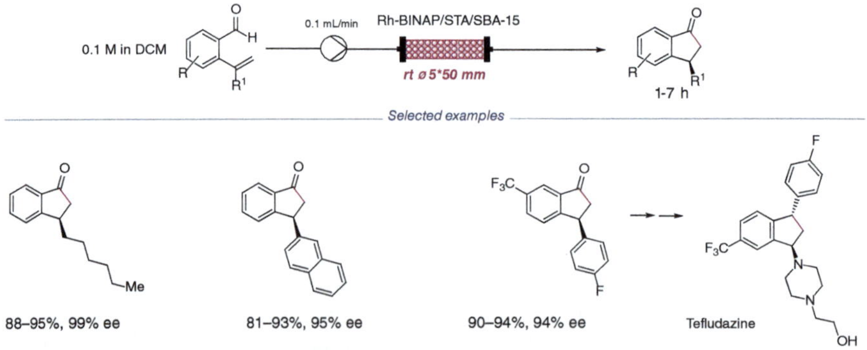

Scheme 4.29 Enantioselective acylation of *ortho*-vinyl benzaldehydes with an immobilized rhodium(I)-catalyst and a BINAP-catalyst.

The applicability of the protocol was showcased by the synthesis of a Tefludazine precursor.

Enantioselective heterogeneous catalysis in flow can also be extended to Lewis-acid catalysis. Whereas enantioselective Lewis-acid catalysis is well-explored in homogeneous conditions, its extension to heterogeneous conditions is facilitated by the use of flow technology. To this end, Kobayashi demonstrated that a heterogeneous Scandium-based Lewis acid is suitable for the catalysis of an enantioselective Friedel–Crafts type carbon–carbon bond formation between indoles and isatins, providing the corresponding addition products in good yield and high enantiomeric excess (Scheme 4.30) [71]. Notably, in contrast to conventional methods, the catalyst is not immobilized through covalent interactions, but an electrostatic acid–base interaction, thus increasing mass transfer efficiency and superior activity and selectivity, showcased by the obtained enantiomeric excess remaining consistently high even after 25 hours of reaction duration.

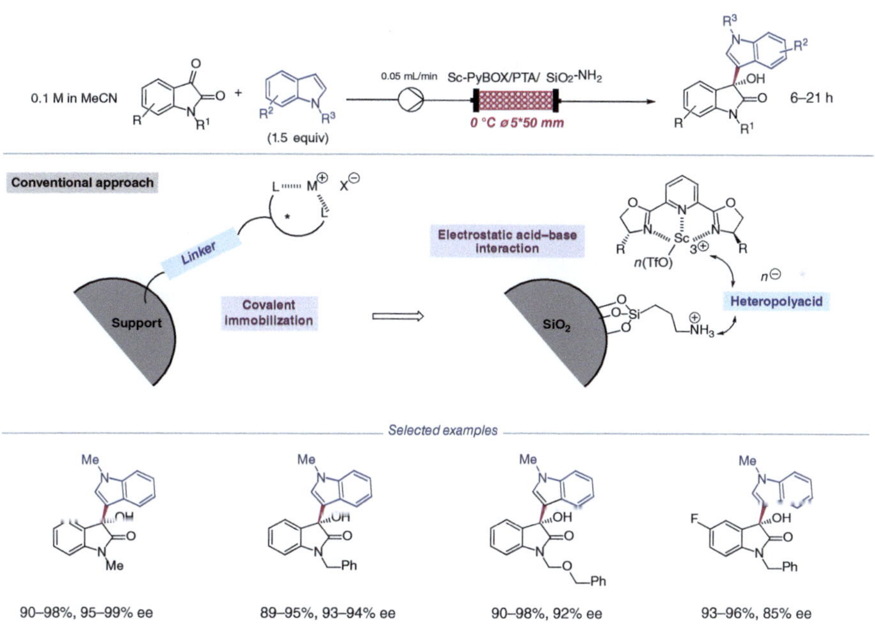

Scheme 4.30 Enantioselective Friedel–Crafts reaction between indoles and isatins.

4.6 Summary and Conclusions

Flow technology as an enabling technology has contributed significantly to advancements in late-stage functionalization with a wide range of advances having been achieved in the last five years. In tandem with photochemical or electrochemical technology, common drawbacks, including limited functional group tolerance, a lack of scalability, or low reproducibility were successfully addressed, allowing more rapid access to natural products, APIs, or hitherto inaccessible motifs for initial hit identification. Within the realm of flash technology, metalation and subsequent functionalization of highly unstable intermediates as well as intermediates bearing susceptible functional groups were achieved, given the possibility to precisely control residence times even in the millisecond timeframe. This is usually precluded from classical batch chemistry, hence drastically improving chemoselectivity issues, or taming intermediates, which had previously been impossible to generate, opening up new chemical space. Finally, homogeneous and heterogeneous catalysis has benefitted from the use of flow technology, by further improving green metrics, reaction efficiency, and chemoselectivity.

References

1 Börgel, J. and Ritter, T. (2020). Late-stage functionalization. *Chem* 6: 1877–1887.

2 Zhang, L. and Ritter, T. (2022). A perspective on late-stage aromatic C—H bond functionalization. *Journal of the American Chemical Society* 144: 2399–2414.

3 Santoro, S., Ferlin, F., Ackermann, L., and Vaccaro, L. (2019). C—H functionalization reactions under flow conditions. *Chemical Society Reviews* 48: 2767–2782.

4 Govaerts, S., Nyuchev, A., and Noel, T. (2020). Pushing the boundaries of C—H bond functionalization chemistry using flow technology. *Journal of Flow Chemistry* 10: 13–71.

5 Alfano, A.I., García-Lacuna, J., Griffiths, O.M. et al. (2024). Continuous flow synthesis enabling reaction discovery. *Chemical Science* 15: 4618–4630.

6 Martins, G.M., Braga, F.C., de Castro, P.P. et al. (2024). Continuous flow reactions in the preparation of active pharmaceutical ingredients and fine chemicals. *Chemical Communications* 60: 3226–3239.

7 Guillemard, L., Kaplaneris, N., Ackermann, L., and Johansson, M.J. (2021). Late-stage C—H functionalization offers new opportunities in drug discovery. *Nature Reviews Chemistry* 5: 522–545.

8 Melchiorre, P. (2022). Introduction: photochemical catalytic processes. *Chemical Reviews* 122: 1483–1484.

9 Bellotti, P., Huang, H.-M., Faber, T., and Glorius, F. (2023). Photocatalytic late-stage C—H functionalization. *Chemical Reviews* 123: 4237–4352.

10 Cambié, D., Bottecchia, C., Straathof, N.J.W. et al. (2016). Applications of continuous-flow photochemistry in organic synthesis, material science, and water treatment. *Chemical Reviews* 116: 10276–10341.

11 Buglioni, L., Raymenants, F., Slattery, A. et al. (2022). Technological innovations in photochemistry for organic synthesis: flow chemistry, high-throughput experimentation, scale-up, and photoelectrochemistry. *Chemical Reviews* 122: 2752–2906.

12 Sambiagio, C. and Noël, T. (2020). Flow photochemistry: shine some light on those tubes! *Trends in Chemistry* 2: 92–106.

13 Donnelly, K. and Baumann, M. (2021). Scalability of photochemical reactions in continuous flow mode. *Journal of Flow Chemistry* 11: 223–241.

14 Colella, M., Gelato, Y., Andresini, M. et al. (2023). Forging C—S bonds on the azetidine ring by continuous flow photochemical addition of thiols and disulfides to azetines. *European Journal of Organic Chemistry* 26: e202300413.

15 Pasca, F., Gelato, Y., Andresini, M. et al. (2024). Synthesis of alcohols: streamlined C1 to Cn hydroxyalkylation through photoredox catalysis. *Chemical Science* 15: 11337–11346.

16 Laudadio, G., Deng, Y., van der Wal, K. et al. (2020). C(sp^3)—H functionalizations of light hydrocarbons using decatungstate photocatalysis in flow. *Science* 369: 92–96.

17 Duhail, T., Bortolato, T., Mateos, J. et al. (2021). Radical α-trifluoromethoxylation of ketones under batch and flow conditions by means of organic photoredox catalysis. *Organic Letters* 23: 7088–7093.

18 Alfano, A.I., Smyth, M., Wharry, S. et al. (2024). Modular synthesis of benzoylpyridines exploiting a reductive arylation strategy. *Organic Letters* 26: 2847–2851.

19 Domański, M., Marcou, G., and Barham, J.P. (2024). An affordable, programmable and interactive continuous flow photoreactor setup for undergraduate organic synthetic teaching labs. *Journal of Flow Chemistry* 14: 349–355.

20 Leisering, S., Ponath, S., Shakeri, K. et al. (2022). Synthesis of 3-epi-hypatulin B featuring a late-stage photo-oxidation in flow. *Organic Letters* 24: 4305–4309.

21 Lockett-Walters, B., Thuillier, S., Baudouin, E., and Nay, B. (2022). Total synthesis of phytotoxic radulanin A facilitated by the photochemical ring expansion of a 2,2-dimethylchromene in flow. *Organic Letters* 24: 4029–4033.

22 Kumar, S. and Singh, A.K. (2022). Micro-photo-flow reactor system for fused N-heteroaryl scaffold synthesis and late-stage functionalization of pyrazolopyridines. *Chemical Communications* 58: 11268–11271.

23 Kaplaneris, N., Akdeniz, M., Fillols, M. et al. (2024). Photocatalytic functionalization of dehydroalanine-derived peptides in batch and flow. *Angewandte Chemie, International Edition* 63: e202403271.

24 Capaldo, L., Ravelli, D., and Fagnoni, M. (2022). Direct photocatalyzed hydrogen atom transfer (HAT) for aliphatic C—H bonds elaboration. *Chemical Reviews* 122: 1875–1924.

25 Sanjosé-Orduna, J., Silva, R.C., Raymenants, F. et al. (2022). Dual role of benzophenone enables a fast and scalable C-4 selective alkylation of pyridines in flow. *Chemical Science* 13: 12527–12532.

26 Luridiana, A., Mazzarella, D., Capaldo, L. et al. (2022). The merger of benzophenone HAT photocatalysis and silyl radical-induced XAT enables both nickel-catalyzed cross-electrophile coupling and 1,2-dicarbofunctionalization of olefins. *ACS Catalysis* 12: 11216–11225.

27 Wen, Z., Pintossi, D., Nuño, M., and Noël, T. (2022). Membrane-based TBADT recovery as a strategy to increase the sustainability of continuous-flow photocatalytic HAT transformations. *Nature Communications* 13: 6147.

28 Wan, T., Capaldo, L., Laudadio, G. et al. (2021). Decatungstate-mediated C(sp^3)—H heteroarylation via radical-polar crossover in batch and flow. *Angewandte Chemie, International Edition* 60: 17893–17897.

29 Trump, L., Lemos, A., Jacq, J. et al. (2020). Development of a general automated flow photoredox 18F-difluoromethylation of N-heteroaromatics in an AllinOne synthesizer. *Organic Process Research and Development* 24: 734–744.

30 Pasca, F., Gelato, Y., Andresini, M. et al. (2025). Continuous flow decarboxylative monofluoroalkylation enabled by photoredox catalysis. *JACS Au* 5: 684–692.

31 Kawasaki, T., Ishida, N., and Murakami, M. (2020). Dehydrogenative coupling of benzylic and aldehydic C—H bonds. *Journal of the American Chemical Society* 142: 3366–3370.

32 Griffiths, O.M., Esteves, H.A., Chen, Y. et al. (2021). Photoredox-catalyzed dehydrogenative Csp^3—Csp^2 cross-coupling of alkylarenes to aldehydes in flow. *The Journal of Organic Chemistry* 86: 13559–13571.

33 Mazzarella, D. (2024). Asymmetric electrochemical synthesis in flow. *Journal of Flow Chemistry* 14: 357–366.

34 Lehnherr, D. and Chen, L. (2024). Overview of recent scale-ups in organic electrosynthesis (2000–2023). *Organic Process Research and Development* 28: 338–366.

35 Kong, W.-J., Finger, L.H., Messinis, A.M. et al. (2019). Flow rhodaelectro-catalyzed alkyne annulations by versatile C—H activation: mechanistic support for rhodium(III/IV). *Journal of the American Chemical Society* 141: 17198–17206.

36 Lin, Y., von Münchow, T., and Ackermann, L. (2023). Cobaltaelectro-catalyzed C—H annulation with allenes for atropochiral and P-stereogenic compounds: late-stage diversification and continuous flow scale-up. *ACS Catalysis* 13 (14): 9713–9723.

37 Huang, C., Qian, X., and Xu, H. (2019). Continuous-flow electrosynthesis of benzofused *S*-heterocycles by dehydrogenative C—S cross-coupling. *Angewandte Chemie, International Edition* 58: 6650–6653.

38 Kooli, A., Wesenberg, L., Beslać, M. et al. (2022). Electrochemical hydroxylation of electron-rich arenes in continuous flow. *European Journal of Organic Chemistry* 2022: e202200011.

39 Long, H., Chen, T.-S., Song, J. et al. (2022). Electrochemical aromatic C—H hydroxylation in continuous flow. *Nature Communications* 13: 3945.

40 Chen, T., Long, H., Gao, Y., and Xu, H. (2023). Continuous flow electrochemistry enables practical and site-selective C—H oxidation. *Angewandte Chemie, International Edition* 62: e202310138.

41 Wang, D., Wang, P., Wang, S. et al. (2019). Direct electrochemical oxidation of alcohols with hydrogen evolution in continuous-flow reactor. *Nature Communications* 10: 2796.

42 Carvalho, M., Demin, S., Martinez-Lamenca, C. et al. (2022). Expedient access to cyanated N-heterocycles by direct flow-electrochemical $C(sp^2)$—H activation. *Chemistry – A European Journal* 28: e202103384.

43 Long, H., Huang, C., Zheng, Y.-T. et al. (2021). Electrochemical C—H phosphorylation of arenes in continuous flow suitable for late-stage functionalization. *Nature Communications* 12: 6629.

44 Schneider, J., Blum, S.P., and Waldvogel, S.R. (2023). Electrochemical synthesis of sulfonamides in single-pass flow. *ChemElectroChem* 10: e202300456.

45 Laudadio, G., Barmpoutsis, E., Schotten, C. et al. (2019). Sulfonamide synthesis through electrochemical oxidative coupling of amines and thiols. *Journal of the American Chemical Society* 141: 5664–5668.

46 Colella, M., Nagaki, A., and Luisi, R. (2020). Flow technology for the genesis and use of (highly) reactive organometallic reagents. *Chemistry – A European Journal* 26: 19–32.

47 Natho, P. and Luisi, R. (2023). Flow chemistry as green technology for the genesis and use of organometallic reagents in the synthesis of key building blocks and APIs – an update. *Tetrahedron Green Chem* 2: 100015.

48 Spennacchio, M., Natho, P., Andresini, M., and Colella, M. (2023). Continuous flow generation of highly reactive organometallic intermediates: a recent update. *Journal of Flow Chemistry* 14: 43–83.

49 Yoshida, J., Takahashi, Y., and Nagaki, A. (2013). Flash chemistry: flow chemistry that cannot be done in batch. *Chemical Communications* 49: 9896–9904.

50 Lee, H.-J., Torii, D., Jeon, Y. et al. (2020). Integrated synthesis using isothiocyanate-substituted aryllithiums by flow chemistry. *Synlett* 31: 1899–1902.

51 Kang, J.-H., Ahn, G.-N., Lee, H. et al. (2022). Scalable subsecond synthesis of drug scaffolds via aryllithium intermediates by numbered-up 3D-printed metal microreactors. *ACS Central Science* 8: 43–50.

52 Kwon, Y., Lee, S., Shin, I., and Kim, W. (2023). Regioselective *ortho*-arylation of polyhalo-substituted (hetero)aryl tosylates using an integrated continuous flow/batch protocol. *Advanced Synthesis and Catalysis* 365: 263–268.

53 Reynard, G., Wimmer, E., Richelet, J. et al. (2023). Chemoselective borylation of bromoiodoarene in continuous flow: synthesis of bromoarylboronic acids. *Journal of Flow Chemistry* 13: 21–29.

54 Kim, D., Lee, H.-J., Shimizu, Y. et al. (2022). Direct C—H metallation of tetrahydrofuran and application in flow. *Nature Synthesis* 1: 558–564.

55 Natho, P., Colella, M., Andresini, M. et al. (2024). Taming 3-oxetanyllithium using continuous flow technology. *Organic Letters* 26: 3032–3036.

56 Graziano, E., Natho, P., Andresini, M. et al. (2024). 1-Oxa-2,6-diazaspiro[3.3]heptane as a new potential piperazine bioisostere – flow-assisted preparation and derivatisation by strain-release of azabicyclo[1.1.0]butanes. *Advanced Synthesis and Catalysis* 366: 3894–3902.

57 Musci, P., Colella, M., Andresini, M. et al. (2022). Flow technology enabled preparation of C3-heterosubstituted 1-azabicyclo[1.1.0]butanes and azetidines: accessing unexplored chemical space in strained heterocyclic chemistry. *Chemical Communications* 58: 6356–6359.

58 Musci, P., Keutz, T., Belaj, F. et al. (2021). Flow technology for telescoped generation, lithiation and electrophilic (C3) functionalization of highly strained 1-azabicyclo[1.1.0]butanes. *Angewandte Chemie, International Edition* 60: 6395–6399.

59 Colella, M., Tota, A., Takahashi, Y. et al. (2020). Fluoro-substituted methyllithium chemistry: external quenching method using flow microreactors. *Angewandte Chemie, International Edition* 59: 10924–10928.

60 Spennacchio, M., Colella, M., Andresini, M. et al. (2023). Unlocking geminal fluorohaloalkanes in nucleophilic fluoroalkylation chemistry: generation and trapping of lithiumfluorocarbenoids enabled by flow microreactors. *Chemical Communications* 59: 1373–1376.

61 Musci, P., Colella, M., Sivo, A. et al. (2020). Flow microreactor technology for taming highly reactive chloroiodomethyllithium carbenoid: direct and chemoselective synthesis of α-chloroaldehydes. *Organic Letters* 22: 3623–3627.

62 von Keutz, T., Cantillo, D., and Kappe, C.O. (2019). Continuous flow synthesis of terminal epoxides from ketones using in situ generated bromomethyl lithium. *Organic Letters* 21: 10094–10098.

63 Graziano, E., Cannillo, D., Spennacchio, M. et al. (2023). A sustainable and chemoselective continuous flow hydrogenation of functionalized 2-azetines to azetidines. *Tetrahedron Green Chem* 1: 100003.

64 Masuda, K., Ichitsuka, T., Koumura, N. et al. (2018). Flow fine synthesis with heterogeneous catalysts. *Tetrahedron* 74: 1705–1730.

65 Das, A. and Chatani, N. (2021). Rh(I)- and Rh(II)-catalyzed C—H alkylation of benzylamines with alkenes and its application in flow chemistry. *Chemical Science* 12: 3202–3209.

66 Zheng, T., Lin, M., and Dong, L. (2022). Continuous flow chemistry synthesis of spirocyclic sultams and isoquinolines through rhodium-catalyzed C—H activation. *European Journal of Organic Chemistry* 2022: e202201170.

67 Saint-Jacques, K., Ladd, C.L., and Charette, A.B. (2022). Access to hexahydroazepinone heterocycles via palladium-catalysed C(sp^3)—H alkenylation/ring-opening of cyclopropanes. *Chemical Communications* 58: 7550–7553.

68 Ferlin, F., Anastasiou, I., Salameh, N. et al. (2022). C(sp^3)—H arylation promoted by a heterogeneous palladium-*N*-heterocyclic carbene complex in batch and continuous flow. *ChemSusChem* 15: e202102736.

69 Ferlin, F., Santoro, S., Ackermann, L., and Vaccaro, L. (2017). Heterogeneous C—H alkenylations in continuous-flow: oxidative palladium-catalysis in a biomass-derived reaction medium. *Green Chemistry* 19: 2510–2514.

70 Saito, Y. and Kobayashi, S. (2024). Continuous-flow enantioselective hydroacylations under heterogeneous chiral rhodium catalysts. *Angewandte Chemie, International Edition* 63: e202313778.

71 Saito, Y. and Kobayashi, S. (2021). Chiral heterogeneous scandium Lewis acid catalysts for continuous-flow enantioselective Friedel–Crafts carbon–carbon bond-forming reactions. *Angewandte Chemie, International Edition* 60: 26566–26570.

5

Radioisotope Insertion Through Late-Stage Functionalization

Emmanuelle Dubost[1,2,3], *Maxime Jay*[1], *and Thomas Cailly*[1,3,4,5]

[1] *Université de Caen Normandie, CERMN UR 4258, 14000 Caen, France*
[2] *Université de Caen Normandie, PhIND UMR-S U1237, 14000 Caen, France*
[3] *Institut Blood and Brain @Caen Normandie (BB@C), 14000 Caen, France*
[4] *Université de Caen Normandie, CYCERON UAR 3408-US50, IMOGERE, 14000 Caen, France*
[5] *CHU Côte de Nacre, Department of Nuclear Medicine, 14000 Caen, France*

5.1 Introduction

Insertion of radioactive isotopes on (bio)molecules has revolutionized research and clinical practice, offering powerful tools for understanding complex biological systems and improving patient care in nuclear medicine. The obtained radiolabeled molecules play a crucial role in various scientific or medical fields. For instance, they are essential in assessing biochemical reactions by allowing to trace the pathways and interactions of molecules within living organisms. Additionally, these labeled molecules are used to measure the *in vivo* distribution of substances, providing essential information for drug development and pharmacokinetic studies. Radiolabeled molecules are also employed in various *in vitro* assays, such as binding assays or radioimmunoassays (RIA). These assays rely on the titration of radiolabeled molecules making them indispensable tools in biochemical and clinical research. In the field of nuclear medicine, the applications of radiolabeled molecules extend to both therapy and diagnosis. Radiotherapeutics, used in radioisotope therapy (RIT) [1, 2], can target and destroy diseased cells, such as cancer cells, with minimal effect on surrounding healthy tissue. In molecular imaging experiments, radiolabeled molecules serve as radiotracers in either positron emission tomography (PET) or single photon emission computed tomography (SPECT) [3, 4] to provide high-resolution images of the physiological and biochemical processes occurring within the body. To date, PET and SPECT are used to diagnose and monitor a wide range of conditions, including cancer, neurological disorders, and cardiovascular diseases.

Several non-metallic radioisotopes, when incorporated into molecular substrates, have diverse and crucial applications in science and medicine (Table 5.1). Carbon-11 is used in PET imaging due to its short half-life and ability to label

Late-Stage Functionalization and Diversification in Organic Synthesis: Methods and Applications, First Edition. Edited by Tatiana Besset.

Table 5.1 Useful non-metallic radioisotopes.

Isotope	Half-life	Main emissions	Main energies (keV)	Applications
Hydrogen-3 (tritium)	12.3 years	β^-	19	Biology, ADME
Carbon-11	20.4 minutes	β^+ γ	960 511	PET
Carbon-14	5730 years	β^-	157	Biology, ADME
Fluorine-18	1.8 hours	β^+ γ	637 511	PET
Iodine-123	13.2 hours	γ	159	SPECT
Iodine-124	4.2 days	β^+ β^+ γ γ	1532 2135 511 603	PET
Iodine-125	59.9 days	X γ	27 35	Biology, research
Iodine-131	8.0 days	β γ	606 364	RIT

various biomolecules. The longer half-life carbon-14 is used to trace biochemical pathways. Iodine-125 is employed in fundamental research, RIA [5, 6] and *in vitro* assays. Iodine-131 is used in the treatment of thyroid cancer and hyperthyroidism. SPECT imaging can be performed with iodine-123 due to its suitable half-life and gamma radiation. Iodine-124 is utilized in PET imaging. Hydrogen-3 or tritium, is used in biochemical research, binding experiments, and *in vivo* fate determination. Fluorine-18 is extensively used in PET imaging.

To date, most of radiolabeling strategies using non-metallic elements rely on a precursor-based approach. In this approach, a functional group, usually introduced early in the synthesis, is replaced in a final radiochemical step to produce the desired radioisotopically labeled derivative (Figure 5.1a). Additionally, a divergent synthetic route is often necessary to produce the non-radioactive reference needed for monitoring the radiochemical process. Several important specifications must be considered when designing precursor-based radiosynthetic strategies:

1) The precursor must be stable and storable and/or easy to obtain and purify.
2) The radiochemical synthesis must achieve high yield or radiochemical yield (RCY) or radiochemical conversion (RCC) [7] , along with high radiochemical purity (RCP) and specific activity (A_S in Bq.g^{-1}) or molar activity (A_m in Bq.mol^{-1}) [8] of the resulting radiochemical.
3) The radiosynthesis duration must align with the half-life of the chosen isotope.
4) The radionuclide should be introduced in the final synthesis step to minimize safety issues and waste production.

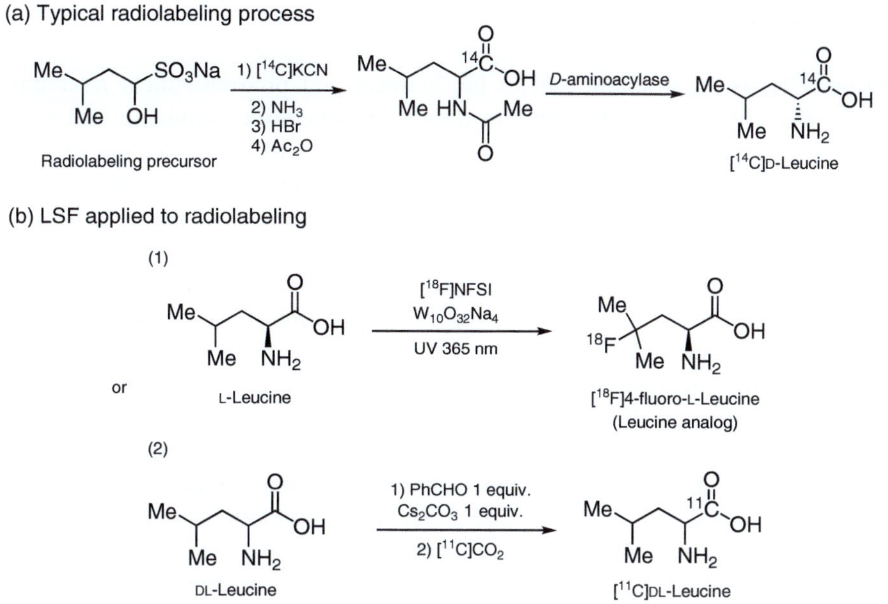

Figure 5.1 Radiolabeling through a precursor-based approach (a) vs. late-stage functionalization (LSF) (b) in the case of leucine. *Source:* Ström et al. [13], Lundgren et al. [14] and Britton et al. [15].

5) The radiosynthesis should ideally be straightforward to implement with minimal equipment. The reaction conditions should be mild, preferably at room temperature, and the resulting radiochemical should be easy to purify.
6) Radiation safety must be taken into account according to local regulation.

In recent decades, these challenging requirements have significantly influenced chemical labeling strategies for increasingly complex (bio)molecules. Considering the difficulties and high costs associated with radiolabeling processes, late-stage functionalization (LSF) [9–12] is emerging as a powerful synthetic strategy to rapidly incorporate non-metallic radioisotopes into complex molecular substrates. This approach meets the required criteria while circumventing the lengthy production of chemical precursors (Figure 5.1b). In the field of radioisotope insertion, according to Ritter's definition [11], LSF approaches can be divided into two classes. The first class involves chemoselective transformation that can be performed on a molecular substrate, resulting in the production of a radiolabeled analog of the starting material. This class includes functional group transformation reactions and C—H bond activation reactions, providing non-isotopically labeled derivatives (Figure 5.1b(1)). The second class, unique to radiochemistry or isotopic labeling, is chemoselective isotopic exchange, where an atom or group of atoms in a molecular substrate is replaced by the same radioactive atom or group, resulting in an isotopically labeled complex molecule (Figure 5.1b(2)). Chapter 6 will present the latest and most innovative LSF strategies for radioisotope insertion. The most relevant examples of tritium, carbon isotopes, fluorine-18, and iodine isotopes insertion will be discussed in dedicated sections.

5.2 Tritium

Tritium (^{3}H) is an essential tool in biochemical and pharmaceutical research. With a half-life of 12.3 years, tritium decays by beta minus emission and produces low-energy beta particles detectable by liquid scintillation. This low energy allows for safe handling in laboratory settings while providing sufficient signal for detection in biological studies. Chemically, tritium behaves similar to hydrogen, allowing incorporation into a wide range of organic molecules without interfering with biological properties. Tritium is used as a tracer in various fields, including biochemistry, molecular biology, and pharmacology. Metabolic pathways, receptor–ligand interactions, and drug distribution and metabolism can be efficiently studied using tritium-labeled tracers. One of the most common applications of tritium labeling is the synthesis of tritiated pharmaceuticals, allowing the assessment of absorption, distribution, metabolism, and excretion (ADME) of drug candidates as well as pharmacokinetics and pharmacodynamics. Tritium is also extensively used in autoradiography to visualize the distribution of tritiated compounds in anatomical cuts. For all these applications, tritiated compounds must be obtained with a sufficient level of isotope incorporation. A_m values of around 37–55 kBq.μmol^{-1} are most of the time sufficient but up to a minimum of 740 MBq.μmol^{-1} can be used in binding assays [16]. Tritium can be produced through several methods. The most common production method involves the irradiation of lithium-6 with neutrons through the reaction $^{6}Li(n,\alpha)^{3}H$. This process is carried out in specialized nuclear reactors dedicated to produce military grade-tritium. Tritium can also be generated as a by-product in heavy water reactors where deuterium (^{2}H) captures neutrons to form tritium. Once produced, tritium is typically stored as tritiated gas (T_2), and this primary source can be further transformed into other tritiated chemicals, such as T_2O, as needed. The latter is then used as a starting material to produce tritium-labeled compounds. The typical synthesis of tritium-labeled compounds involves the incorporation of tritium into precursor molecules [17], and the chemical routes employed are generally optimized through deuteration. This can be achieved through various methods, including catalytic hydrogenation, where T_2 is used to replace hydrogen atoms in organic molecules. Another common method is the hydrogen isotope exchange (HIE), where tritium atoms are exchanged with hydrogen atoms in specific positions of the molecule. Advances in tritium labeling techniques have led mainly to the development of highly efficient and selective HIE methods for incorporating tritium into complex molecules in efficient LSF processes.

5.2.1 Transition Metal-Catalyzed HIE

Transition-metal-catalyzed HIE is one of the modalities which has been extensively developed in the last four decades in tritium LSF, and several transition-metal-based catalysts have been used in this approach.

5.2.1.1 Iridium-Based Catalysts

Homogeneous iridium catalysis is a method of choice to perform late-stage tritiation. In 1979, Crabtree's works led to the discovery of the iridium(I) homogeneous hydrogenation catalyst, $[Ir(COD)(PCy_3)(Py)]PF_6$ [18]. This kind of catalyst has been studied for HIE applications, leading to few structural modifications in order to improve the deuterium and tritium labeling properties, as described by Kerr [19] or Tamm group (Figure 5.2) [20].

This technology allows the late-stage labeling of a large number of compounds through an *ortho* C—H activation, using functionalities such as sulfonamides, carbonyls, or nitrogen-rich heterocycles as directing groups. Using this approach, several molecules of interest have been tritiated [21, 22]. These catalysts are air-stable and usable even at low pressure (25 mbar of T_2) (Figure 5.3) [22].

Since few years, heterogeneous iridium catalysis has been widely developed. Indeed, these catalysts are cheaper than most of homogeneous iridium catalysts, stable to air and humidity, and can be easily separated from the reaction mixture at the end of the reaction. Pieters group has developed transition-metal nanoparticles and has shown their advantages compared to homogeneous catalysis in HIE. $[Ir(COD)\text{-}(OMe)]_2$, a commercially available and air-stable iridium precatalyst, has been shown to generate *in situ* both Ir complexes and nanoclusters

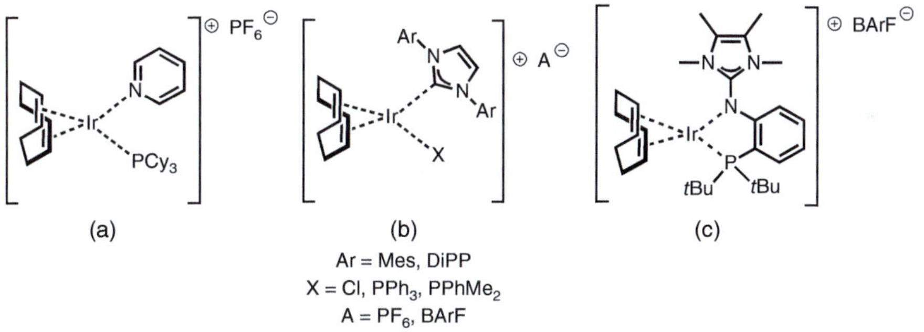

Figure 5.2 Catalysts developed by (a) Crabtree, (b) Kerr, and (c) Tamm.

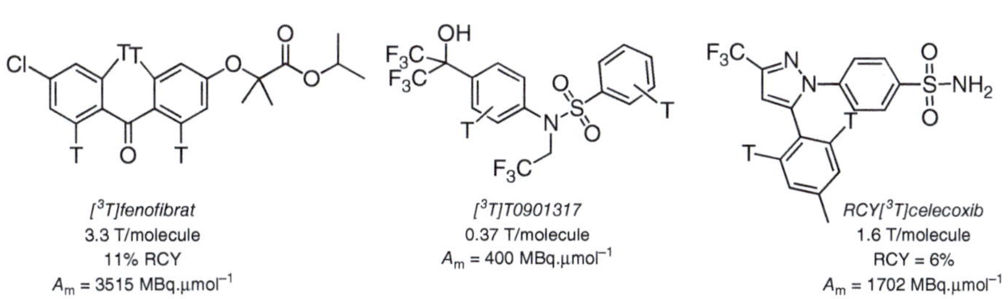

Figure 5.3 Examples of pharmaceutics labeled using homogeneous iridium catalyst. *Source:* Marek et al. [21] and Derdau et al. [22].

catalytically actives for HIE. This approach allows a multi-site tritiation of several pharmaceuticals mainly on C(sp²)—H, pyridine, and/or benzene, with high functional group tolerance (for instance, halogens, sulfurs, and nitriles) and high A_m (Scheme 5.1) [23].

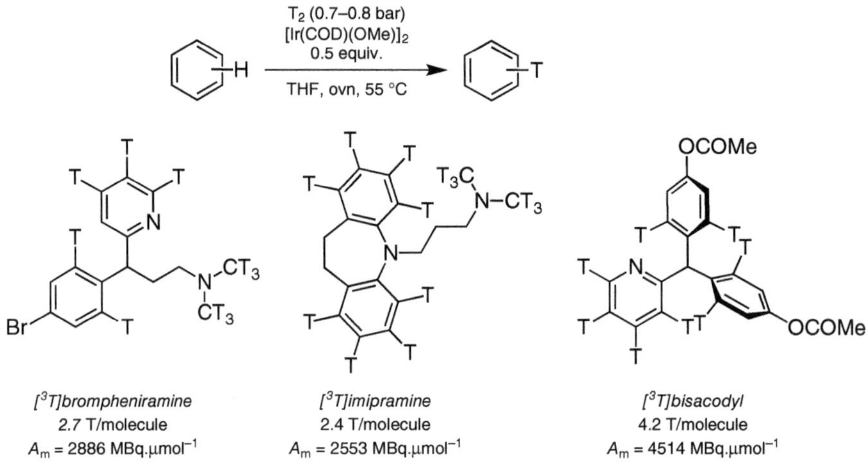

Scheme 5.1 $[Ir(COD)\text{-}(OMe)]_2$-catalyzed HIE catalyzed. *Source:* Adapted from Pieters [23].

Iridium nanoparticles (IrNP) have been prepared by the same group from $[Ir(COD)\text{-}(OMe)]_2$ in combination with *N*-heterocyclic carbene ligand 1,3-dicyclohexylimidazole under 5 bar of H_2 pressure [24]. Used as HIE nanocatalyst, these nanoparticles gave really good results concerning deuteration of anilines. Applied onto a volixibat precursor, its tritiated analog has been obtained with a high selectivity on the *ortho*-position of the aniline motif, along with a high A_m (Scheme 5.2).

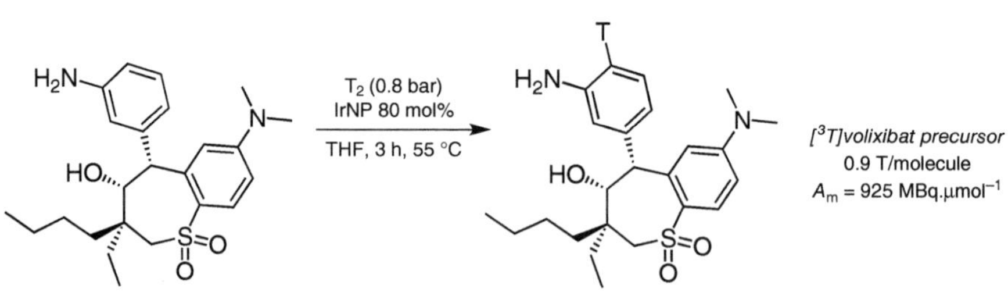

Scheme 5.2 Nanocatalyzed tritiation of volixibat pharmacophore. *Source:* Adapted from Pieters, Chaudret and Derdau [24].

5.2.1.2 Iron-Based Catalysts

The first example of undirected catalysis has been provided by Chirik (Scheme 5.3) [25]. Using an iron(0) catalyst, their methodology allows the labeling of aromatic position without the need of directing group, where the labeling is driven by steric hindrance. This methodology has been successfully applied to five pharmaceuticals

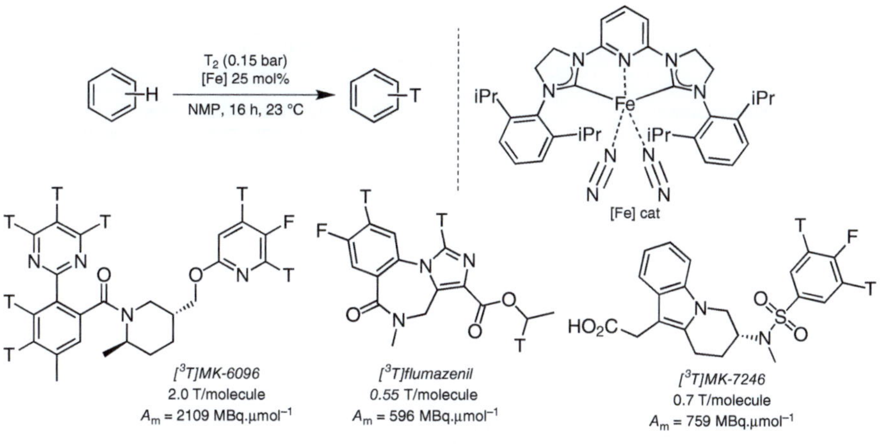

Scheme 5.3 Undirected iron-catalyzed labeling of aromatic positions. *Source:* Adapted from Chirik [25].

and proven to give similar to better results compared to iridium-mediated tritiation. However, even powerful, this catalyst is air- and moisture-sensitive and requires several steps of synthesis.

5.2.1.3 Ni-Catalyzed Tritiation

Chirik's group has studied the effect of nickel catalysis onto tritiation through site-selective HIE. In 2018, the first study has been reported using α-diimine nickel hydride complex, either generated *in situ* or used as precatalyst (Scheme 5.4). This methodology allowed interestingly high HIE on different substrates, in particular on nitrogen rich molecules and with a high functional group tolerance. This study included the tritiation of nine pharmaceuticals, obtained with high A_m [26]. Based on these results, Chirik's group has developed a second-generation nickel hydride catalyst in order to improve its HIE performance by increasing both the number of labeled sites and their isotopic enrichment [27]. Applied to the same pharmaceuticals, this catalyst has been proven to provide labeled compounds with higher A_m.

5.2.1.4 Palladium-Based Catalysts

In 2018, Hoover's team has described the first directed palladium(II)-mediated tritiation of molecules of interest (Scheme 5.5) [28]. Their procedure use the *in situ* formation of a palladacycle using an excess of $Pd(OAc)_2$ which is then placed under T_2 atmosphere in the presence of JohnPhos. This methodology has been successfully applied to the late-stage tritiation of nine complex pharmaceuticals, labeled mainly on the *ortho*-position of carbonyls and benzene rings, with high A_m.

In 2021, Pfeifer et al. proposed an approach-based heterogeneous palladium catalysis using polyvinylpyrrolidone (PVP) nanoparticles (Scheme 5.6) [29]. Produced by placing a palladium precursor complex in reductive atmosphere in the presence of PVP polymer in THF, the use of these nanoparticles with T_2 allows the insertion of tritium atom on benzylic positions onto pharmaceuticals with high A_m and a very good selectivity.

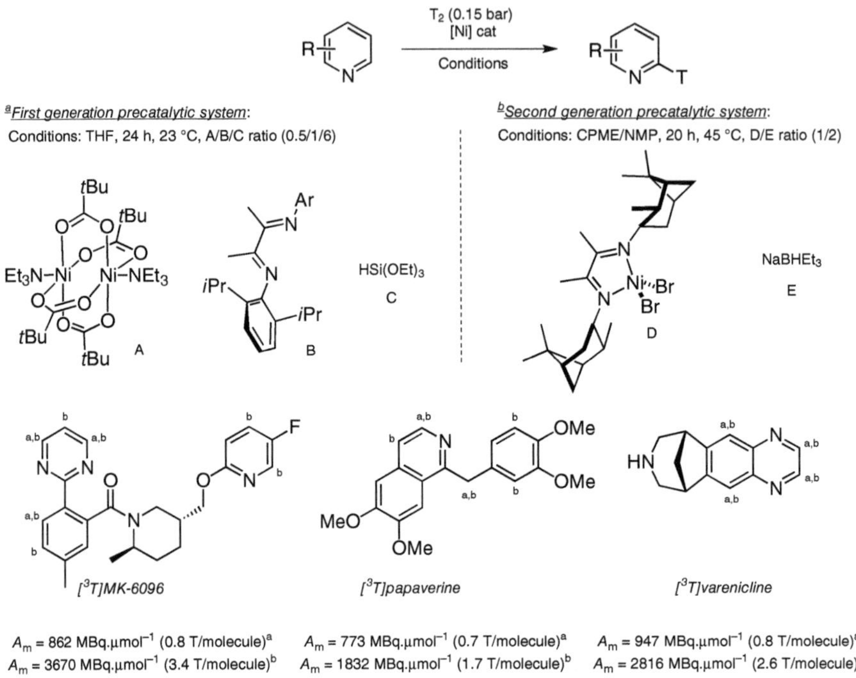

Scheme 5.4 Site-selective HIE using α-diimine nickel hydride complexes. *Source:* Chirik [26, 27].

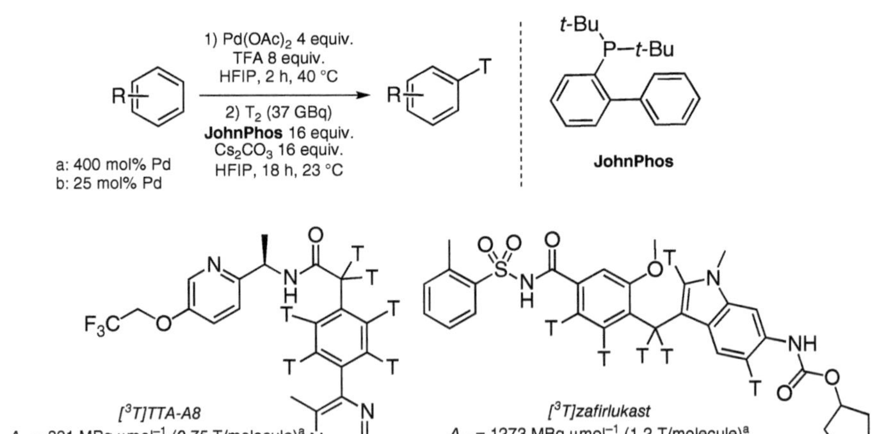

Scheme 5.5 Palladium-mediated tritiation of aromatics. *Source:* Adapted from Hoover [28].

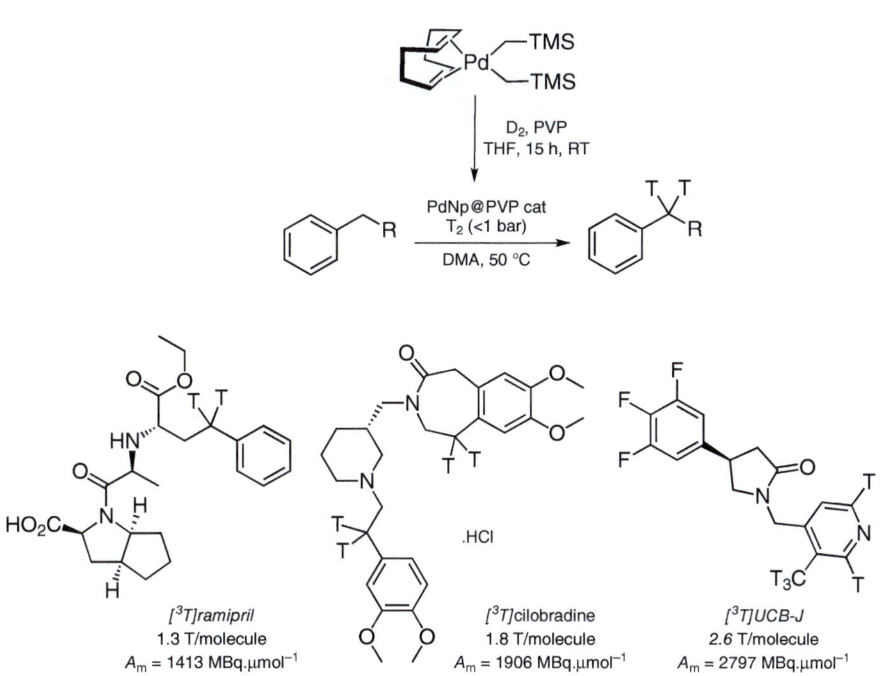

Scheme 5.6 Heterogeneous palladium-catalyzed labeling of benzylic positions. *Source:* Adapted from Pfeifer [29].

A different approach has been proposed in 2021 by Ritter et al. (Scheme 5.7) [30]. According to their methodology, a thianthrenium salt is prepared from the compound of interest using thianthrenium-*S*-oxide through S_EAr. The formed intermediate is filtered and directly used in palladium-catalyzed tritiation. The use of this intermediate allows a very selective incorporation of tritium with high A_m, using mild conditions, even with sensitive functions.

5.2.1.5 Ruthenium Catalysis

In 2018, Pieters et al. presented the first example of the use of ruthenium catalysis in sulfur-directed C(sp^3)—H activation – tritiation. Using commercially available Ru/C as a catalyst and T_2 as a tritium source, the late-stage labeling of pergolide, a dopamine receptor agonist used in Parkinson's disease, has been performed selectively on the α-position of sulfur atom with a good A_m (555 MBq · μmol^{-1}) (Scheme 5.8) [31]. A similar approach has been described by Ackermann's group using ruthenium-catalyzed C—H activation of benzoic acids in combination with a T_2O source (Scheme 5.8) [32].

In their efforts to propose efficient tritiation methodologies using metallic nanoparticles, Pieters et al. investigated the effect of incorporated stabilizers on the ruthenium nanoparticles catalytic behavior (Scheme 5.9). Using PVP, they have been able to radiolabel nitrogen-rich scaffolds as benzimidazole, triazole, and carbazole [33] while *N*-heterocyclic carbene (NHC-ICy) led to the efficient radiolabeling of nucleobases [34].

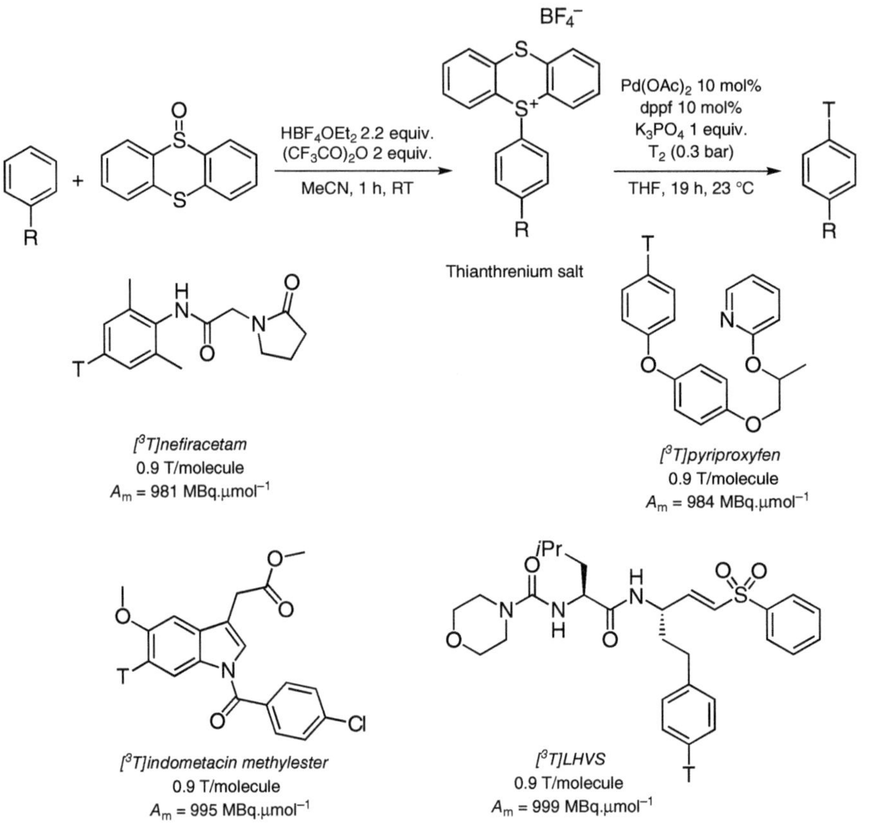

Scheme 5.7 Selective tritiation of aromatic using thianthrenium salts as intermediate. *Source:* Ritter [30] / Springer Nature / CC BY 4.0.

5.2.1.6 Rhodium Catalysis

In 2022, Pieters et al. proposed a methodology, using *in situ* formed rhodium nanoparticles from commercially available and air-stable rhodium dimer [35]. This catalyst allows the selective C—H activation of C(sp^3)—H, leading to radiolabeled analogs of two radiopharmaceuticals (Scheme 5.10).

5.2.2 Photocatalysis

In 2017, McMillan's group published the application of a photocatalytic methodology to successfully incorporate tritium into molecules of interest. This methodology uses an iridium-based photocatalyst and *in situ*-generated T_2O as the tritium source (Scheme 5.11) [36]. It allows the introduction of tritium on tertiary aliphatic amine α-position with high selectivity and high A_m. This methodology has been applied on 18 drug molecules, highlighting its broad scope.

In 2022, König et al. described the selective activation of C(sp^3)—H bond on amine α position, using photoredox-catalyzed deprotonation (Scheme 5.12) [37]. Developed for deuteration of molecules of interest, this methodology has been

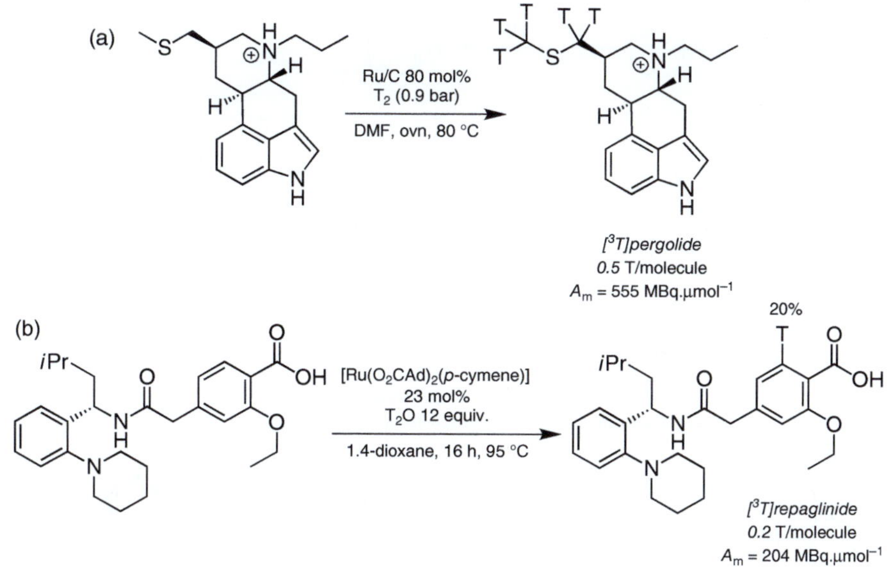

Scheme 5.8 Tritiation through ruthenium-catalyzed C—H activation. *Source:* (a) Adapted from Pieters [31] and (b) Ackermann [32] / John Wiley & Sons.

applied on the last-stage isotopic exchange of linezolid, giving the proof of concept of the use of this approach with a controlled amount of T_2O (5 equiv.).

5.2.3 Alkali-Based HIE

Mainly used in deuterium labeling, alkali-based HIE have been scarcely used with tritium as this methodology often requires large excess of T_2O. Nevertheless, a recent example has been published in 2023 by Yang H. and Guan B.-T. group, where a selective labeling of benzylic position have been reported, using a bis-trimethylsilylamido catalyst to promote a kinetic deprotonative equilibrium, allowing the HIE with T_2 (Scheme 5.13) [38].

5.3 Carbon Isotopes

Carbon-11 and carbon-14 play significant roles in the field of radiolabeling. Each of these isotopes has unique properties associated with specific applications, ranging from *in vivo* PET imaging to ecosystem studies. To date, most of the late-stage carbon isotopes insertion have been developed using an isotope-exchange approach.

Carbon-11 has a half-life of 20.3 minutes and decays by positron emission. The emitted positrons annihilate with electrons producing two 511 keV gamma rays 180° to each other. These properties make carbon-11 particularly useful in PET imaging allowing rapid imaging protocols with minimal radiation exposure to patients. Carbon-11 is produced in a cyclotron through the proton bombardment

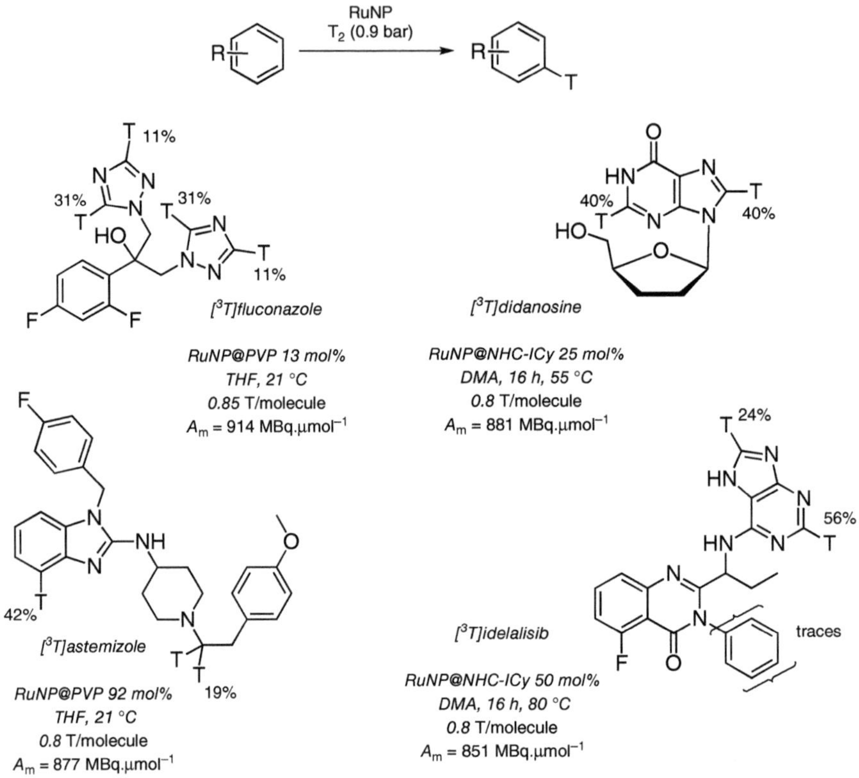

Scheme 5.9 Nanocatalyzed tritiation of nitrogen-rich heterocycles. *Source:* Pieters [33, 34].

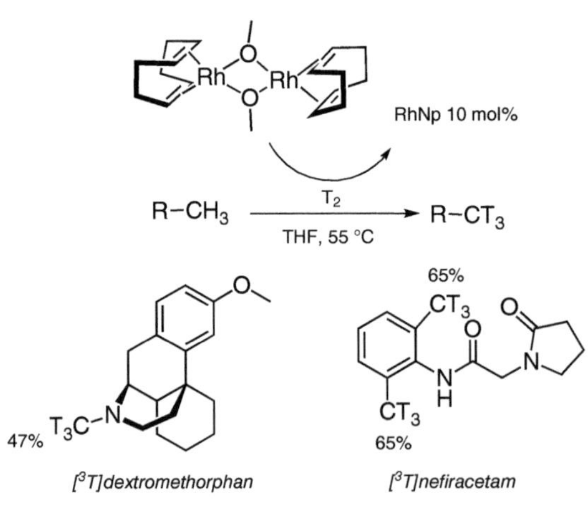

Scheme 5.10 Tritiation using rhodium nanoparticles-catalyzed $C(sp^3){-}H$ activation. *Source:* Pieters [35] / with permission of American Chemical Society.

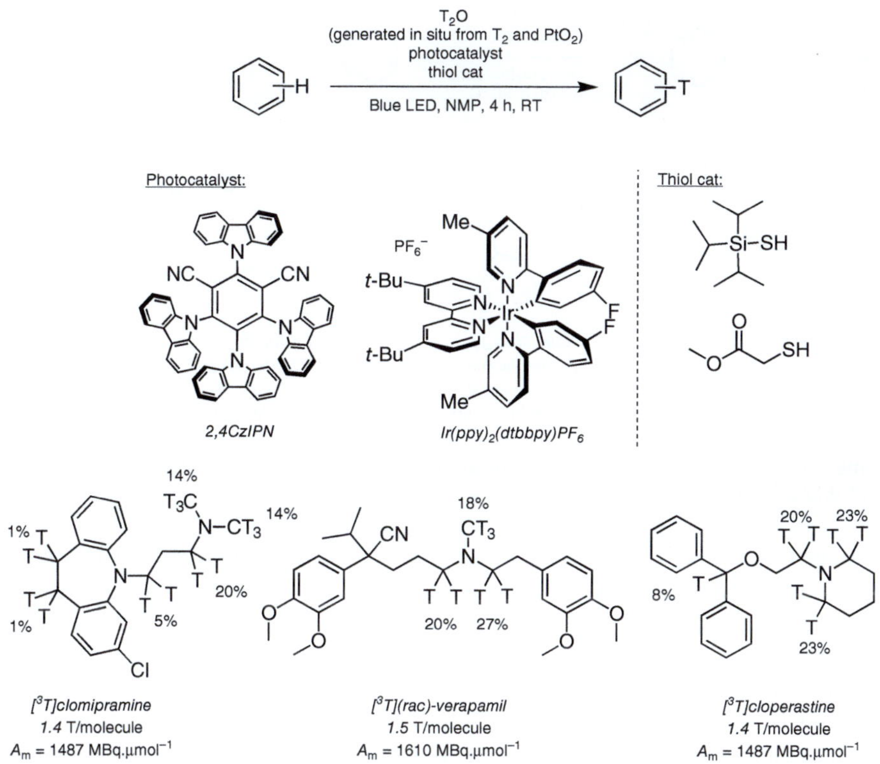

Scheme 5.11 Radiolabeling of aliphatic amine through photocatalysis. *Source:* McMillan [36] / American Association for the Advancement of Science.

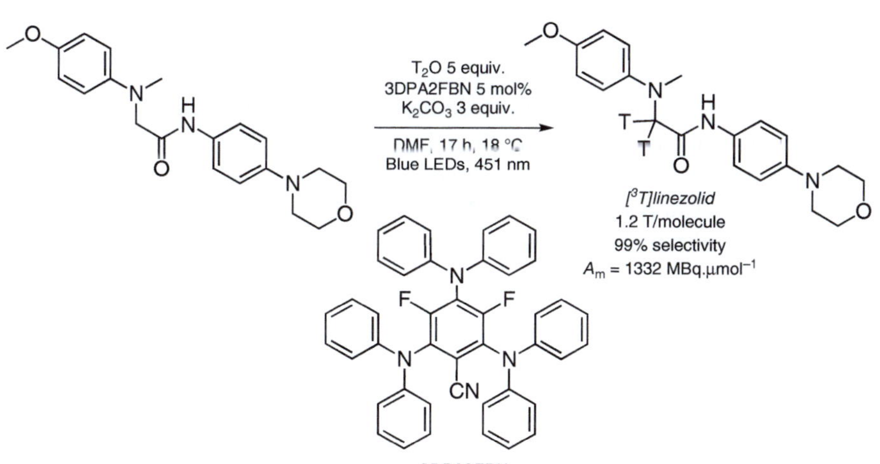

Scheme 5.12 Selective tritiation on amine α-position using activation of C(sp^3)−H bond. *Source:* König [37] / with permission of American Chemical Society.

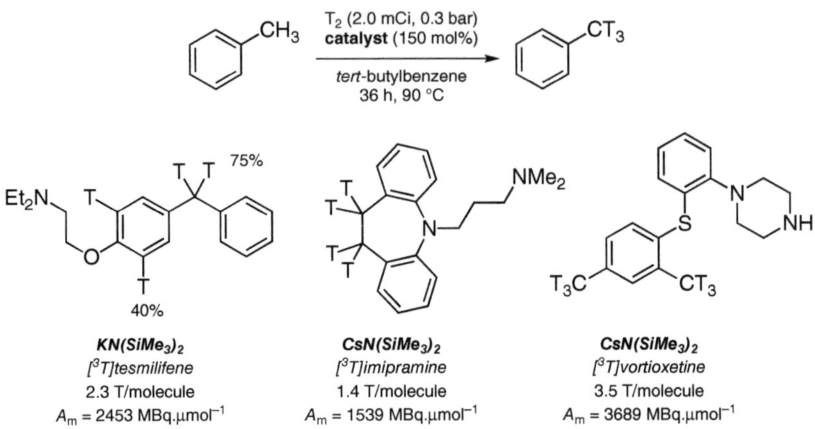

Scheme 5.13 Selective deprotonation–tritiation of benzylic position. *Source:* Yang and Guan [38] / John Wiley & Sons.

of nitrogen-14 through the $^{14}N(p,\alpha)^{11}C$ reaction. The generated carbon-11 then interacts with trace amounts of a deliberately added O_2 or H_2 added in the cyclotron to form either $[^{11}C]CO_2$ or $[^{11}C]CH_4$ with high A_m. $[^{11}C]CO_2$ or $[^{11}C]CH_4$ can be used directly or, more commonly, converted into more reactive intermediates, such as $[^{11}C]CH_3I$ or $[^{11}C]CO$ [39, 40]. The rapid and efficient production of carbon-11 derivatives is crucial due to their short half-life, necessitating their immediate use in synthesis and application. Tracer production is usually achieved through a series of automated reactions using advanced synthesis modules and labeling precursors.

Carbon-14 has a much longer half-life of 5730 years and decays by beta minus emission. This long half-life is ideal for studying long-term biological processes without significant decay-related signal loss. As a consequence, carbon-14 is used in environmental studies or in biochemistry where ^{14}C-labeled compounds help elucidate metabolic pathways. In ecology, carbon-14 helps understanding at carbon cycling in ecosystems, and carbon-14 dating is a powerful tool for determining the age of biological specimens though this latter technique is not related to radiochemistry. Carbon-14 is also used in pharmacokinetics to study drug ADME over prolonged periods. Carbon-14 is produced through neutron activation of nitrogen-14, $^{14}N(n,p)^{14}C$, in nuclear reactors. The only carbon-14 chemical source produced in these reactions is $[^{14}C]BaCO_3$. This starting material is often transformed into $[^{14}C]CO_2$ [41].

5.3.1 Isotopic Exchange

5.3.1.1 Decarboxylation/Recarboxylation Strategies

In 2018, Gauthier's group reported a palladium-catalyzed carbon isotope exchange on aliphatic and benzoic acids using $[^{14}C]CO$ (Scheme 5.14) [42]. This approach was made possible by the use of specialized glassware, COware [43, 44], designed to release stoichiometric amounts of $[^{14}C]CO$. $[^{14}C]COgen$, previously synthesized from 9-methylfluorene and $[^{14}C]CO_2$, is introduced into chamber 1

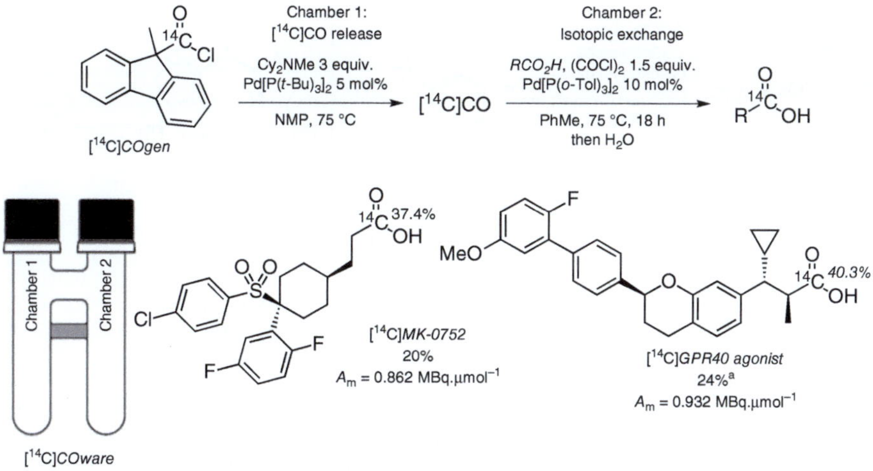

Scheme 5.14 Palladium-catalyzed carbon isotope exchange on aliphatic and benzoic acids with [^{14}C]CO. ^{a}Pd$_2$dba$_3$ and JohnPhos were used to generate the catalyst in chamber 2. *Source:* Gauthier [42] / with permission of American Chemical Society.

along with a palladium source and Cy$_2$NMe, releasing [^{14}C]CO. The isotopic exchange occurs in chamber 2 in the presence of a second palladium source and oxalyl chloride. Mechanistically, the reaction proceeds through the formation of an acyl chloride, followed by oxidative insertion of palladium, CO elimination/exchange, and migratory insertion. The resulting [^{14}C]acyl chloride is then hydrolyzed. The article demonstrates the method primarily with ^{13}C, and a few examples with ^{14}C, achieving late-stage ^{14}C isotopic exchange with MK-0752 and a GPR40 agonist with high A_m.

In 2019, Cantat and Audisio described the direct use of the universal precursor [^{14}C]CO$_2$ to promote isotopic exchange at carboxylic acid sites (Scheme 5.15) [45]. In this method, cesium benzoate salts are heated with a catalytic amount of CuBr and a bisoxazoline (BOX) ligand, along with three equivalents of [^{14}C]CO$_2$, to achieve dynamic carbon isotope exchange. This approach was elegantly demonstrated in the synthesis of ^{14}C-enriched Probenecid, Flumequine, and Febuxostat, all achieving high A_m.

In 2020, Audisio applied this latter approach to the carbon isotope exchange of phenylacetic acids, but this time reported a transition-metal-free process (Scheme 5.16) [46]. Inspired from a similar study starting potassium carboxylates and [^{13}C]CO$_2$ [47], cesium carboxylate salts were simply heated at 150 °C with three equivalents of [^{14}C]CO$_2$. The addition of K$_3$PO$_4$ was found to promote decarboxylation in some cases. This method was successfully used for the late-stage insertion of ^{14}C into pharmaceutical substrates. The introduction of short half-life [^{11}C]CO$_2$ was also evaluated using the same procedure, yielding modest RCY.

A related procedure was published by Baran in 2019, utilizing a redox-active ester (RAE) as the starting material (Scheme 5.17) [48]. These RAEs are synthesized in a single step from the corresponding aliphatic carboxylic acids using *N*-hydroxyphthalimide (NHPI), *N*,*N*′-diisopropylcarbodiimide (DIC), and

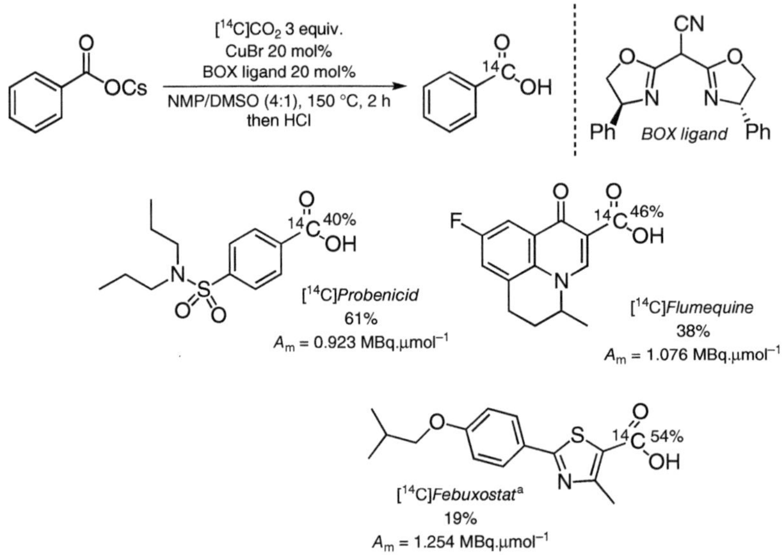

Scheme 5.15 Dynamic carbon isotope exchange on (het)arylcarboxylates. [a]Reaction performed at 190 °C. *Source:* Adapted from Cantat and Audisio [45].

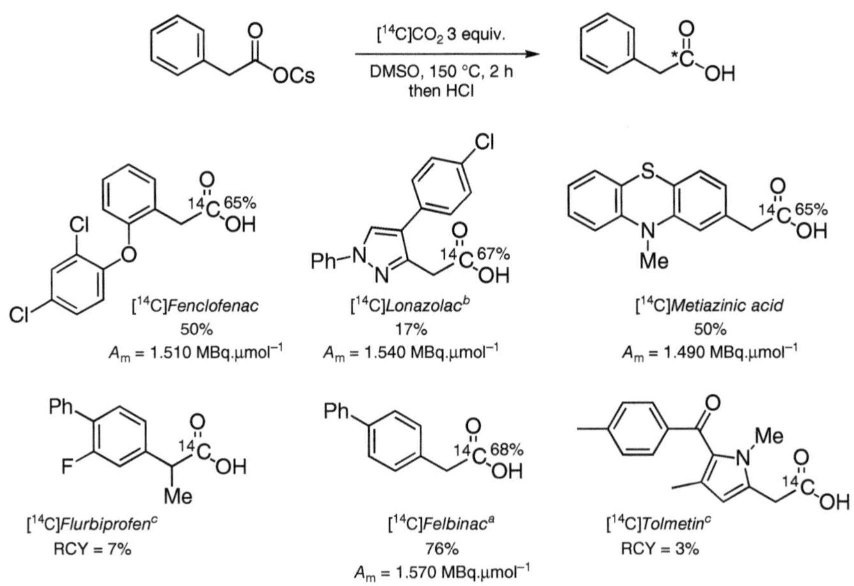

Scheme 5.16 Transition-metal-free carbon isotope exchange of phenyl acetic acids. [a]Reaction time was set to 4hours; [b]reaction performed overnight with K_3PO_4; and [c]reaction time was set to 10 minutes, and RCY is given decay-corrected. *Source:* Audisio [46] / with permission of American Chemical Society.

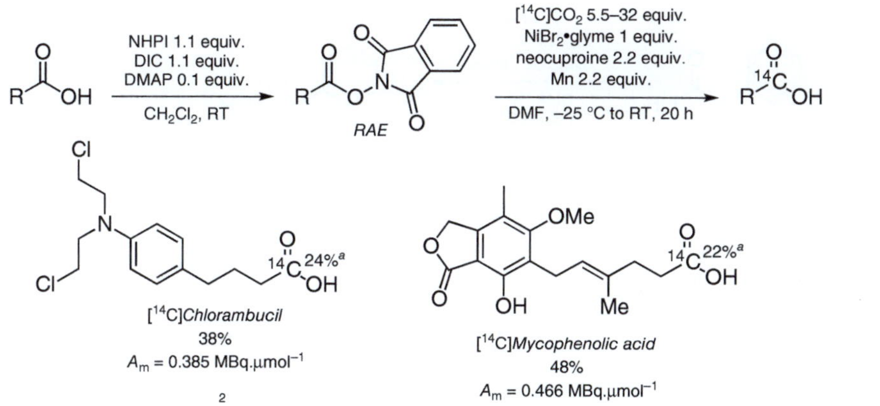

Scheme 5.17 Baran's nickel-mediated reductive decarboxylation–[^{14}C]recarboxylation of aliphatic carboxylic acids. [a]Isotopic enrichment estimated from the same experiment with [^{13}C]CO. *Source:* Kingston et al. [48] / with permission of American Chemical Society.

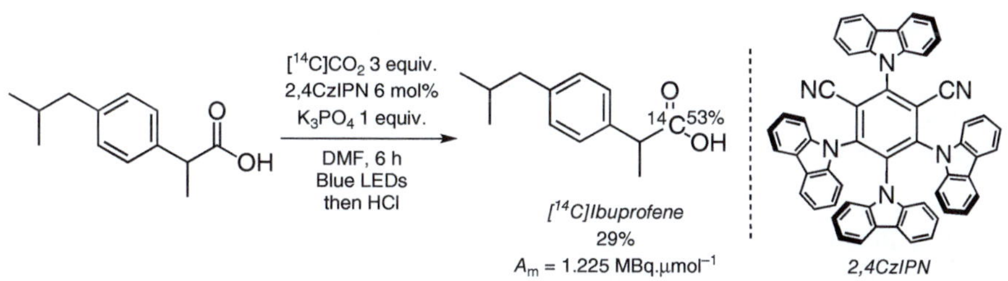

Scheme 5.18 Ibuprofen photochemical carbon isotope exchange with [^{14}C]CO_2. *Source:* Audisio [50] / with permission of American Chemical Society.

4-dimethylaminopyridine (DMAP). Once the RAE is obtained, isotopic exchange occurs at −25 °C with an excess of [^{14}C]CO_2, using $NiBr_2$ and neocuproine as the catalytic system and manganese as the reductant. Mechanistically, the reaction proceeds through a nickel-mediated reductive decarboxylation–[^{14}C]recarboxylation. The method tolerates sensitive functional groups, including ketones, unprotected alcohols, and alkenes, but it is incompatible with tertiary carboxylic acids. This approach is effectively illustrated in the article by the ^{14}C-labeling of chlorambucil and mycophenolic acid, achieving satisfactory A_m. Similar conditions were established and published simultaneously by Audisio's group [49].

These decarboxylation–recarboxylation approaches were further explored through a photochemical method. Audisio directly labeled [^{14}C]ibuprofen using [^{14}C]CO_2, blue LEDs, K_3PO_4, and 2,4CzIPN, achieving high A_m (Scheme 5.18) [50]. Rotstein and Lundgren applied this strategy for ^{11}C-labeling of various substrates using Cs_2CO_3 as the base (Scheme 5.19) [51]. Utilizing [^{11}C]CO_2, several anti-inflammatory drugs were labeled with RCC and RCY. However, the measured A_m for Fenoprofen was too low for this technique to be suitable for PET imaging tracer production.

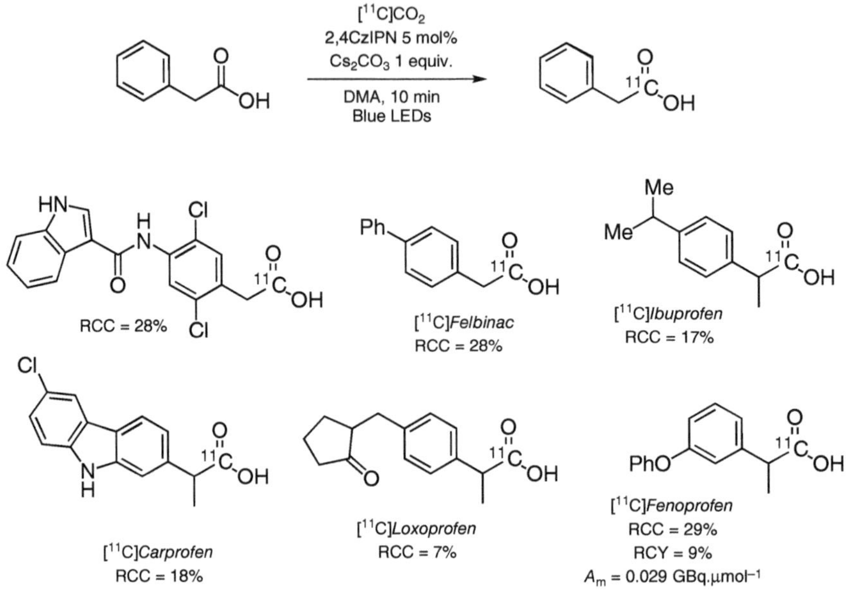

Scheme 5.19 Phenylacetic acids photochemical carbon isotope exchange with $[^{11}C]CO_2$. RCC and RCY are decay-corrected. *Source:* Rotstein and Lundgren [51] / with permission of American Chemical Society.

In 2022, Rotstein and Lundgren described a carbon isotope exchange procedure that enables the direct labeling of unprotected α-amino acids using either $[^{11}C]CO_2$ or $[^{14}C]CO_2$ produced from $[^{14}C]BaCO_3$ (Scheme 5.20) [14, 52]. In this process, an imine is formed *in situ* with a catalytic amount of benzaldehyde, followed by deprotonation with Cs_2CO_3 to form an enolate, which traps either $[^{11}C]CO_2$ or $[^{14}C]CO_2$. Subsequent decarboxylation and imine hydrolysis yield the labeled derivatives. This strategy was applied to a variety of amino acids; however, the A_m remains modest, and racemization occurs systematically during the process.

5.3.1.2 Nitrile Group Exchange

Nitrile groups can also undergo exchange using nickel catalysis with $[^{14}C]Zn(CN)_2$ as the cyanide source. Reilly and Strotman first described such a process in 2021 (Scheme 5.21) [53], which, although primarily demonstrated with carbon-13, has been applied to the ^{14}C-labeling of belzutifan, resulting in modest A_m. The reaction involves nickel insertion into the C—CN bond using an air-stable Ni^0 source and a Lewis acid (BPh_3) to promote the departure and exchange of the nitrile group. Around the same time, Audisio's group reported a related process under basic conditions (Scheme 5.22) [54], employing DMAP and the air-sensitive $Ni(COD)_2$ as the catalyst. Although these processes are mechanistically similar, the conditions proposed by Audisio significantly improved the A_m and was applied to Perampanel, Vilazodone, Febuxostat, and Finerenone.

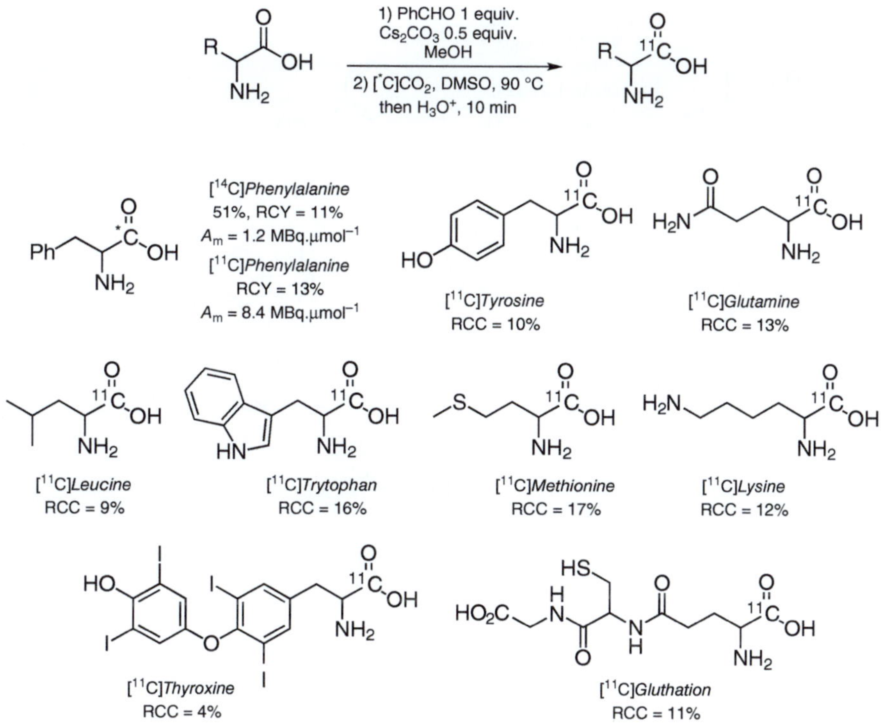

Scheme 5.20 Aldehyde-catalyzed carboxylate exchange in α-amino acids [^{11}C]CO_2 or [^{14}C]CO_2. RCC is decay-corrected. *Source:* Rotstein and Lundgren [14, 52].

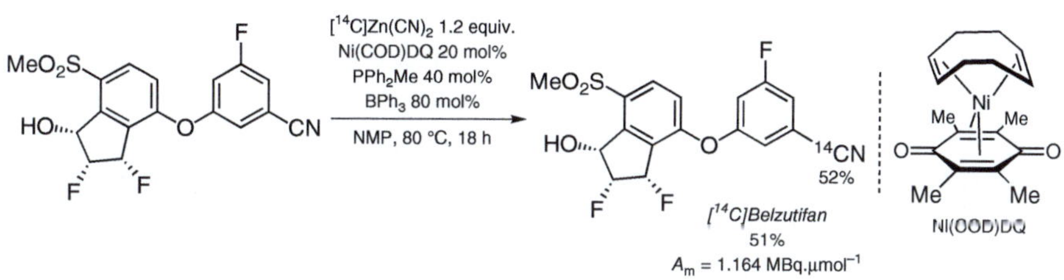

Scheme 5.21 Ni-catalyzed carbon isotope exchange of aryl nitriles. *Source:* Reilly and Strotman [53] / with permission of American Chemical Society.

5.3.2 C—F Activation

In 2023, Hoyosa reported C—F bond activation approach applied to [^{11}C]cyanation (Scheme 5.23) [55]. In this one-pot process, C—F activation is first performed in the presence of Ni(COD)$_2$, tricyclohexylphosphine (PCy$_3$), and LiCl, resulting in the formation of arylnickel(II) species. The resulting solution is then treated with [^{11}C]HCN, yielding the corresponding [^{11}C]cyano-derivatives. The process was demonstrated on several biologically relevant substrates, affording the ^{11}C-labeled derivatives with good RCY and A_m. Functional groups bearing labile hydrogen had to be protected during the reaction.

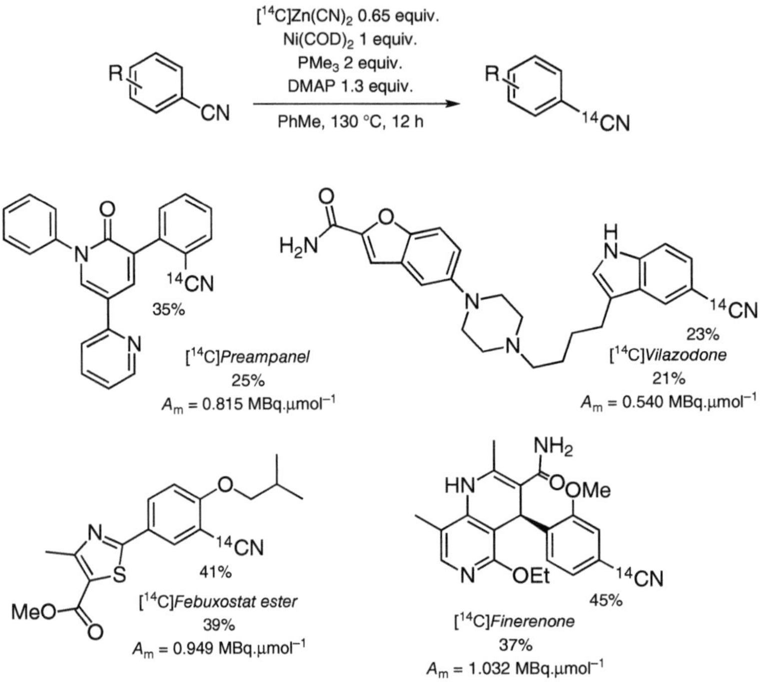

Scheme 5.22 Ni-catalyzed carbon isotope exchange of aryl nitriles. *Source:* Audisio [54] / with permission of American Chemical Society.

5.4 Fluorine-18

Like carbon-11, fluorine-18 decays through positron emission but with a longer half-life of 109.7 minutes. These properties make fluorine-18 a crucial tool in both clinical diagnostics and research applications through PET imaging. Indeed, like carbon-11, the positron emission produces positrons that annihilate with electrons in the body, resulting in the emission of two 511 keV gamma photons. Detection of these photons by PET scanners creates detailed functional images of the body's internal processes. Fluorine-18 is mainly produced through the well-established bombardment of $[^{18}O]H_2O$ with protons in a cyclotron in the $^{18}O(p,n)^{18}F$ reaction. The resulting fluorine-18 is then extracted from the cyclotron as aqueous $[^{18}F]$fluoride with high A_m ~40 000 GBq.µmol^{-1}. To make aqueous $[^{18}F]$fluoride sufficiently nucleophilic, it must first be activated. This activation is accomplished through a multistep procedure that begins with capturing the $[^{18}F]$fluoride on an ion-exchange column. It is then eluted with a solution of acetonitrile and water that includes a counterion. Commonly, tetraalkylammonium ions or potassium ions complexed with the aminopolyether cryptand Kryptofix 2.2.2 (K_{222}) are used as fluoride counterions. Carbonate or hydrogenocarbonate is preferred as counterions for these cations due to their non-nucleophilic and relatively non-basic nature, which helps prevent competitive substitution and elimination reactions. To overcome the high solvation energy of fluoride in water, azeotropic distillation

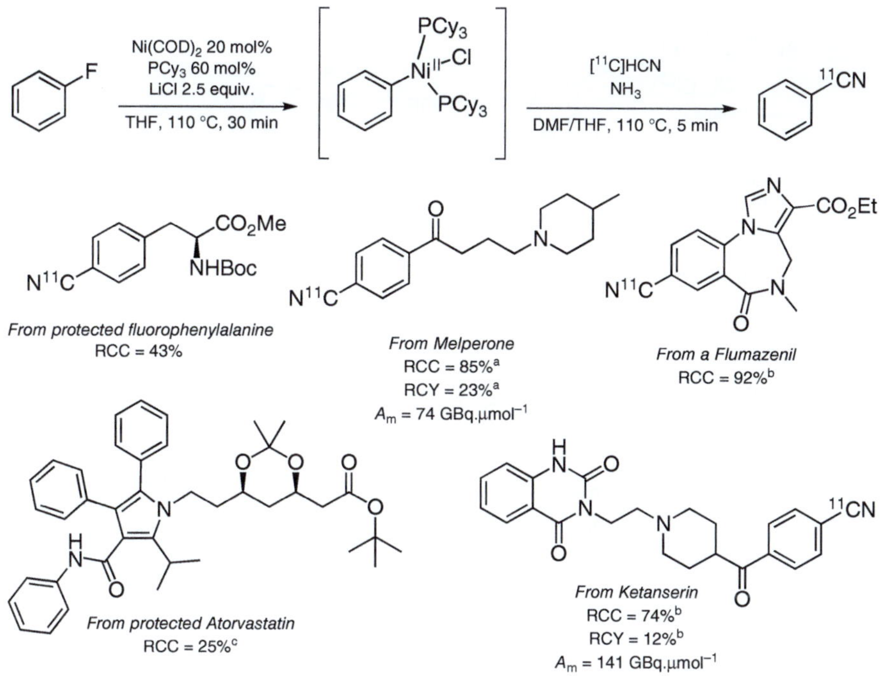

Scheme 5.23 ^{11}C-cyanation of aryl fluorides through C–F bond activation. RCC and RCY are decay-corrected. ^{a}First step was performed at 80 °C for one hour; ^{b}first step was performed at 100 °C; and ^{c}first step was performed at 60 °C. *Source:* Hosoya [55] / John Wiley & Sons.

with acetonitrile is employed after elution. This technique minimizes the water content, resulting in a much more reactive [^{18}F]fluoride. The remaining residue is then dissolved in an organic solvent for subsequent ^{18}F-labeling processes. These procedures have been automated to reliably produce reactive and high A_m [^{18}F]fluoride. To date, most of the ^{18}F-radiotracers syntheses are achieved using radiolabeling precursors [56–58]. While ^{19}F/^{18}F isotopic exchanges have been investigated, they remain challenging strategies in the context of C—F bond formation. These approaches generally produce labeled derivatives with low A_m [59, 60]. The formation of Si—F, B—F, or S—F bonds through ^{19}F/^{18}F isotopic exchanges offers improved A_m but often requires the introduction of dedicated functional groups, moving away from the principles of LSF [61]. In this field, sulfur [^{18}F]fluoride exchange ([^{18}F]SuFEx) has emerged as a pivotal reaction, offering an efficient pathway for incorporating fluorine-18 into complex molecules [62]. To date, the most efficient late-stage fluorine-18 insertions are based on either functional group transformation or C—H activation processes.

5.4.1 Functional Group Transformation

In 2016, Ritter's group presented an unusual concerted nucleophilic aromatic substitution reaction applied to deoxyradiofluorination [63] (Scheme 5.24). In this

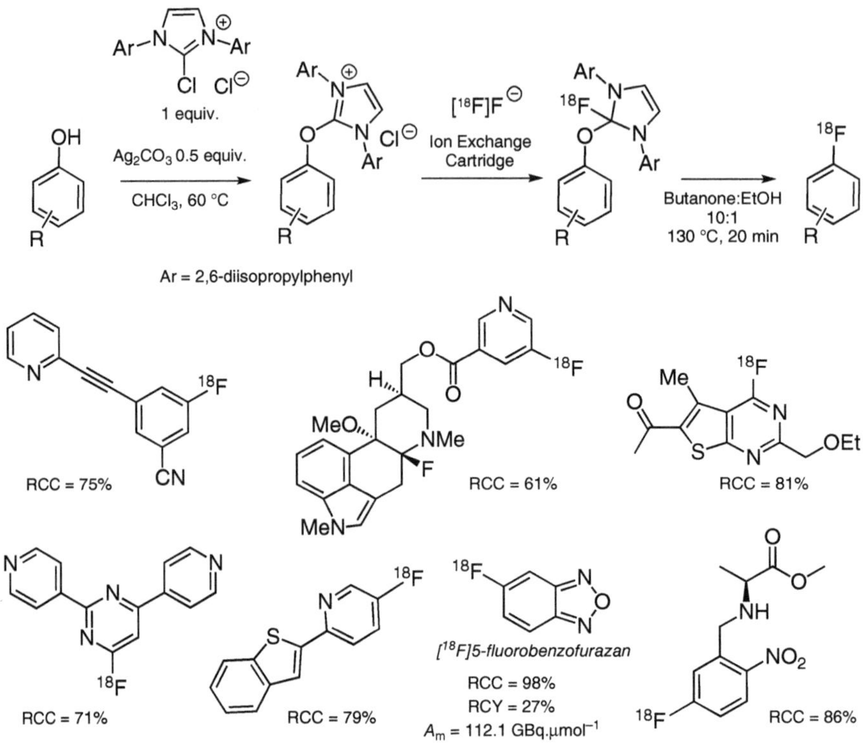

Scheme 5.24 Ritter's concerted nucleophilic aromatic substitution with ^{18}F. RCC is decay-corrected, and RCY is non-decay-corrected. *Source:* Adapted from Neumann et al. [63].

process, a phenol is first reacted with a chloroimidazolium salt to form an uronium ion. A ^{18}F-tetrahedral intermediate is then formed using an anion exchange cartridge charged with ^{18}F-fluoride. This elution procedure eliminates the need for azeotropic drying of ^{18}F-fluoride, and subsequent heating of the resulting solution directly provides ^{18}F-labeled aryl fluorides. Unlike typical nucleophilic aromatic substitution (S_NAr) reactions, which proceed through a Meisenheimer intermediate and are limited to electron-poor arenes, this reaction is not restricted in this way. The methodology was applied to a variety of polyfunctional phenols, achieving high RCC. The isolation and purification of [^{18}F]5-fluorobenzofurazan resulted in a RCY of 27% with high A_m, suitable for PET imaging.

In 2017, the same group published a modification of this approach that enables the conversion of highly electron-rich phenols into ^{18}F-labeled arenes [64] (Scheme 5.25). Using CpRu(COD)Cl, the activation of the phenol ring was achieved through the formation of an η^6 π-coordination ruthenium complex. The presence of the ruthenium reduces the electron density of the phenol, thereby enhancing its reactivity towards nucleophiles. The reaction was tested on a wide variety of substrates, yielding spectacular results on sensitive polyfunctional scaffolds. Notably, the reaction demonstrated tolerance to basic amines and unprotected alcohols.

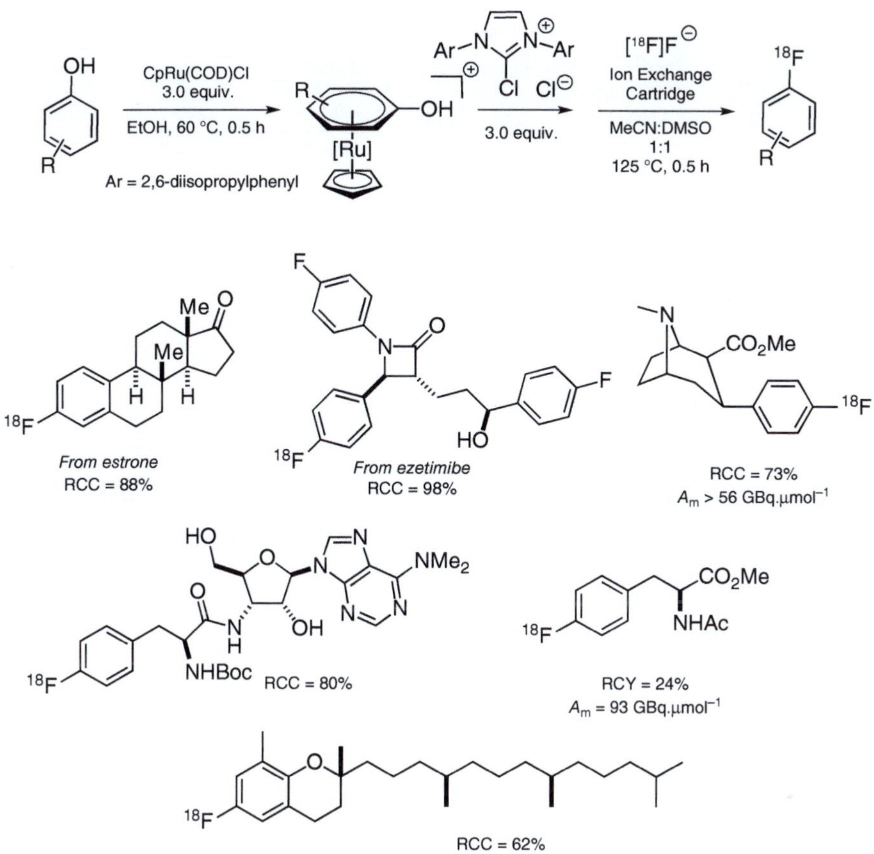

Scheme 5.25 Ritter's ^{18}F-deoxyfluorination of phenols via Ru π-complexes. RCC is decay-corrected, and RCY is non-decay-corrected. *Source:* Adapted from Beyzavi et al. [64].

In 2024, Tredwell, Sap, and Gouverneur reported that pyridinium salts derived from anilines represent a novel precursor class for S_NAr reactions with [^{18}F]fluoride (Scheme 5.26) [65]. These pyridinium salts are prepared from the corresponding aniline and pyrylium tetrafluoroborate. The described process is efficient when using isolated pyridinium salts but results in lower conversion in a one-pot process. The resulting ^{18}F-fluorodeamination was demonstrated with the synthesis of a large panel of electron-poor (hetero)aryl substrates. The protocol tolerates functionalities commonly found in drugs, leads to high A_m, and is amenable to scale-up and automation.

5.4.2 C—H Activation

In 2014, Groves' group described an efficient C—H benzylic [^{18}F]fluorination method (Scheme 5.27) [66]. This approach uses a catalytic amount of manganese(III) salt (Mn(salen)OTs) to introduce fluorine-18 into several bioactive molecules, typically achieving high RCC. The method was successfully applied to

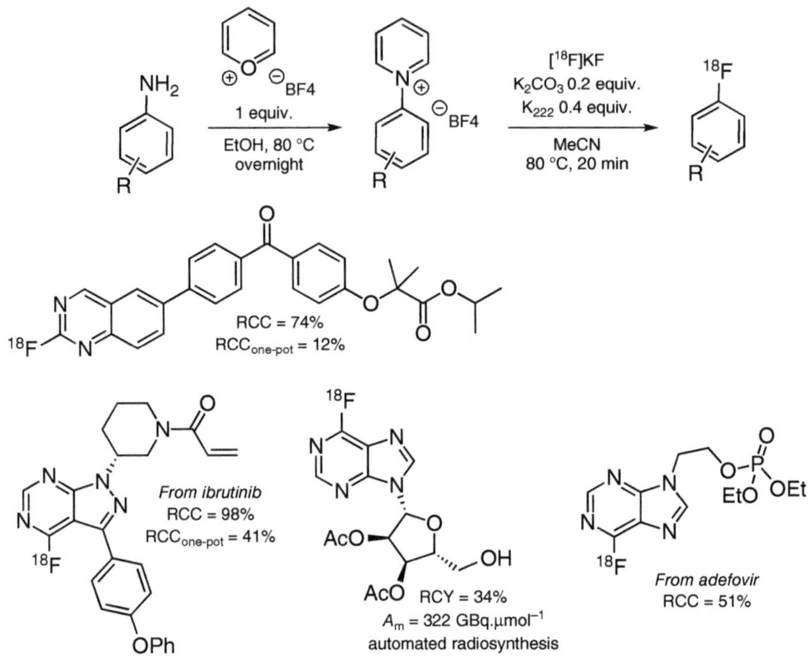

Scheme 5.26 Gouverneur et al. ^{18}F-fluorodeamination. RCC is decay-corrected and provided for the last step unless otherwise stated, RCY is non-decay-corrected. *Source:* Adapted from Ford et al. [65].

isolate a fluorinated analog of Celestolide with an A_m suitable for PET imaging. According to the proposed mechanism, the tosyl group of Mn(salen)OTs is substituted by [^{18}F]fluoride, followed by oxidation with PhIO to form a manganese(V) intermediate. This intermediate generates a radical at a benzylic site of the substrate, which then undergoes fluorination.

In 2017, Britton and coworkers reported a related process that enables the direct [^{18}F]fluorination of aliphatic amino acids (Scheme 5.28) [15]. In this method, the alkyl radical is generated using sodium decatungstate under UV irradiation, and an electrophilic fluorine-18 source, [^{18}F]NFSI, is required. However, the synthesis of [^{18}F]NFSI from [^{18}F]F_2 in a preliminary step complicates the process and limits its ability to achieve sufficient A_m for PET imaging.

Scott and Sanford reported the first approach toward CH radiofluorination of arenes in 2017 (Scheme 5.29) [67]. This copper(I)-mediated reaction proceeds in a two-step sequence from arenes bearing electron-donating groups (EDG). Initially, a crude iodonium solution is prepared from the chosen substrate with [hydroxy(tosyloxy)iodo]benzene and trimethylsilyl trifluoromethanesulfonate (TMSOTf) as an activator. This crude solution is then reacted with a [^{18}F]KF · 18-crown-6 complex in the presence of $(MeCN)_4$CuOTf. The addition of a base (iPr_2NEt) and quinaldic acid significantly increases the conversion. The reaction has a broad scope, tolerating heterocycles and sensitive functional groups, though protic functional groups (e.g., alcohols, amines, and amides) require

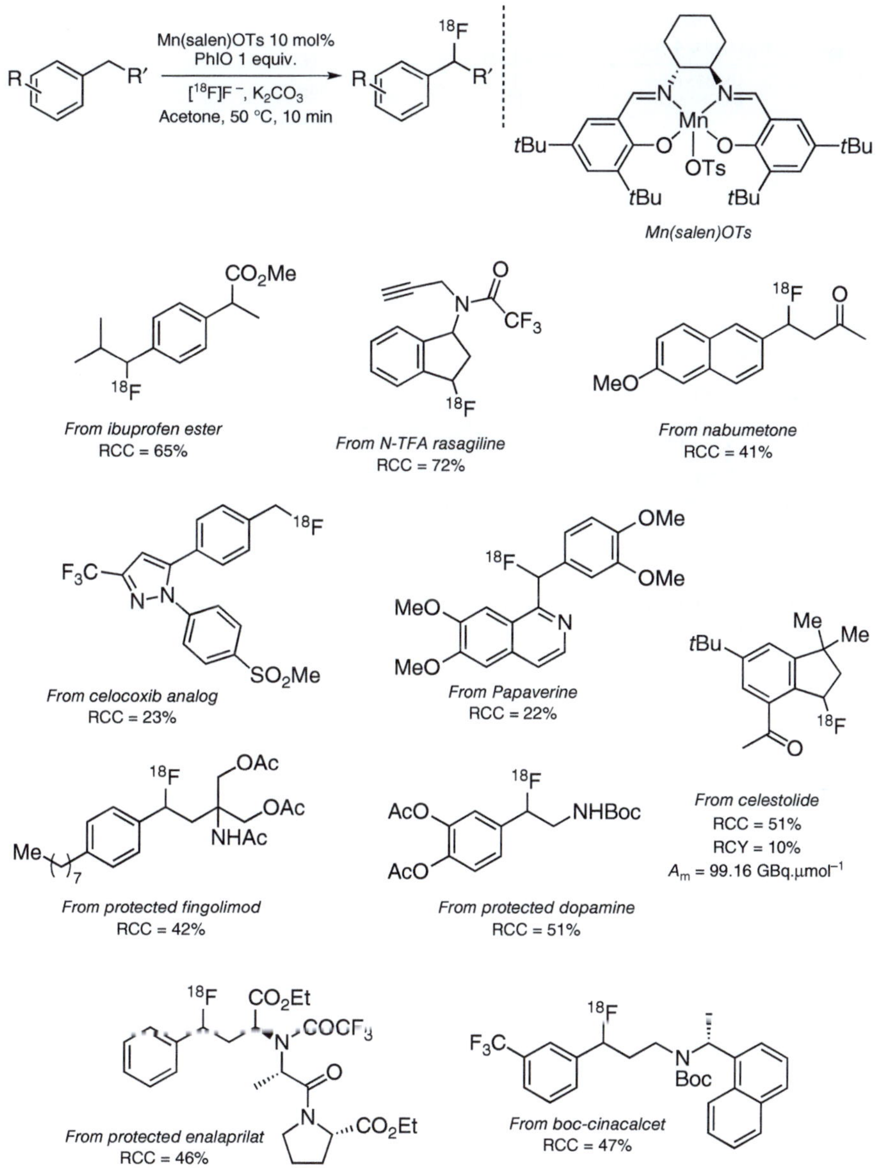

Scheme 5.27 Benzylic C—H [^{18}F]fluorination. RCC is decay-corrected, and RCY is non-decay-corrected. *Source:* Adapted from Groves [66].

protection prior to the reaction. The method was exemplified on two bioactive substrates, achieving high A_m with *N*-benzylnimesulide.

A similar approach was reported by Ritter in 2020, involving the formation of sulfonium salts. The process begins with a selective *para* C—H thianthrenation of arenes using hindered sulfoxides (Scheme 5.30) [68]. After isolating the formed sulfonium salts, reaction with [^{18}F]fluoride at high temperature provides ^{18}F-labeled

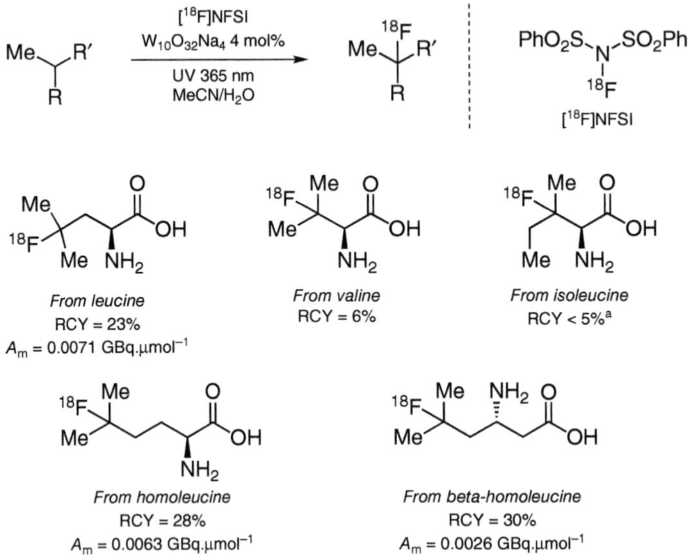

Scheme 5.28 C—H [^{18}F]fluorination of aliphatic amino acids. RCY is decay-corrected. ^{a}A mixture 3- and 4-fluoro isomers is obtained. *Source:* Adapted from Britton and coworkers [15].

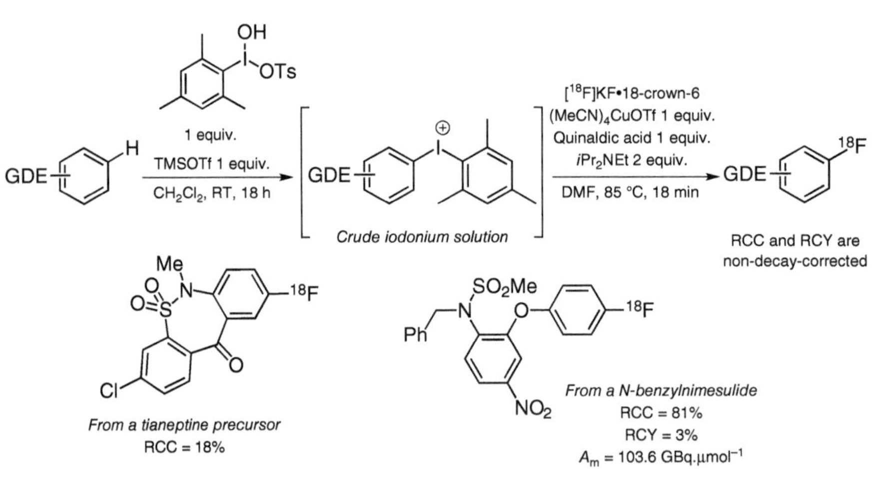

Scheme 5.29 Cu-mediated C—H [^{18}F]fluorination of electron-rich arenes. *Source:* Adapted from Sanford and Scott [67].

derivatives through S_NAr. Protic functional groups (e.g., alcohols and carboxylic acids) have been protected during the process. However, the method demonstrates high functional group tolerance, accommodating ketones, nitriles, esters, and heterocycles. Reported RCC are high, although isolation was performed only once for Dicamba methyl ester and no RCY was provided. Nevertheless, a high A_m was observed.

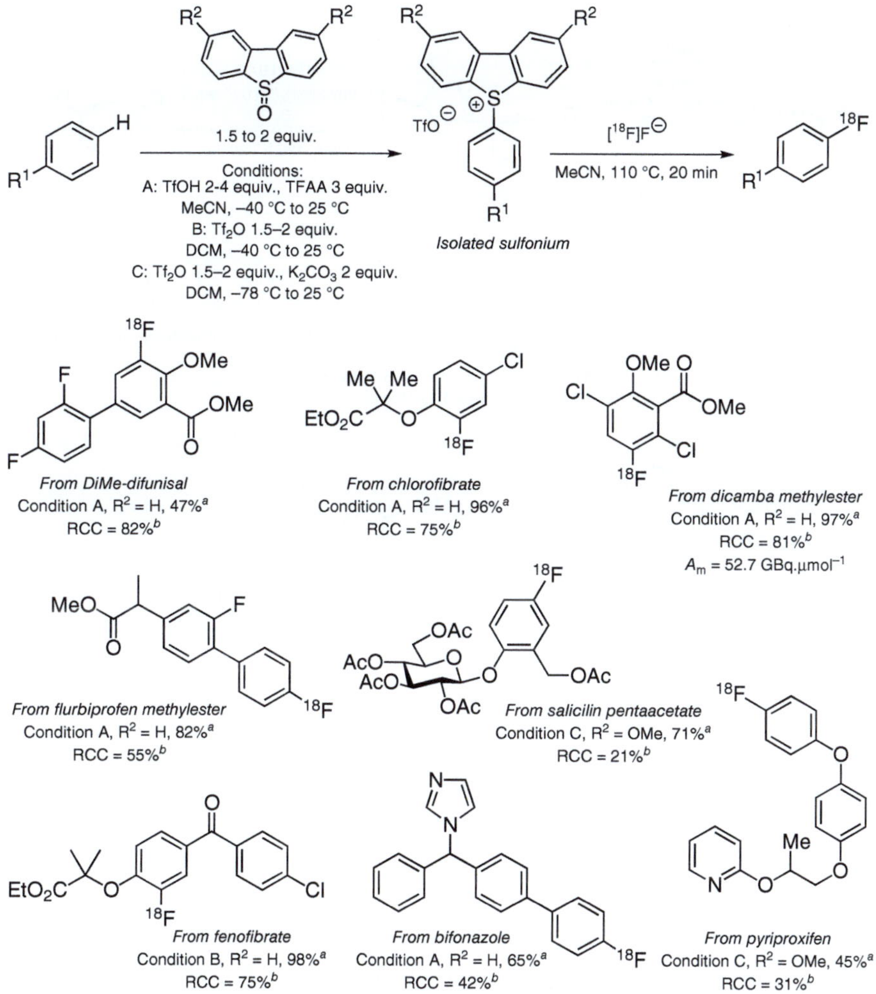

Scheme 5.30 Late stage aromatic ^{18}F-fluorination via aryl sulfonium salts. [a]Isolated sulfonium yield; [b]decay-corrected RCC for the last step; and [c]labeling performed in DMSO with Kryptofix 222, K_2CO_3. *Source:* Adapted from Ritter [68].

In 2021, Scott and Sanford proposed a one-pot *meta*-selective C—H borylation/[^{18}F]fluorination of arenes (Scheme 5.31) [69]. The sequence begins with a C—H borylation using pinacol borane, an iridium(I) catalyst, and 3,4,7,8-tetramethyl-1,10-phenanthroline (tmphen), predominantly targeting the *meta* position of the substrates. The resulting crude pinacol ester solution is then mixed with [^{18}F]tetrabutylammonium fluoride ([^{18}F]TBAF) and a copper(II) catalyst and heated to yield the desired ^{18}F-labeled derivatives. This reaction demonstrates a broad scope, though, again, protic functional groups must be protected. The reported A_m and RCY are suitable for PET *in vivo* imaging.

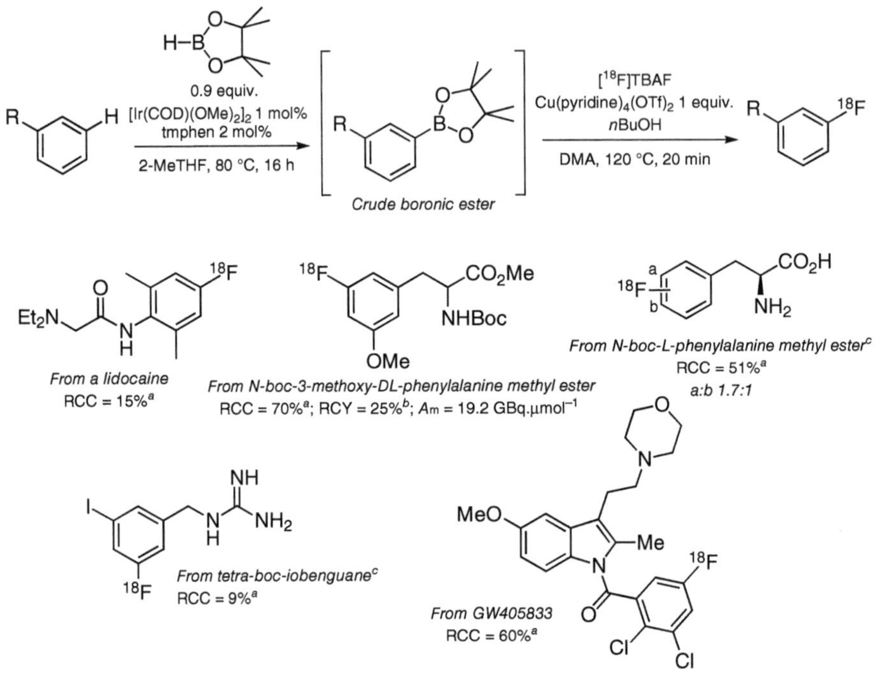

Scheme 5.31 *Meta*-selective C—H borylation/radiofluorination of arenes. [a]RCC is non-decay-corrected; [b]RCY is decay-corrected; [c]an acidic deprotection step was added to the process; and [d]$Cu(pyridine)_4(OTf)_2$ was replaced by Cu(imidazo[1,2-*b*]pyridazine)$_4$(OTf)$_2$. *Source:* Adapted from Sanford and Scott [69].

5.4.3 C—F Activation

In 2015, Hosoya published the first example of C—F bond activation applied to late-stage radioisotope insertion (Scheme 5.32) [70]. This transformation involves a $^{19}F/^{18}F$ exchange on protected dihydrofluvastatin. The process begins with a Ni^0-catalyzed C—F bond borylation, yielding a pinacol boronic ester. A copper-mediated radiofluorodeboronation is then achieved using previously described conditions [71]. However, in this work, no isolation of the compound is carried out, and the A_m is only estimated.

5.5 Iodine Radioisotopes

Among the 37 known isotopes of iodine, 4 of them, iodine-123, -125, -124, and -131, offer valuable radioactive properties to the field of biology, medicine and fundamental research. With a half-life of 13.2 hours, iodine-123 decays by electron capture and emits gamma rays. Its half-life and gamma emission properties make it ideal for diagnostic imaging in nuclear medicine through SPECT where it provides clear images while minimizing radiation exposure to the patient. Iodine-123 is commonly produced in a cyclotron via the proton bombardment of xenon-124

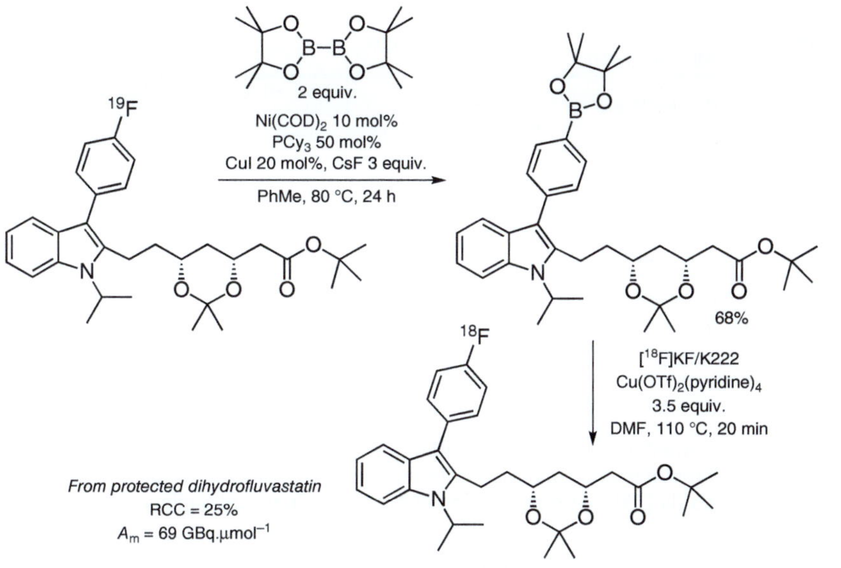

Scheme 5.32 $^{19}F/^{18}F$ exchange through Ni(0)-catalyzed C—F bond activation. RCC is non-decay-corrected. *Source:* Adapted from Niwa et al. [70].

through the $^{124}Xe(p,np)^{123}Xe \rightarrow {}^{123}I$ reaction, where xenon-123 decays to iodine-123. Iodine-125 decays by electron capture, emitting low-energy gamma rays with a half-life of 59.4 days. Iodine-125 is used in RIA, where its gamma emissions help detect biological substances. It is also employed in sealed sources to treat cancer (brachytherapy), where its low-energy emissions minimize damage to surrounding tissues. Most of the iodine investigational radiochemistry is also developed from this isotope due to its physical properties and availability. Iodine-125 is produced in a nuclear reactor through neutron irradiation of xenon-124. The reaction $^{124}Xe(n, \gamma)^{125}Xe \rightarrow {}^{125}I$ involves the capture of neutrons by xenon-124, which decays to iodine-125. Iodine-124 is produced in cyclotron through the proton bombardment of tellurium-124 in the $^{124}Te(p, n)^{124}I$ nuclear reaction. Iodine-124 has a half-life of 4.2 days and decays by positron emission and electron capture. Its positron emission makes it useful for PET imaging; nevertheless, side emission of highly energetic gamma rays limits its applications. Iodine-131 has a half-life of 8.0 days and decays by beta and gamma emissions. Its beta emissions are used in therapeutic applications through RIT, while its gamma emissions were historically used to perform SPECT imaging. Iodine-131 is generated through the neutron irradiation of tellurium-130 in the $^{130}Te(n, \gamma)^{131}Te \rightarrow {}^{131}I$ reaction. All iodine isotopes are commonly provided as aqueous [*I]NaI solutions, and most radiochemical strategies to date focus on labeling Csp^2 sites, to provide metabolically stable molecules, using radiolabeling precursors with oxidizing agents to produce electrophilic iodine species. Established methods in this field include the radioiododestannylation reaction and electrophilic aromatic substitution of tyrosine, which are used to label small molecules and proteins, respectively [72]. Late-stage iodine isotopic exchanges have been known for decades but usually require harsh conditions and

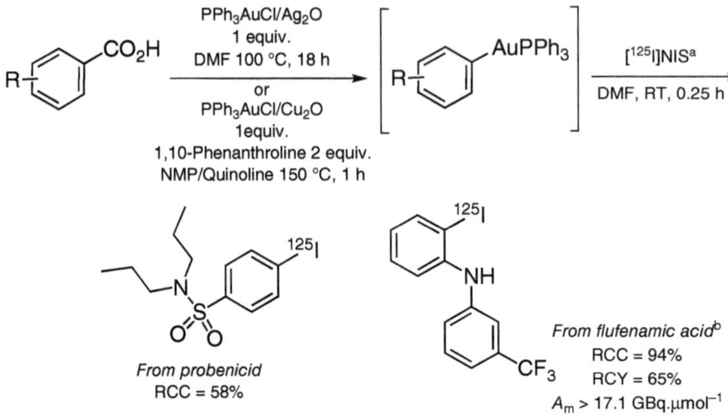

Scheme 5.33 Gold(I)-mediated radio-iododecarboxylation. [a]Prepared by mixing [^{125}I]NaI and NCS at RT for 15 minutes and [b]obtained from the N-Boc derivative after deprotection with H_2SO_4. *Source:* Cailly [75] / with permission of American Chemical Society.

result in low A_m [73, 74]. Recently, a few LSF approaches have been developed, based on functional group transformations and C—H activation strategies.

5.5.1 Functional Group Transformation

In 2023, Cailly's group unveiled a novel gold(I)-mediated radioiododecarboxylation of arenes (Scheme 5.33) [75]. This approach features decarboxylation in the presence of $AuPPh_3Cl$, resulting in the formation of an arylgold(I) intermediate. Subsequently, a crude solution of this intermediate is treated with a [^{125}I]*N*-iodosuccinimide ([^{125}I]NIS) solution, prepared *in situ* from [^{125}I]NaI and NCS. The reaction exhibits a broad scope, accommodating sensitive functional groups such as aldehydes, ketones, and esters. Moreover, heteroarenes, cinnamic acid, and phenylpropiolic acid proved to be valuable substrates for this transformation. Notably, the method successfully generated radioiodinated analogs of probenecid and flufenamic acid with convincing RCY and A_m.

5.5.2 C—H Activation Reactions

In 2018, Cailly's group described a palladium-mediated C—H radioiodination of arenes. The reaction involves a two-step sequence (Scheme 5.34) [76]. First, an arene with a C—H activation-promoting group is reacted at room temperature for 24 hours with a stoichiometric amount of palladium(II) acetate, yielding a crude palladacycle solution. This solution is then mixed with [1^{25}I]NIS, prepared *in situ* from [^{125}I]NaI and NCS. The ^{125}I-labeled derivatives are obtained after 15 minutes through oxidative insertion and reductive elimination. The process is efficiently demonstrated with *N*-acylsulfonamide substrates, and 15 other directing groups are identified in the article as capable of promoting the transformation. Additionally, a modification using a catalytic amount of palladium and a palladacycle formation time of 15 minutes have also been published [77].

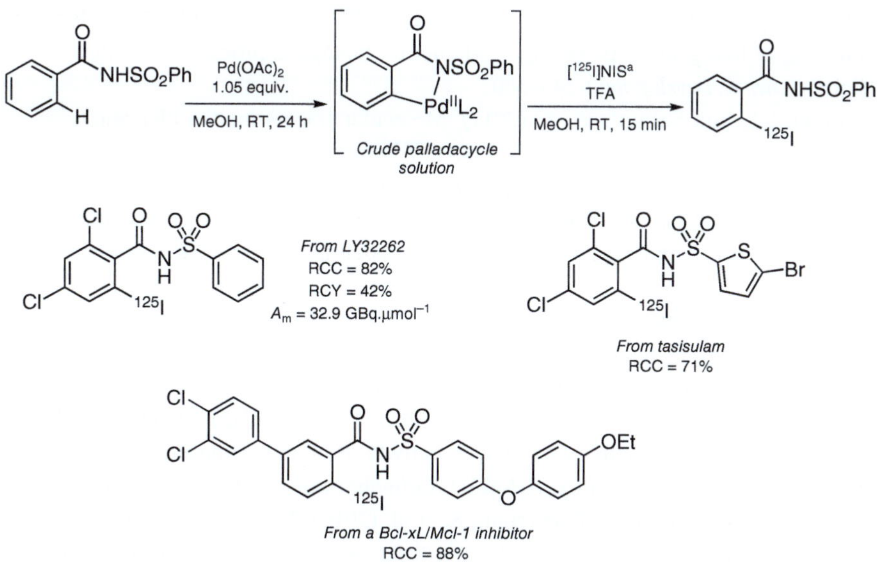

Scheme 5.34 Palladium-mediated site-selective C—H radioiodination. [a]Prepared by mixing [^{125}I]NaI and NCS at RT for 15 minutes. *Source:* Adapted from Cailly [76].

5.6 Summary and Conclusions

This chapter presents the most advanced LSF strategies currently applicable in the field of radioisotope insertion. As the reader may have noticed, most of the articles cited in this chapter are from the past decade and have been produced by a few research groups. The insertion of radioisotopes using late-stage approaches is a relatively new area of chemistry pioneered by a small community. Moreover, considering the nature of the isotopes, much of the work discussed focuses on publications involving tritium and carbon-14, where isotopic exchange predominates. These isotopic labeling methods are now well accepted by the scientific community, and some of these LSF strategies have already become references in the field. For fluorine-18, carbon-11, and iodine isotopes, the community focused on the preparation and discovery of PET and SPECT radiotracers remains largely reliant on precursor-based approaches and LSF is rarely used to develop new diagnostic tools. In the future, alongside the established LSF processes, the discovery of new efficient LSF strategies for isotope incorporation may lead to a paradigm shift in the conception of radiotracers, especially in non-isotopic labeling.

References

1 Chatal, J.-F. and Hoefnagel, C.A. (1999). Radionuclide therapy. *Lancet* 354: 931–935.

2 Volkert, W.A. and Hoffman, T.J. (1999). Therapeutic radiopharmaceuticals. *Chemical Reviews* 99: 2269–2292.

3 Massoud, T.F., Massoud, T.F., Gambhir, S.S., and Gambhir, S.S. (2003). Molecular imaging in living subjects: seeing fundamental biological processes in a new light. *Genes & Development* 545–580.
4 Pimlott, S.L. and Sutherland, A. (2011). Molecular tracers for the PET and SPECT imaging of disease. *Chemical Society Reviews* 40: 149–162.
5 Hermanson, G.T. (2013). *Bioconjugate Techniques*. Elsevier.
6 Wilbur, D.S. (1992). Radiohalogenation of proteins: an overview of radionuclides, labeling methods and reagents for conjugate labeling. *Bioconjugate Chemistry* 3: 433–470.
7 Herth, M.M., Ametamey, S., Antuganov, D. et al. (2021). On the consensus nomenclature rules for radiopharmaceutical chemistry – reconsideration of radiochemical conversion. *Nuclear Medicine and Biology* 93: 19–21.
8 Luurtsema, G., Pichler, V., Bongarzone, S. et al. (2021). EANM guideline for harmonisation on molar activity or specific activity of radiopharmaceuticals: impact on safety and imaging quality. *EJNMMI Radiopharmacy and Chemistry* 6: 34.
9 Cernak, T., Dykstra, K.D., Tyagarajan, S. et al. (2016). The medicinal chemist's toolbox for late stage functionalization of drug-like molecules. *Chemical Society Reviews* 45: 546–576.
10 Moir, M., Danon, J.J., Reekie, T.A., and Kassiou, M. (2019). An overview of late-stage functionalization in today's drug discovery. *Expert Opinion on Drug Discovery* 14: 1137–1149.
11 Börgel, J. and Ritter, T. (2020). Late-stage functionalization. *Chem* 6: 1877–1887.
12 Guillemard, L., Kaplaneris, N., Ackermann, L., and Johansson, M.J. (2021). Late-stage C—H functionalization offers new opportunities in drug discovery. *Nature Reviews Chemistry* 5: 522–545.
13 Malmquist, J. and Ström, P. (2012). Multiple labeling of a potent CX3CR1 antagonist for the treatment of multiple sclerosis. *Journal of Labelled Compounds and Radiopharmaceuticals* 55: 387–392.
14 Bsharat, O., Doyle, M.G.J., Munch, M. et al. (2022). Aldehyde-catalysed carboxylate exchange in α-amino acids with isotopically labelled CO_2. *Nature Chemistry* 14: 1367–1374.
15 Nodwell, M.B., Yang, H., Čolović, M. et al. (2017). ^{18}F-fluorination of unactivated C—H bonds in branched aliphatic amino acids: direct synthesis of oncological positron emission tomography imaging agents. *Journal of the American Chemical Society* 139: 3595–3598.
16 Voges, R., Heys, J.R., and Moenius, T. (2009). Chapter 2, Strategies for target preparation. In: *Preparation of Compounds Labeled with Tritium and Carbon-14*, 25–45. Wiley.
17 Saljoughian, M. (2002). Synthetic tritium labeling: reagents and methodologies. *Synthesis* 2002: 1781–1801.
18 Crabtree, R. (1979). Iridium compounds in catalysis. *Accounts of Chemical Research* 12: 331–337.
19 Timofeeva, D.S., Lindsay, D.M., Kerr, W.J., and Nelson, D.J. (2020). A quantitative empirical directing group scale for selectivity in iridium-catalysed hydrogen isotope exchange reactions. *Catalysis Science & Technology* 10: 7249–7255.

20 Jess, K., Derdau, V., Weck, R. et al. (2017). Hydrogen isotope exchange with iridium(I) complexes supported by phosphine-imidazolin-2-imine P,N ligands. *Advanced Synthesis and Catalysis* 359: 629–638.

21 Yongsong, T., Brož, B., Tureček, F., and Marek, A. (2022). Tritium hydrogen-isotope exchange with electron-poor tertiary benzenesulfonamide moiety; application in late-stage labeling of T0901317. *Journal of Labelled Compounds and Radiopharmaceuticals* 65: 36–44.

22 Morawietz, P., Weck, R., Scholte, A.A. et al. (2022). Significantly improved radiochemical yields in gaseous tritium reactions by iridium(I)-catalyzed hydrogen isotope exchange. *Green Chemistry* 24: 4824–4829.

23 Daniel-Bertrand, M., Garcia-Argote, S., Palazzolo, A. et al. (2020). Multiple site hydrogen isotope labelling of pharmaceuticals. *Angewandte Chemie, International Edition* 59: 21114–21120.

24 Valero, M., Bouzouita, D., Palazzolo, A. et al. (2020). NHC-stabilized iridium nanoparticles as catalysts in hydrogen isotope exchange reactions of anilines. *Angewandte Chemie, International Edition* 59: 3517–3522.

25 Pony Yu, R., Hesk, D., Rivera, N. et al. (2016). Iron-catalysed tritiation of pharmaceuticals. *Nature* 529: 195–199.

26 Yang, H., Zarate, C., Palmer, W.N. et al. (2018). Site-selective nickel-catalyzed hydrogen isotope exchange in N-heterocycles and its application to the tritiation of pharmaceuticals. *ACS Catalysis* 8: 10210–10218.

27 Zarate, C., Yang, H., Bezdek, M.J. et al. (2019). Ni(I)-X complexes bearing a bulky α-diimine ligand: synthesis, structure, and superior catalytic performance in the hydrogen isotope exchange in pharmaceuticals. *Journal of the American Chemical Society* 141: 5034–5044.

28 Yang, H., Dormer, P.G., Rivera, N.R., and Hoover, A.J. (2018). Palladium(II)-mediated C—H tritiation of complex pharmaceuticals. *Angewandte Chemie, International Edition* 57: 1883–1887.

29 Pfeifer, V., Zeltner, T., Fackler, C. et al. (2021). Palladium nanoparticles for the deuteration and tritiation of benzylic positions on complex molecules. *Angewandte Chemie, International Edition* 60: 26671–26676.

30 Zhao, D., Petzold, R., Yan, J. et al. (2021). Tritiation of aryl thianthrenium salts with a molecular palladium catalyst. *Nature* 600: 444–449.

31 Gao, L., Perato, S., Garcia-Argote, S. et al. (2018). Ruthenium-catalyzed hydrogen isotope exchange of C(sp^3)—H bonds directed by a sulfur atom. *Chemical Communications* 54: 2986–2989.

32 Müller, V., Weck, R., Derdau, V., and Ackermann, L. (2020). Ruthenium(II)-catalyzed hydrogen isotope exchange of pharmaceutical drugs by C—H deuteration and C—H tritiation. *ChemCatChem* 12: 100–104.

33 Pfeifer, V., Certiat, M., Bouzouita, D. et al. (2020). Hydrogen isotope exchange catalyzed by Ru nanocatalysts: labelling of complex molecules containing N-heterocycles and reaction mechanism insights. *Chemistry – A European Journal* 26: 4988–4996.

34 Palazzolo, A., Feuillastre, S., Pfeifer, V. et al. (2019). Efficient access to deuterated and tritiated nucleobase pharmaceuticals and oligonucleotides using hydrogen-isotope exchange. *Angewandte Chemie, International Edition* 58: 4891–4895.

35 Levernier, E., Tatoueix, K., Garcia-Argote, S. et al. (2022). Easy-to-implement hydrogen isotope exchange for the labeling of N-heterocycles, alkylkamines, benzylic scaffolds, and pharmaceuticals. *JACS Au* 2: 801–808.

36 Loh, Y.Y., Nagao, K., Hoover, A.J. et al. (2017). Photoredox-catalyzed deuteration and tritiation of pharmaceutical compounds. *Science* 358: 1182–1187.

37 Murugesan, K., Donabauer, K., Narobe, R. et al. (2022). Photoredox-catalyzed site-selective generation of carbanions from C(sp^3)—H bonds in amines. *ACS Catalysis* 12: 3974–3984.

38 Du, H., Fan, J., Wang, Z. et al. (2023). Cesium amide-catalyzed selective deuteration of benzylic C—H bonds with D 2 and application for tritiation of pharmaceuticals. *Angewandte Chemie, International Edition* 62: e202214461.

39 Scott, P.J.H. (2009). Methods for the incorporation of carbon-11 to generate radiopharmaceuticals for PET imaging. *Angewandte Chemie, International Edition* 48: 6001–6004.

40 Antoni, G. (2015). Development of carbon-11 labelled PET tracers-radiochemical and technological challenges in a historic perspective. *Journal of Labelled Compounds and Radiopharmaceuticals* 58: 65–72.

41 Babin, V., Taran, F., and Audisio, D. (2022). Late-stage carbon-14 labeling and isotope exchange: emerging opportunities and future challenges. *JACS Au* 2: 1234–1251.

42 Gauthier, D.R., Rivera, N.R., Yang, H. et al. (2018). Palladium-catalyzed carbon isotope exchange on aliphatic and benzoic acid chlorides. *Journal of the American Chemical Society* 140: 15596–15600.

43 Friis, S.D., Taaning, R.H., Lindhardt, A.T., and Skrydstrup, T. (2011). Silacarboxylic acids as efficient carbon monoxide releasing molecules: synthesis and application in palladium-catalyzed carbonylation reactions. *Journal of the American Chemical Society* 133: 18114–18117.

44 Friis, S.D., Lindhardt, A.T., and Skrydstrup, T. (2016). The development and application of two-chamber reactors and carbon monoxide precursors for safe carbonylation reactions. *Accounts of Chemical Research* 49: 594–605.

45 Destro, G., Loreau, O., Marcon, E. et al. (2019). Dynamic carbon isotope exchange of pharmaceuticals with labeled CO_2. *Journal of the American Chemical Society* 141: 780–784.

46 Destro, G., Horkka, K., Loreau, O. et al. (2020). Transition-metal-free carbon isotope exchange of phenyl acetic acids. *Angewandte Chemie, International Edition* 59: 13490–13495.

47 Kong, D., Moon, P.J., Lui, E.K.J. et al. (2020). Direct reversible decarboxylation from stable organic acids in dimethylformamide solution. *Science* 369: 557–561.

48 Kingston, C., Wallace, M.A., Allentoff, A.J. et al. (2019). Direct carbon isotope exchange through decarboxylative carboxylation. *Journal of the American Chemical Society* 141: 774–779.

49 Tortajada, A., Duan, Y., Sahoo, B. et al. (2019). Catalytic decarboxylation/carboxylation platform for accessing isotopically labeled carboxylic acids. *ACS Catalysis* 9: 5897–5901.

50 Babin, V., Talbot, A., Labiche, A. et al. (2021). Photochemical strategy for carbon isotope exchange with CO_2. *ACS Catalysis* 11: 2968–2976.

51 Kong, D., Munch, M., Qiqige, Q. et al. (2021). Fast carbon isotope exchange of carboxylic acids enabled by organic photoredox catalysis. *Journal of the American Chemical Society* 143: 2200–2206.

52 Doyle, M.G.J., Mair, B.A., Sib, A. et al. (2024). *A Practical Guide for the Preparation of C1-Labeled α-Amino Acids Using Aldehyde Catalysis with Isotopically Labeled* CO_2. US: Springer.

53 Reilly, S.W., Lam, Y.H., Ren, S., and Strotman, N.A. (2021). Late-stage carbon isotope exchange of aryl nitriles through Ni-catalyzed C—CN bond activation. *Journal of the American Chemical Society* 143: 4817–4823.

54 Feng, M., De Oliveira, J., Sallustrau, A. et al. (2021). Direct carbon isotope exchange of pharmaceuticals via reversible decyanation. *Journal of the American Chemical Society* 143: 5659–5665.

55 Zhang, Z., Niwa, T., Watanabe, K., and Hosoya, T. (2023). 11 C-cyanation of aryl fluorides via nickel and lithium chloride-mediated C—F bond activation. *Angewandte Chemie, International Edition* 62: 1–5.

56 Jacobson, O., Kiesewetter, D.O., and Chen, X. (2015). Fluorine-18 radiochemistry, labeling strategies and synthetic routes. *Bioconjugate Chemistry* 26: 1–18.

57 Tredwell, M. and Gouverneur, V. (2012). 18F labeling of arenes. *Angewandte Chemie, International Edition* 51: 11426–11437.

58 Preshlock, S., Tredwell, M., and Gouverneur, V. (2016). 18 F-labeling of arenes and heteroarenes for applications in positron emission tomography. *Chemical Reviews* 116: 719–766.

59 Blom, E., Karimi, F., and Långström, B. (2009). [18F]/19F exchange in fluorine containing compounds for potential use in 18F-labelling strategies. *Journal of Labelled Compounds and Radiopharmaceuticals* 52: 504–511.

60 Ryzhikov, N.N., Gomzina, N.A., Fedorova, O.S. et al. (2004). Preparation of [^{10}F]flumazenil, a potential radioligand for PET imaging of central benzodiazepine receptors, by isotope exchange. *Radiochemistry* 46: 290–294.

61 Wang, T., Lv, S., Mou, Z. et al. (2023). Isotope exchange-based ^{18}F-labeling methods. *Bioconjugate Chemistry* 34: 140–161.

62 Zheng, Q., Xu, H., Wang, H. et al. (2021). Sulfur [18F]fluoride exchange click chemistry enabled ultrafast late-stage radiosynthesis. *Journal of the American Chemical Society* 143: 3753–3763.

63 Neumann, C.N., Hooker, J.M., and Ritter, T. (2016). Concerted nucleophilic aromatic substitution with $19F^-$ and $18F^-$. *Nature* 534: 369–373.

64 Beyzavi, H., Mandal, D., Strebl, M.G. et al. (2017). 18 F-deoxyfluorination of phenols via Ru π-complexes. *ACS Central Science* 3: 944–948.

65 Ford, J., Ortalli, S., Chen, Z. et al. (2024). Expedient access to ^{18}F-fluoroheteroarenes via deaminative radiofluorination of aniline-derived pyridinium salts. *Angewandte Chemie, International Edition* 63: e202404945.

66 Huang, X., Liu, W., Ren, H. et al. (2014). Late stage benzylic C—H fluorination with [^{18}F]fluoride for PET imaging. *Journal of the American Chemical Society* 136: 6842–6845.

67 McCammant, M.S., Thompson, S., Brooks, A.F. et al. (2017). Cu-mediated C—H ^{18}F-fluorination of electron-rich (hetero)arenes. *Organic Letters* 19: 3939–3942.

68 Xu, P., Zhao, D., Berger, F. et al. (2020). Site-selective late-stage aromatic ^{18}F fluorination via aryl sulfonium salts.pdf. *Angewandte Chemie, International Edition* 59: 1956–1960.

69 Wright, J.S., Sharninghausen, L.S., Preshlock, S. et al. (2021). Sequential Ir/Cu-mediated method for the meta-selective C—H radiofluorination of (hetero)arenes. *Journal of the American Chemical Society* 143: 6915–6921.

70 Niwa, T., Ochiai, H., Watanabe, Y., and Hosoya, T. (2015). Ni/Cu-catalyzed defluoroborylation of fluoroarenes for diverse C—F bond functionalizations. *Journal of the American Chemical Society* 137: 14313–14318.

71 Tredwell, M., Preshlock, S.M., Taylor, N.J. et al. (2014). A general copper-mediated nucleophilic 18F fluorination of arenes. *Angewandte Chemie, International Edition* 53: 7751–7755.

72 Dubost, E., McErlain, H., Babin, V. et al. (2020). Recent advances in synthetic methods for radioiodination. *Journal of Organometallic Chemistry* 85: 8300–8310.

73 Chezal, J., Papon, J., Labarre, P. et al. (2008). Evaluation of radiolabeled (hetero)aromatic analogues of *N*-(2-diethylaminoethyl)-4-iodobenzamide for imaging and targeted radionuclide therapy of melanoma. *Journal of Medicinal Chemistry* 51: 3133–3144.

74 Giglio, B.C., Wang, H., Yan, X., and Li, Z. (2019). Synthesis and initial evaluation of radioactive 5-I-α-methyl-tryptophan: a Trp based agent targeting IDO-1. *MedChemComm* 10: 814–816.

75 Bloux, H., Khouya, A.A., Sopkova-de Oliveira Santos, J. et al. (2023). Gold(I)-mediated radioiododecarboxylation of arenes. *Organic Letters* 25: 8100–8104.

76 Dubost, E., Babin, V., Benoist, F. et al. (2018). Palladium-mediated site-selective C—H radio-iodination. *Organic Letters* 20: 6302–6305.

77 Dubost, E., Babin, V., Benoist, F. et al. (2019). Improvements of C—H radio-iodination of *N*-acylsulfonamides toward implementation in clinics. *Synthesis* 51: 4393–4400.

Part II

Diversity-Oriented Synthesis

6

Late-Stage Functionalization Strategies Using C—O Bond Activation

Rémi Blieck

Sorbonne Université, CNRS, Institut Parisien de Chimie Moléculaire, IPCM, 4 place Jussieu 75252, F-75005, Paris, France

6.1 Introduction

Late-stage functionalization is an extremely dynamic field of research in organic synthesis, with the potential to profoundly transform chemical synthesis and, in particular, drug discovery [1]. Focusing on the selective modification of complex molecules at advanced stages of synthesis, including the selective derivatization of natural products, this strategy faces major challenges in terms of functional group tolerance, efficiency, and chemoselectivity. However, many recent advances in the field demonstrate that late-stage functionalization represents a fundamental shift in the way chemists address the synthesis and modification of highly functionalized compounds.

Most of the strategies developed in late-stage functionalization focus on selective C—H functionalization [2], but this approach is certainly not the only one that is the subject of intense research. This chapter therefore illustrates the late-stage functionalization of compounds of interest by C—O bond activation. Although this chapter cannot be an exhaustive overview of the extensive range of examples developed in this area of research, it highlights the broad spectrum of approaches available to date and their efficacy. A classification by functional group converted has been adopted to illustrate this wide variety, and following the recent definition of late-stage functionalization [1], reactions requiring the pre-installation of activating groups to enable the selective reaction have been considered out of scope.

6.2 Late-Stage Activation of Free Alcohols

Alcohols are ubiquitous in compounds of interest. This functional group is present in a wide variety of natural products and is one of the 10 most common functional groups in bioactive compounds [3], playing a pivotal role in the pharmacological properties of drugs, with a significant influence on solubility and interactions

Late-Stage Functionalization and Diversification in Organic Synthesis: Methods and Applications,
First Edition. Edited by Tatiana Besset.

with biological targets. In addition to this widespread occurrence of alcohols in pharmaceuticals, the possibility of directly using alcohols for late diversification, by activating the C—O bond, encourages a great number of research to adopt this strategy. Numerous distinct mechanisms have been extensively used for the late-stage functionalization of sophisticated products with a free alcohol through activation of the C—O bond, including hydrogen borrowing mechanism, radical activation, and transition metal-catalyzed coupling reactions.

6.2.1 *N*- and *C*-Alkylation by Hydrogen Borrowing Mechanism

One of the most represented strategies for the use of alcohols for late-stage functionalization by C—O bond activation is based on hydrogen borrowing mechanism. Selective functionalization is then achieved by the formation of a reactive carbonyl intermediate from the free alcohol, which can subsequently be involved in a wide variety of reactions, including alkylation, allylation, cyclization, and cross-coupling reactions.

The scope of application of this mechanism is particularly broad, and it has been applied to the late-stage functionalization and diversification of a large number of compounds. Among the reported examples, the possibility of activating the C—O bond of methanol allows late *N*-methylation without the need for toxic precursors such as diazomethane or methyl iodide [4]. When used with ruthenium catalysis, this method has been particularly efficient for the synthesis of atomoxetine [5], naftifine [6], and venlafaxine [7]. In addition, the same strategy is efficient for late-stage *N*-alkylation with primary and secondary alcohols, using nickel- [8], manganese- [9] and cobalt-based catalytic systems [10] (Scheme 6.1). Furthermore, this hydrogen borrowing mechanism is efficient for several *C*-alkylations, as demonstrated in the ruthenium-catalyzed α-alkylation of methylene ketones, leading to donepezil [11], the cobalt-catalyzed derivatization of cholesterol [12], and the iridium-catalyzed derivatization of vitamin E [13]. A comprehensive overview of the applications of hydrogen borrowing strategies for the synthesis of drugs and late-stage diversifications from free alcohols has recently been reported [14], revealing the wide range of applications available.

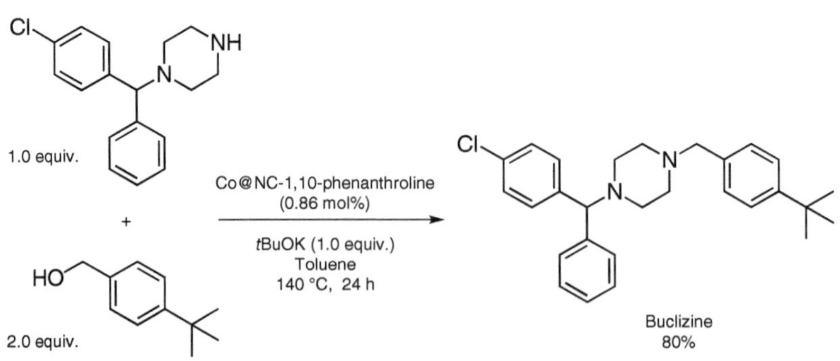

Scheme 6.1 Co-catalyzed *N*-alkylation with primary alcohols.

6.2.2 Radical Activation of Alcohols

Since 2022, *in situ* activation of free alcohols by *N*-heterocyclic carbene salts has demonstrated its high efficacy for the late-stage derivatization of compounds and has been adapted to a great variety of transition metal-catalyzed couplings with diverse nucleophiles. First developed for the deoxygenative arylation of alcohols [15] and directly applied to the late-stage derivatization of Taxol and simvastatin (Scheme 6.2), this strategy has been rapidly adapted to cross-coupling with alkenes [16], carboxylic acids [17], alkyl bromides [18], aromatic amines and carbamates [19], and alcohols [20] and to phosphonylation [21], trifluoromethylation [22], and sulfination reactions [23]. The high flexibility of this approach has been illustrated through late-stage functionalization of pharmaceuticals.

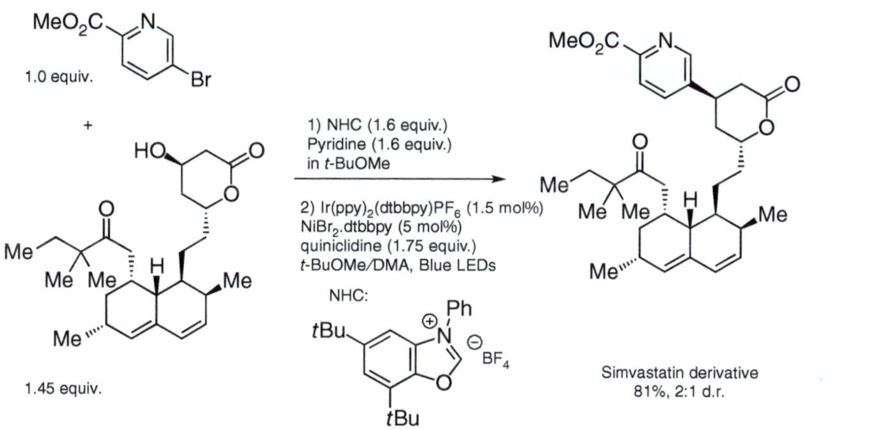

Scheme 6.2 Deoxygenative arylation of free alcohols using *N*-heterocyclic carbene salts.

Complementary to this approach, other radical activations of alcohols have been reported. In particular, transition metal catalysis based on titanium catalysis has been exploited for the selective deoxygenative radical alkylation of tertiary alcohols [24], including gomisin A (Scheme 6.3), and copper catalysis is efficient for the aryl radical-activated deoxygenative difluoromethylation of alcohols, providing access to bioisosteres [25]. Radical species also demonstrated their efficiency in alcohols activation, by using phosphoranyl radicals [26] or boryl radicals [27] as activators under photoredox catalysis.

6.2.3 Deoxyhalogenation of Alcohols

The halogenation of alcohols is another efficient route for the late-stage diversification of free alcohols. In particular, as a direct pathway for the synthesis of fluorine-containing analogs of compounds of interest, the late-stage deoxyfluorination is more developed than the other deoxyhalogenations [28]. Until recently, the classical fluorinating reagents were highly efficient but lacked sufficient functional

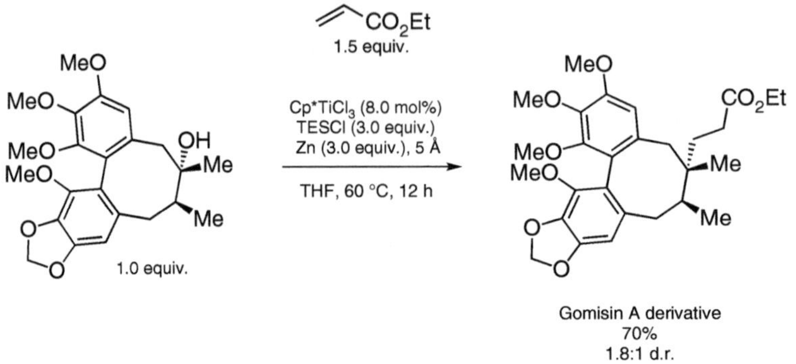

Scheme 6.3 Ti-catalyzed deoxygenative alkylation of tertiary alcohols.

group tolerance to be used in late-stage approach, but since 2011, several late-stage deoxyfluorination reactions have been reported. In the case of C(sp^2)—O bond activation, the use of PhenoFluor has demonstrated its efficiency in the deoxyfluorination of phenolic derivatives [29], including bio-relevant compounds (Scheme 6.4).

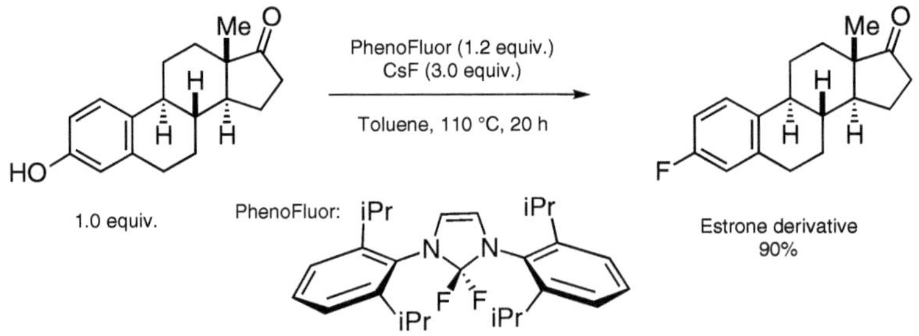

Scheme 6.4 Deoxyfluorination of phenols using PhenoFluor.

Analogs of PhenoFluor have been developed, demonstrating higher stability toward hydrolysis [30] and greater reactivity under milder conditions [31].

Complementary to the functionalization of C(sp^2)—O bonds, PhenoFluor has also been applied to the deoxyfluorination of C(sp^3)—O bonds. The method has demonstrated its compatibility with a wide range of bioactive compounds, including alkaloids, with high selectivity [32]. Applied to oligomycin A (Scheme 6.5), a single hydroxyl group was converted, in the presence of other tertiary- or β,β′-disubstituted secondary alcohols.

Following these pioneering examples, several fluorinating reagents have been developed: PyFluor [33], CpFluor [34], AlkylFluor [35], SulfoxFluor [36], SelectFluor/R_3P/ICH_2CH_2I [37], and methanesulfonic acid [38], all proved to be efficient in the context of late-stage transformations and were applied to the conversion of challenging compounds. Circumventing the use of complex fluorinating agents, the group of Watson demonstrated in 2020 the possibility of using CuF_2 as fluorine source [39] together with diisopropylcarbodiimide (DIC) to perform a selective deoxyfluorination, efficient, for example, on Lumefantrine.

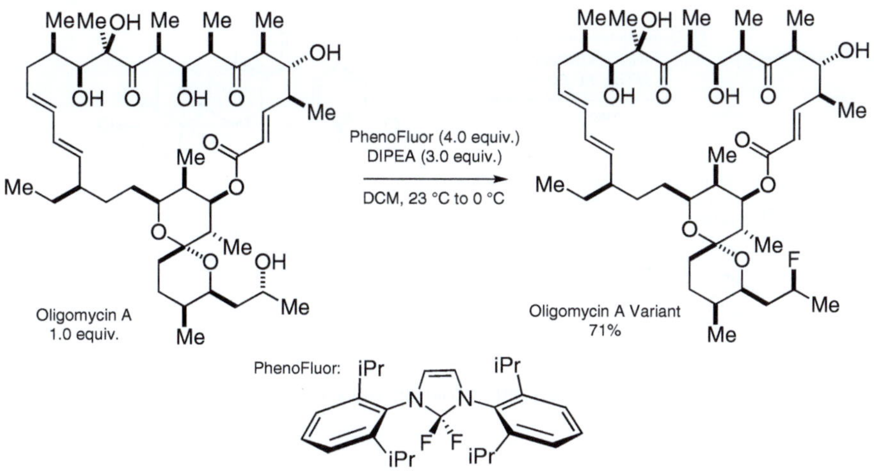

Scheme 6.5 Deoxyfluorination of aliphatic alcohols.

Alternatively, the Appel mechanism has been suitably adapted for deoxychlorination [40], deoxybromination [41], and deoxyiodination [42] in the context of late-stage derivatization of alcohols.

6.2.4 Transition Metal-Catalyzed Arylation and Alkylation of Free Alcohols

Although considered inert in many transition metal-catalyzed coupling reactions and often used in the form of activated sulfonate derivatives, free alcohols have been exploited for the late-stage diversification of compounds of interest, particularly in arylation and allylation reactions.

6.2.4.1 Arylation

Among the efficient catalysts, the use of nickel has allowed an efficient late diversification of alcohols by arylation. In 2011, a Suzuki–Miyaura coupling between naphthol derivatives and various aryl boroxines gave a direct access to biaryl compounds [43]. Nickel-catalyzed coupling between free alcohols and aryl (pseudo)halides has extended the scope of applications to a wide range of alcohols, including both aromatic and aliphatic alcohols [44]. In particular, an enantioconvergent deoxygenative reductive coupling [45] has been used for the asymmetric synthesis of bioactive compounds (Scheme 6.6).

6.2.4.2 Alkylation

Using an efficient catalytic system based on palladium and calcium, the dehydrative cross-coupling of allylic alcohols has been described as a late-stage allylation method. It has been applied to alkynes [46], tolerating the presence of several functional groups, to keto-ester and malonate-ester derivatives [47], and to allylic or alkenyl C—H bonds (Scheme 6.7) [48].

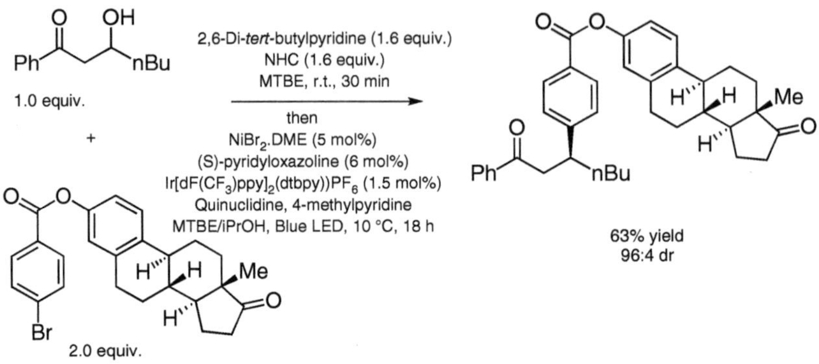

Scheme 6.6 Deoxygenative Ni-catalyzed Suzuki–Miyaura cross-coupling.

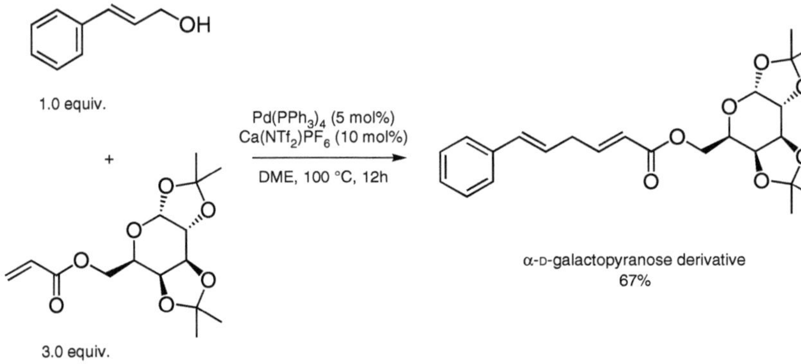

Scheme 6.7 Pd-catalyzed alkylation of alcohols.

Among the catalytic systems that have demonstrated their efficiency in late-stage alkylation with free alcohols, titanium can also be mentioned for the vinylation of tertiary alcohols [49], along with iridium for the alkynylation of allylic alcohols [50].

6.2.5 Alcohols Oxidation

The selective oxidation of alcohols to the corresponding ketones or aldehydes has been widely described and is an important means of diversifying compounds of interest. Efficient methods are based on metal-free visible light-induced oxidation [51], CO_2-mediated oxidation [52], and organometallic catalyzed transfer dehydrogenation [53]. Catalytic acceptorless dehydrogenation of alcohols, including aliphatic alcohols [54], has also been applied for late-stage oxidation (Scheme 6.8).

6.2.6 Deoxyborylation of Alcohols

Given that access to a borylated derivative of any compound of interest provides direct gateways to numerous derivatization pathways, the late-stage

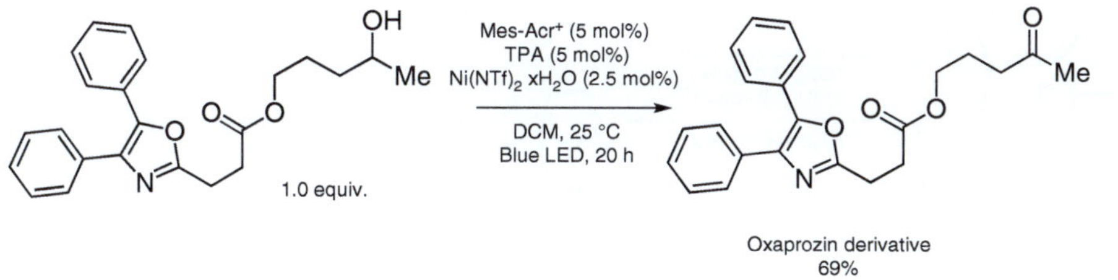

Scheme 6.8 Ni-catalyzed acceptorless dehydrogenation of aliphatic alcohols.

borylation has been widely studied [55], mainly via C—H bond activation. The direct and selective deoxyborylation of free alcohols has also been developed with a simple procedure and has recently been applied to late-stage derivatization. The one-pot deoxyborylation of phenolic derivatives has been reported using SO_2F_2 as *in situ* activator and was applied in particular to the selective deoxyborylation of (*S*)-10-hydroxycamptothecin [56]. Regarding the deoxyborylation of aliphatic alcohols, an electrochemical approach allowed the selective borylation of alcohols and carbonyl derivatives [57] and was efficiently applied to several bioactive compounds and analogs (Scheme 6.9).

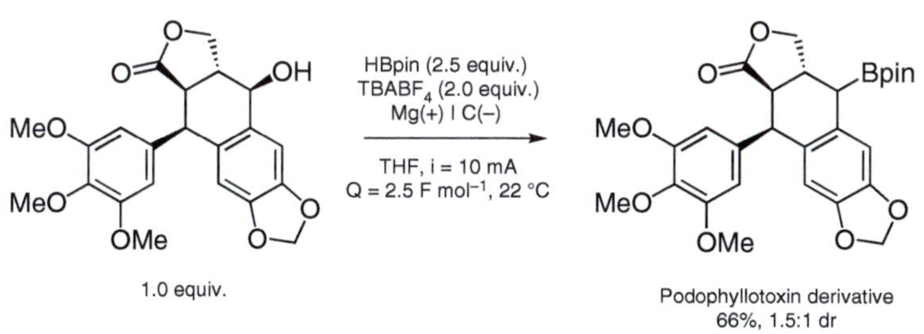

Scheme 6.9 Electrochemical deoxyborylation of alcohols.

6.2.7 Deoxyazidation of Alcohols

Until recently, the most efficient method for azide formation from alcohols relied on Mitsunobu conditions [58] but was rarely used in a late-stage approach. The other efficient deoxyazidations of alcohols suffered from harsh conditions and limited scope [59], preventing the use of this reaction in the context of late-stage functionalization. Yet, in 2022, two distinct approaches made it possible to achieve the late-stage deoxyazidation of free alcohols for a rapid diversification. One approach involves the use of a hypervalent azido-iodine(III) reagent in the presence of triphenylphosphine and was efficient for the azidation of a broad scope of aliphatic alcohols [60]. The other approach relies on the use of simple NaN_3 in the presence of a fluorinated sulfoximine and demonstrated a high functional group tolerance (Scheme 6.10) [61].

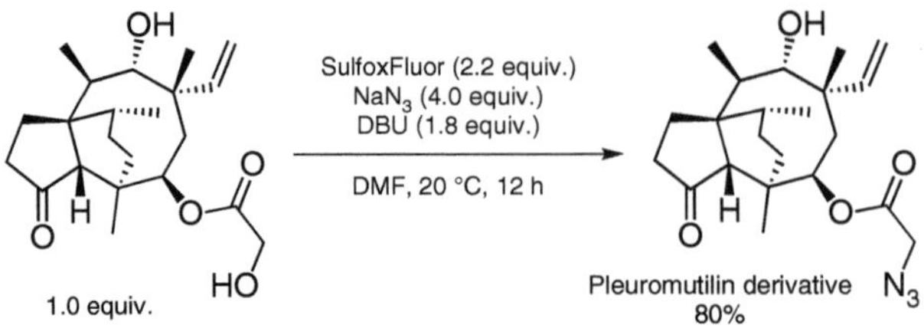

Scheme 6.10 Deoxyazidation of alcohols using NaN_3.

6.3 Activation of Carboxylic Acids

Carboxylic acids are among the most represented functional groups in compounds of interest [3], and their reactivities have also been exploited in late-stage derivatization. Thus, numerous examples can be found in the scientific literature exploiting the presence of carboxylic acids, in decarboxylative or non-decarboxylative processes, for late-stage couplings. In addition, the late-stage reduction of acids to the corresponding alcohols, aldehydes, or alkanes has also been an effective tool for derivatizing these compounds.

6.3.1 Decarboxylative Couplings

6.3.1.1 Decarboxylative Arylation

One of the most classical reactions for a decarboxylative arylation of carboxylic acids is the Minisci-type reaction, efficient with a wide range of radical precursors and classically developed on carboxylic acids [62]. This process has demonstrated its efficiency with metal-catalyzed approaches, notably with silver [63], and major advances have been observed with light irradiation, enabling its use for late-stage functionalization [64]. A pioneering example described the efficiency of an iridium-based catalyst [65], in the presence of an excess of oxidant, and applied it to the alkylation of a fasudil analog with carboxylic acid (Scheme 6.11). Cerium also demonstrated its efficiency for the decarboxylative carbamoylation of azoles, even in late-stage functionalization, using an electrophotocatalytic process with oxamic acid derivatives [66].

Complementary to the use of metallic species, organic photocatalysts have also been efficient for late-stage applications of this reaction. Organic photocatalysts that have demonstrated their effectiveness include 4CzIPN [67], acridinium salts [68], and hypervalent iodine [69].

Notably, in the absence of light and metal catalyst, a late-stage Minisci reaction was reported efficient under mild conditions with an excess of persulfate oxidant [70]. To demonstrate the functional group tolerance of the system, it has been applied to the alkylation of ligands and bioactive compounds, such as caffeine (Scheme 6.12).

Another approach developed for the decarboxylative arylation of carboxylic acid relies on nickel-catalyzed cross-coupling with aryl halides [71]. This highly efficient

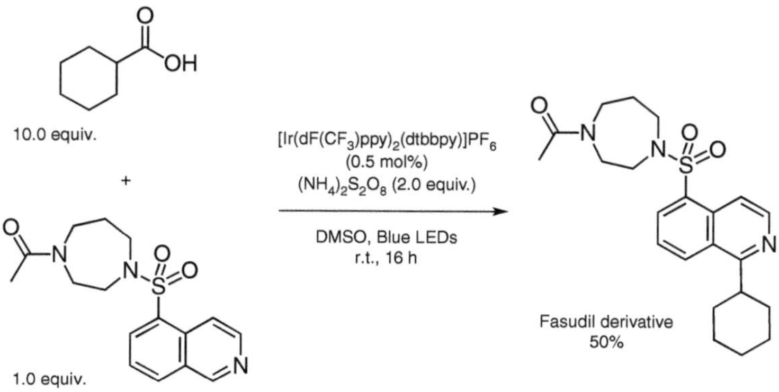

Scheme 6.11 Photocatalyzed decarboxylative arylation of carboxylic acids.

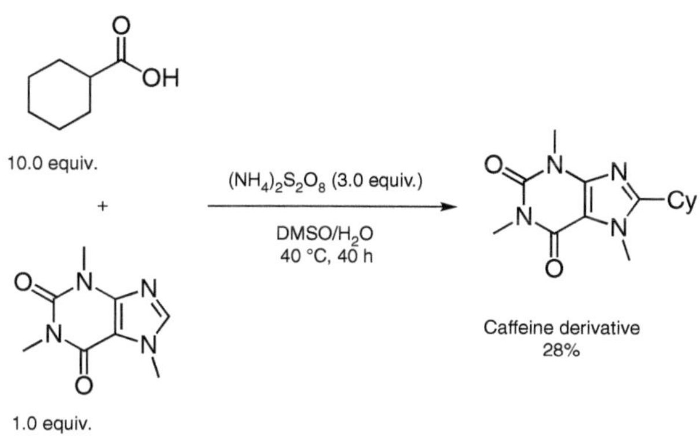

Scheme 6.12 Minisci reaction using excess of persulfate oxidant.

approach was quickly adapted to a broad scope of carboxylic acids, including late-stage use of α-aminocarboxylic acids [72], α-fluorocarboxylic acids [73], and tertiary acids [74] (Scheme 6.13).

6.3.1.2 Decarboxylative Alkylation, Alkenylation, and Alkynylation

Besides arylation, several other reactions that allow the selective formation of C—C bonds have been adapted for late-stage functionalization via the decarboxylation of acids.

A photocatalyzed decarboxylative alkynylation [75] has thus been shown to be efficient on natural terpenoid derivatives (Scheme 6.14). Remarkably, the same process carried out under high pressure of carbon monoxide allows a carbonylative variation of the reaction, giving access to the corresponding ynones.

Decarboxylative alkenylations have also been developed for direct use of carboxylic acids, adopting strategies that combine transition metal catalysis and photocatalysis [76]. Again, catalytic systems have been adapted for late-stage purposes: a cobalt-catalyzed system, using an acridinium salt as photocatalyst [77], was indeed efficient for the late-stage alkenylation of indomethacin (Scheme 6.15),

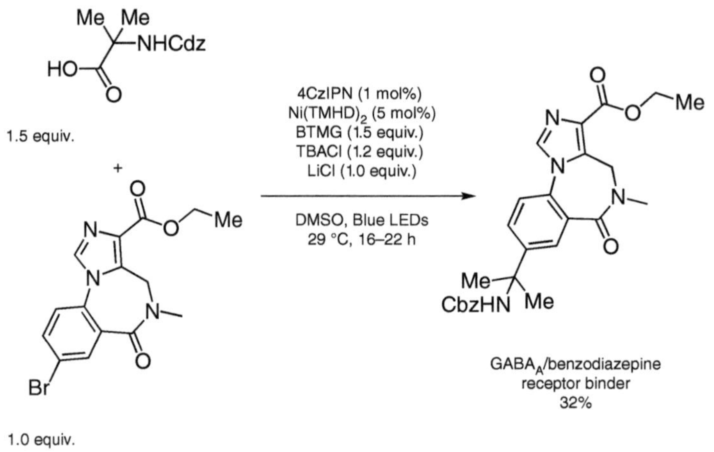

Scheme 6.13 Ni-catalyzed decarboxylative arylation of carboxylic acids.

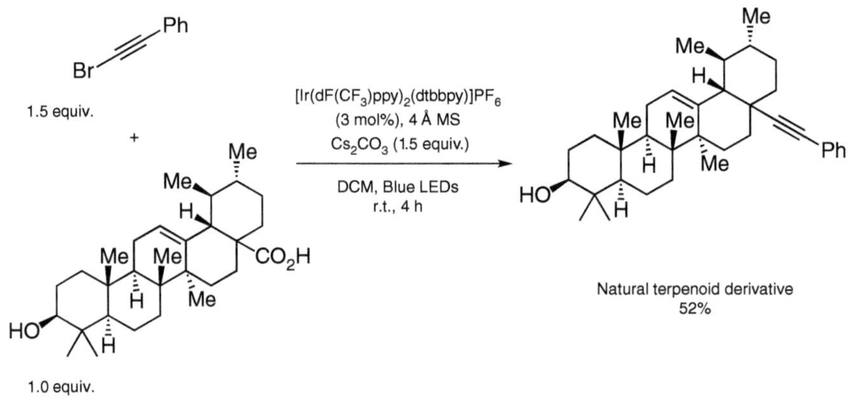

Scheme 6.14 Photocatalyzed decarboxylative alkynylation of carboxylic acids.

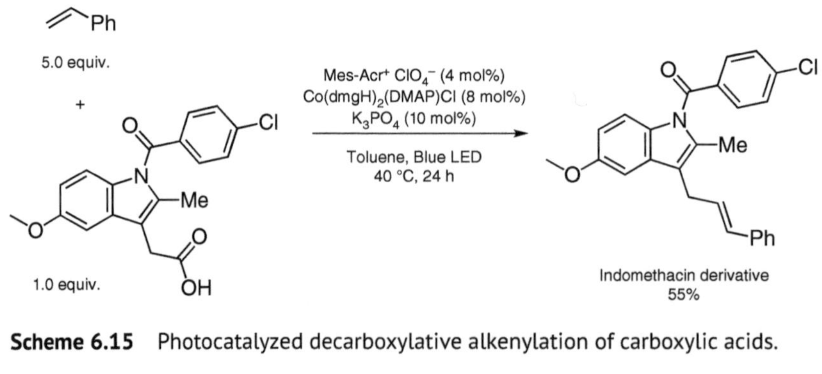

Scheme 6.15 Photocatalyzed decarboxylative alkenylation of carboxylic acids.

and a cerium–nickel dual transition metal electrophotocatalytic system demonstrated its efficiency on a broad scope of substrates [78].

In addition, the decarboxylative alkylations of numerous carboxylic acids have been shown to be efficient in the context of late-stage functionalization, including the photocatalyzed addition to vinyl boronic esters [79], cyclopropanation [80], trifluoromethylation (Scheme 6.16) [81], and cyclobutylation [82].

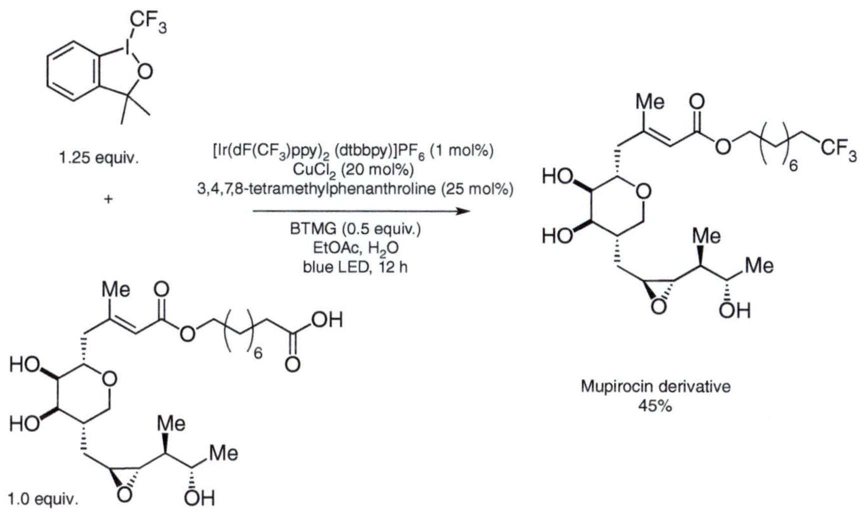

Scheme 6.16 Photocatalyzed decarboxylative alkylation of carboxylic acids.

6.3.1.3 Decarboxylative Cyanation

Decarboxylative cyanation of free carboxylic acids remains scarce in the literature, as preactivation to redox-active ester intermediates is usually required for an efficient cyanation [83]. Recently, however, enantioselective approaches have been reported, demonstrating the possibility to perform decarboxylative cyanation of carboxylic acids, without preactivation, using either photoelectrocatalysis [84] or hypervalent iodine(III) reagents [85]. This strategy has been applied for the selective cyanation of several 2-arylpropionic acid derivatives (Scheme 6.17).

6.3.1.4 Decarboxylative C—Heteroatom Bond Formation

Complementary to C—C bond formation, late-stage decarboxylative strategies have been applied to the formation of a wide range of C—heteroatom bonds, mainly through photocatalytic processes.

In terms of C—O bond formation, carboxylic acids, including 2-propionic acid derivatives [86], have been involved in decarboxylative hydroxylation using molecular oxygen. Similarly, in the presence of free alcohols, decarboxylative processes gave access to ethers from carboxylic acids [87] and to esters from phenylglyoxalic acid derivatives [88]. In the latter case, the potential for late-stage derivatization was demonstrated for several drugs and natural products (Scheme 6.18).

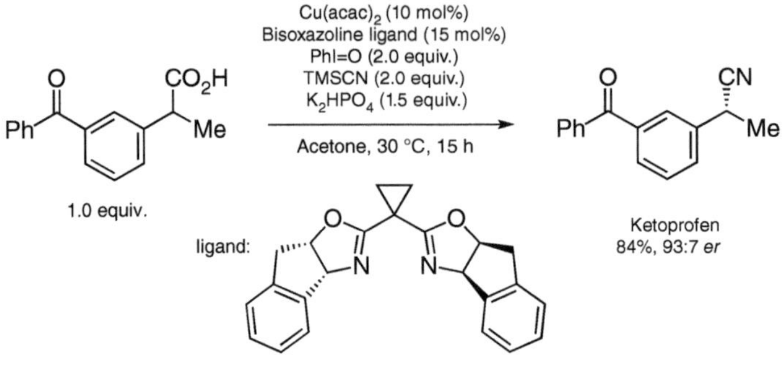

Scheme 6.17 Decarboxylative cyanation of carboxylic acids.

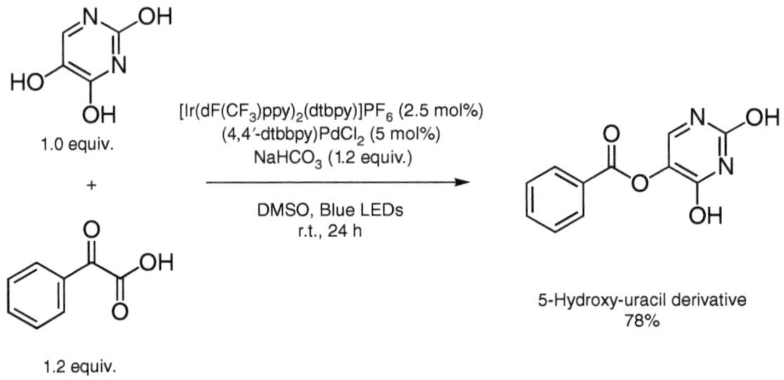

Scheme 6.18 Decarboxylative esterification of phenylglyoxalic acids.

In addition, examples can be found for the decarboxylative oxygenation of acids with tetramethylpiperidine 1-oxyl, efficient on several bioactive compounds, by a combination of iron catalysis and photocatalysis [89], as well as a copper-catalyzed nitrooxylation of aliphatic acids, also suitable for late-stage derivatization [90].

For C—N bond formations, similar reactions, including those using highly functionalized amines or free carboxylic acids, have demonstrated their efficiency without any pre-activation. Again, the combined use of photocatalysis and organometallic catalysis, particularly copper [91], has demonstrated its ability to perform decarboxylative *N*-alkylation (Scheme 6.19). The late-stage decarboxylative hydrazination of carboxylic acids with di-*tert*-butylazodicarboxylate, using a cerium photocatalyst, has been reported on advanced structures [92], and the decarboxylative amidation of carboxylic acids has also been performed in the absence of transition metal or light irradiation through the opening of dioxazolones [93].

Complementing these C—O and C—N bond formations, examples of using non-activated alcohols for decarboxylative borylation [94], xanthylation [95], sulfonylation [96], and fluorination [97] can be found.

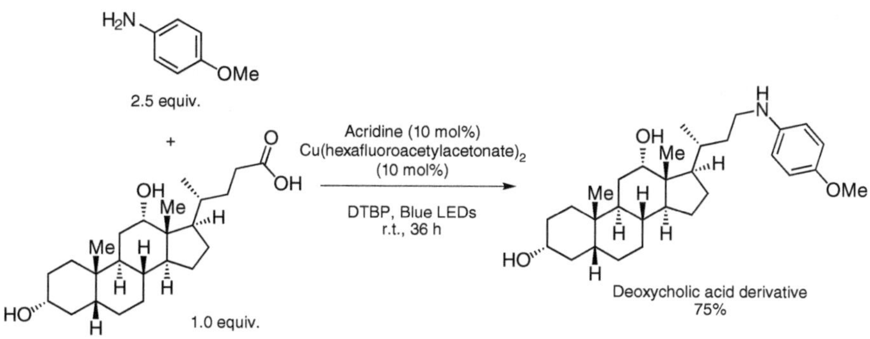

Scheme 6.19 Decarboxylative *N*-alkylation of carboxylic acids.

6.3.2 Non-decarboxylative Couplings

In addition to decarboxylative processes, late-stage conversion of free carboxylic acids to the corresponding ketones, amides, or esters has also proven to be efficient for the rapid and selective diversification of compounds of interest.

6.3.2.1 Conversion to Ketones

Direct conversion of carboxylic acids to ketones can usually be achieved by the addition of organolithium or Grignard reagents, but this approach cannot be applied for late-stage derivatization as it suffers from a low functional group tolerance and poor selectivity. One way around this problem is the use of stoichiometric organozinc, organocopper, or organoboron derivatives, but these approaches usually require activation of the carboxylic acid derivatives.

On the other hand, the use of transition metal catalysis and photocatalysis allowed the selective conversion of carboxylic acids to ketones, without pre-activation, and was efficient in late-stage diversification, in the presence of a wide variety of coupling partners.

Thus, an efficient process using nickel catalysis together with photoactivation has been reported with alkyltrifluoroborates [98], glycoside derivatives [99], and alcohol derivatives [100] even in late-stage applications (Scheme 6.20).

In addition, photocatalysis in combination with either phosphines (Scheme 6.21) [101] or *N*-heterocyclic carbenes [102] has also been used in late-stage conversion of carboxylic acids to ketones.

6.3.2.2 Amidation

Due to the prevalence of the amide pattern in natural products and compounds of interest, several methods have been disclosed to perform selective and efficient late-stage amidation of free carboxylic acids. Again, photocatalytic processes have proven to be highly effective, with [103] or without [104] the use of an additional metal catalyst and have been adapted for late-stage derivatization (Scheme 6.22).

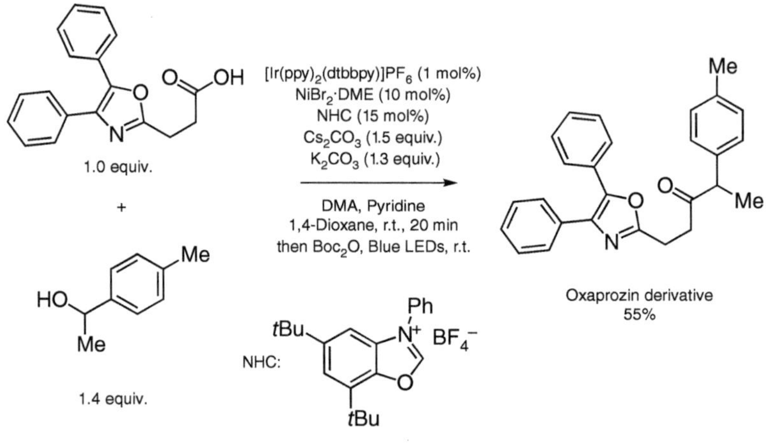

Scheme 6.20 Ni-catalyzed conversion of carboxylic acids to ketones under photoactivation.

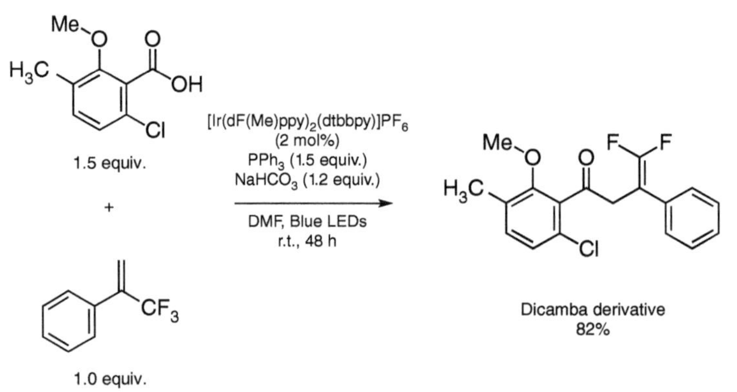

Scheme 6.21 Photocatalyzed conversion of carboxylic acids to ketones with phosphines.

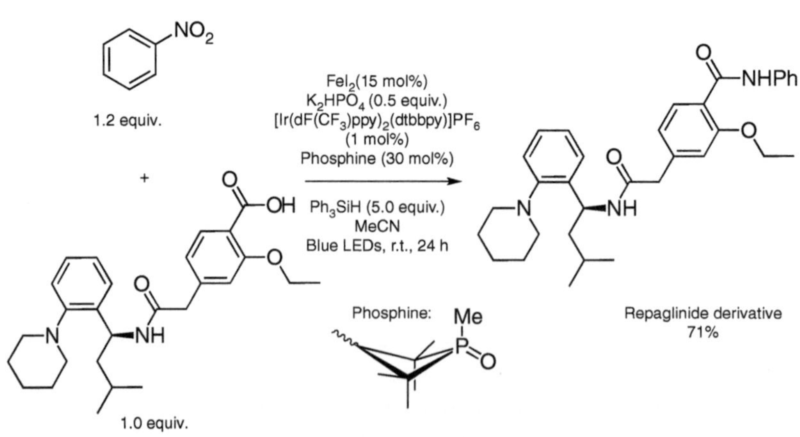

Scheme 6.22 Photocatalyzed amidation of carboxylic acids.

In addition, the use of organocatalysts such as DMAP also allows the selective coupling between free carboxylic acids and *N*-cyano-*N*-phenyl-*p*-tosylsulfonamide [105], with a good functional group tolerance.

The late-stage synthesis of amides from carboxylic acids has also been reported efficiently via a silver-catalyzed radical cross-coupling with isocyanides [106] through a decarboxylation/oxidation sequence (Scheme 6.23).

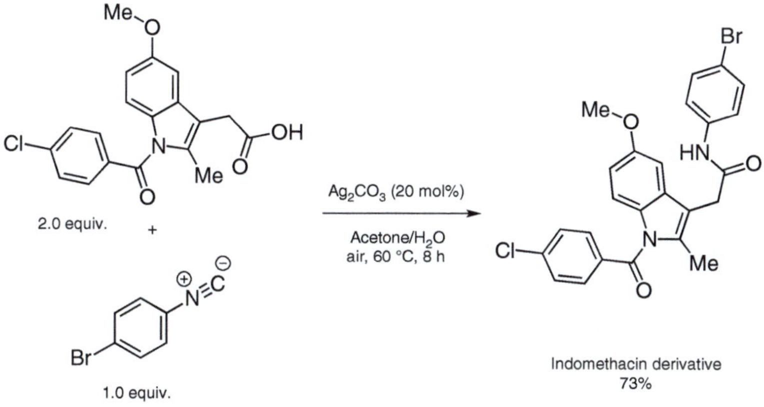

Scheme 6.23 Ag-catalyzed radical amidation of carboxylic acids.

In addition to the formation of amides, the late-stage conversion of carboxylic acids to the corresponding imides was performed by a copper-catalyzed transformation [107]. This process, previously reported for simple molecules, was indeed efficient for the conversion of several bioactive compounds (Scheme 6.24).

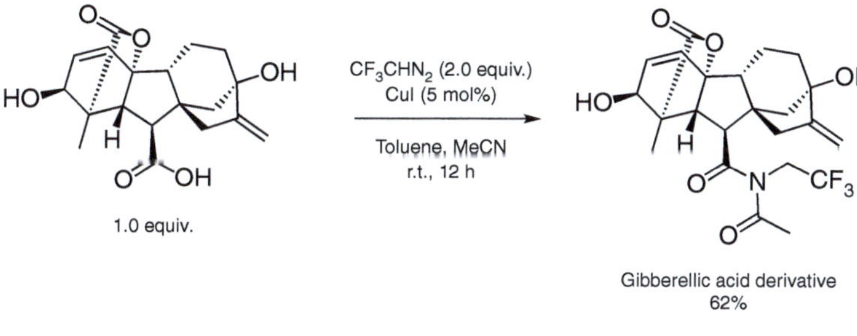

Scheme 6.24 Cu-catalyzed imidation of carboxylic acids.

6.3.2.3 Esterification

As for amides, esters are prevalent in active compounds, encouraging the development of numerous strategies for direct esterification of carboxylic acids. Among the methods that are effective on complex substrates are esterifications with alcohols [108], with aryl halides [109], by C—H bond activation [110], and with allenamides [111]. The latter was particularly efficient and selective by metal-free hydrocarboxylation (Scheme 6.25).

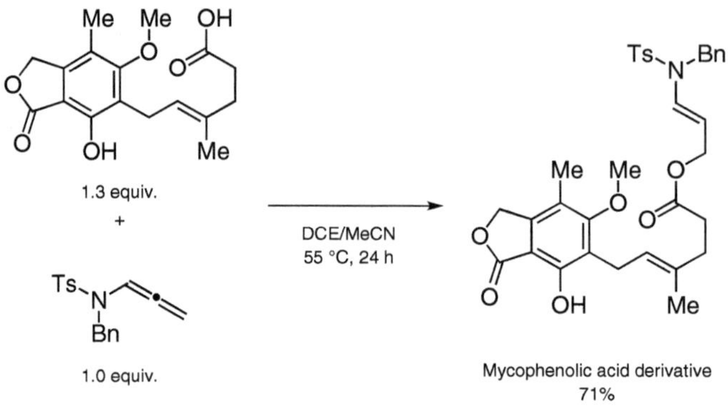

Scheme 6.25 Metal-free esterification of carboxylic acids with allenes.

6.3.2.4 Fluorination

Another interesting direct conversion of carboxylic acids is the formation of acyl fluorides. These are stable acid derivatives with valuable reactivity. If traditional fluorination reagents have shown good efficiency, they have also suffered from harsh reaction conditions and low selectivity, limiting their applications in late-stage derivatization. However, recent examples have been applied under mild conditions to highly complex structures, enabling late-stage application. These include the use of $(Me_4N)SCF_3$ (Scheme 6.26) [112], trifluoromethyltrifluoromethanesulfonate [113], and cesium fluoride with trichloroisocyanuric acid [114].

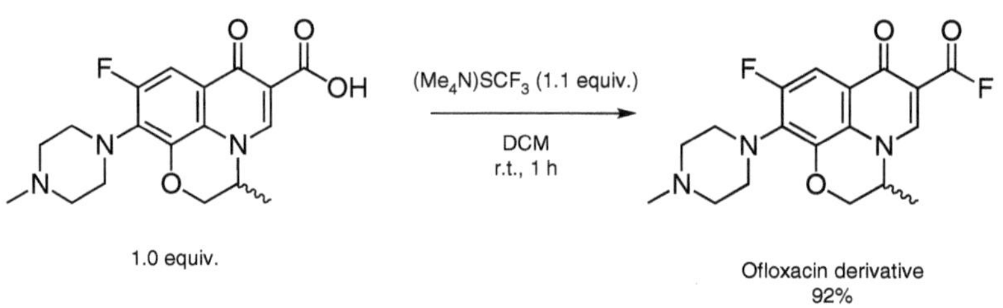

Scheme 6.26 Selective fluorination of carboxylic acids.

It should be noted that in addition to the fluorination reaction, the selective formation of trifluoromethyl thioesters from carboxylic acids has been achieved [115] and applied in a late-stage strategy (Scheme 6.27).

6.3.3 Reductions

The conversion of free carboxylic acids to the corresponding reduced species is another active field of research, as totally selective methods remain rare. Indeed, application to late-stage derivatization usually implies the presence of other functional groups more prone to react with anionic of neutral hydride sources than carboxylic acids.

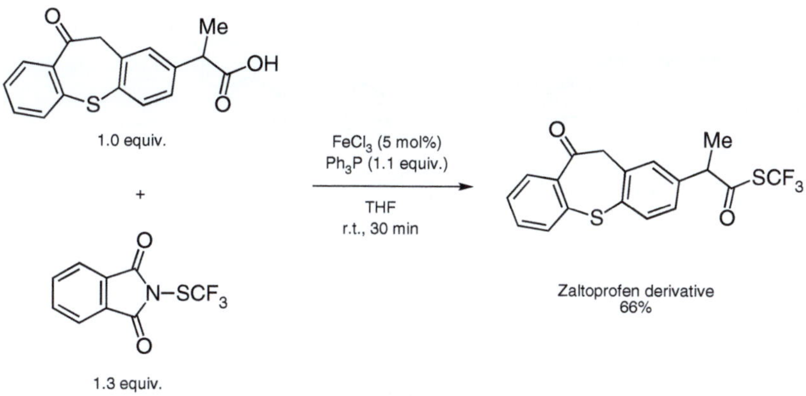

Scheme 6.27 Fe-catalyzed trifluoromethyl thiolation of carboxylic acids.

The use of phosphoranyl radicals, efficient for the activation of free alcohols [26], has also demonstrated its efficacy for the selective conversion of carboxylic acids to the corresponding acyl radicals by direct C—O bond activation, which then leads to the formation of the corresponding aldehydes (Scheme 6.28) or cyclized products. Strategies based on photocatalytic approaches have also been efficient for late-stage hydrodecarboxylation [116] and dehydrodecarboxylation of acids [117].

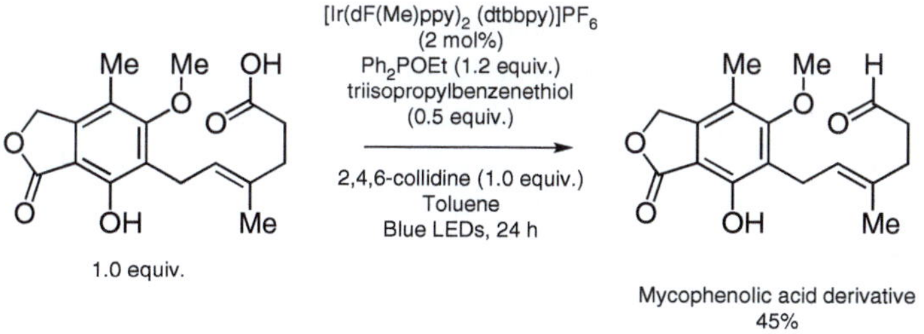

Scheme 6.28 Selective reduction of carboxylic acids using phosphoranyl radicals.

In addition to these reductions performed under photoredox conditions, the palladium-catalyzed hydrodecarboxylation of aromatic acids has been reported with high chemoselectivity, although requiring a high reaction temperature [118]. The possibility of using this approach in late-stage derivatization has been demonstrated for complex structures (Scheme 6.29).

Selective conversion of alkyl carboxylic acids to the corresponding alcohols is also possible in the absence of a metal catalyst by a base-promoted reduction with pinacolborane [119].

6.4 Activation of Ketones and Aldehydes

Compared to the other functional groups discussed in this chapter, the occurrence of ketones and aldehydes is more limited [3]. However, the conversion of these

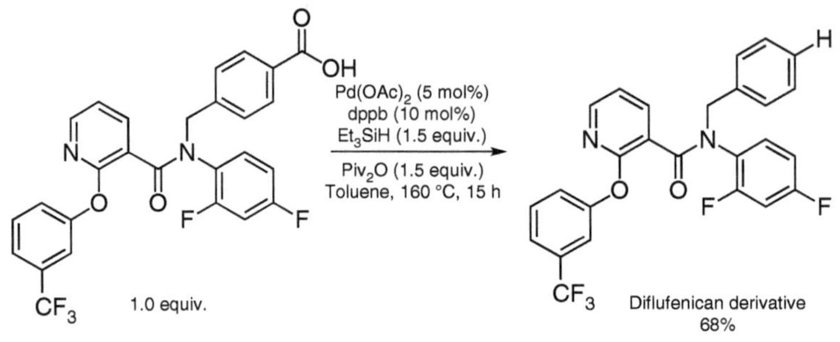

Scheme 6.29 Pd-catalyzed hydrodecarboxylation of carboxylic acids.

functional groups is an important area of research that opens up new perspectives, including in late-stage strategies.

6.4.1 Alkylation and Arylation

The conversion of ketones to tertiary alcohols by nucleophilic addition on the carbon atom is a very well-described and long-established process [120]. However, the highly nucleophilic organometallic reagents required, used in stoichiometric proportions, are a limitation for this method, limiting functional-group tolerance and therefore its applications in late-stage strategies.

Nevertheless, catalytic methods have been used to circumvent this problem, such as a copper-catalyzed asymmetric alkylation of ketones with enynes, reported in 2016 [121]. The high functional-group tolerance was particularly demonstrated in the late-stage derivatization of compounds of interest with high selectivity (Scheme 6.30). This pioneering example has paved the way for the development of numerous catalytic systems, such as the enantioselective nickel-catalyzed late-stage arylation of active compounds with (hetero)aryl boronates [122]. In addition to these organometallic-catalyzed approaches, photocatalyzed strategies have also allowed the selective late-stage alkylation of ketones [123].

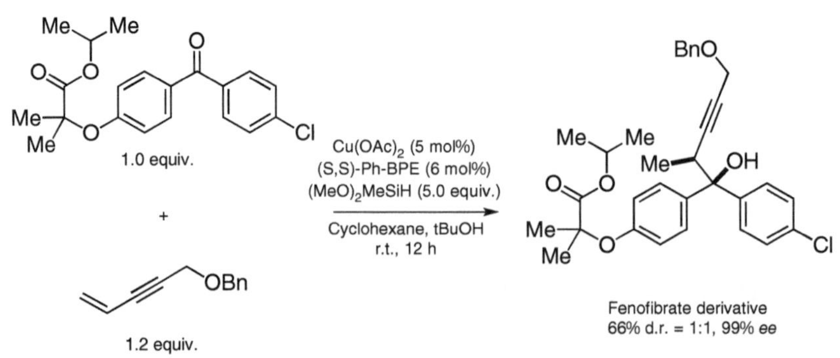

Scheme 6.30 Cu-catalyzed alkylation of ketones.

The ability of ketones and aldehydes to undergo multicomponent reactions has also been exploited in late-stage derivatization, and a representative example is the photocatalyzed hydroaminoalkylation of olefins, leading to tertiary amines [124].

6.4.2 Deoxygenative Couplings

Besides these classical nucleophilic additions, ketones have been involved in deoxygenative couplings, allowing them to be used as alkylative reagents [125] or to undergo efficient borylations [126]. In the first example, a radical activation under photoredox conditions allowed the C—H functionalization of heteroarenes with simple ketones and was compatible with the use of sophisticated compounds (Scheme 6.31).

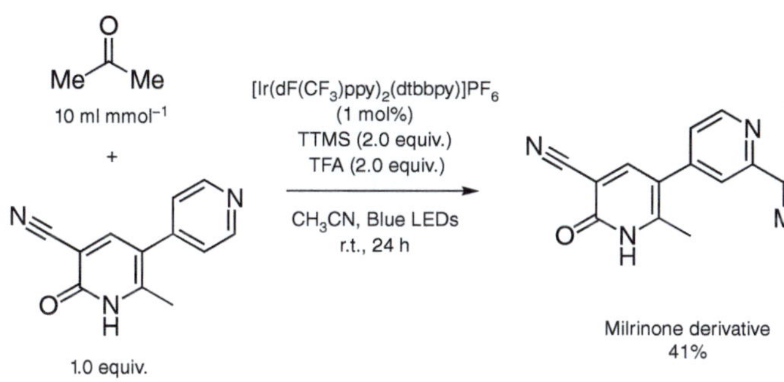

Scheme 6.31 Photocatalyzed deoxygenative arylation of ketones.

For aldehydes, specific deoxygenative cross-coupling has been developed, such as the deoxygenative carbonyl cross-metathesis. Among recent examples, suitable for late-stage applications, a gem-di-metal catalysis efficiently converted aldehyde to alkenes, with good stereocontrol depending on the use of either iron or chromium catalyst [127]. More recently, NHC catalysis in the presence of carbonate also demonstrated its efficacy to perform the cross-coupling of aldehydes to corresponding ketones [128], with good functional group tolerance (Scheme 6.32).

6.4.3 Rearrangement and Chain-Walking

Another elegant approach to the late-stage diversification of ketones is the insertion of either oxygen or nitrogen atoms. Regarding the oxygen insertion, Baeyer–Villiger oxidation is well established [129], while nitrogen insertion, especially on complex structures, has been developed more recently. In 2019, the use of nitromethane as nitrogen donor in the conversion of ketones and aldehydes to the corresponding amides [130] was shown to be efficient, even on bioactive compounds. Similar insertions have been observed using chiral amino-indanol [131] and aminodiphenylphosphinates [132] to generate γ-lactams from cyclobutanones.

Through a strategy similar to the Willgerodt–Kindler rearrangement, the chain-walking isomerization of ketones allowed the late-stage modification of several steroids [133], giving a direct access to rare derivatives (Scheme 6.33).

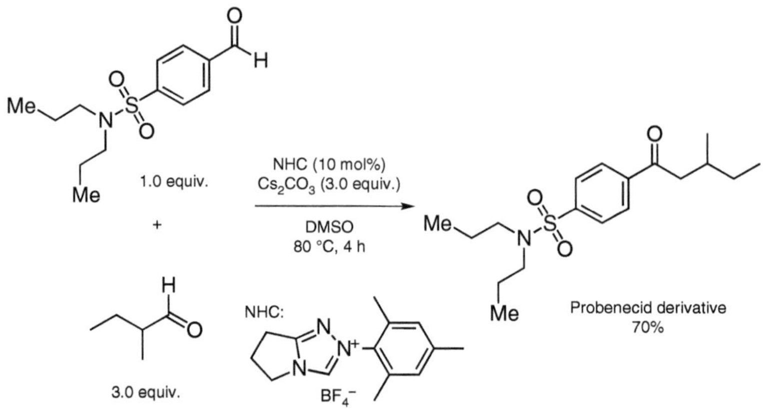

Scheme 6.32 Deoxygenative cross-coupling of aldehydes.

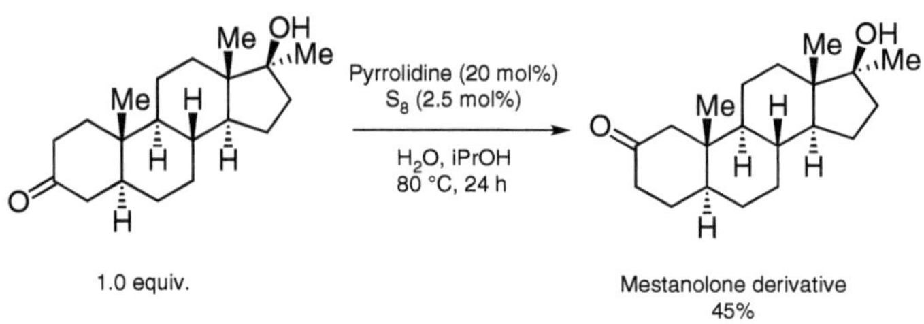

Scheme 6.33 Chain-walking isomerization of ketones.

6.5 Activation of Esters

Ester group is another ubiquitous functional group [3], with wide availability and great stability, which come at the price of a relative inertness, making its efficient conversion a major challenge. Among the solution developed to achieve late-stage C—O bond activation of ester, the use of transition metal catalysis based mainly on nickel and palladium has demonstrated its efficacy. Base-promoted conversion of esters to amide was also a fruitful strategy.

6.5.1 Metal-Catalyzed C–C Bond Formation

Esters have demonstrated their efficiency as reactive electrophilic species through the activation of C—O bond, especially in the case of alkylation [134]. It is noteworthy that the occurrence of a decarbonylative process or the selective activation of the C(acyl)—O bond can be controlled by the ligand, using either nickel [135] or palladium catalysis [136]. This ligand-controlled selectivity has also been applied in the case of late-stage derivatization (Scheme 6.34).

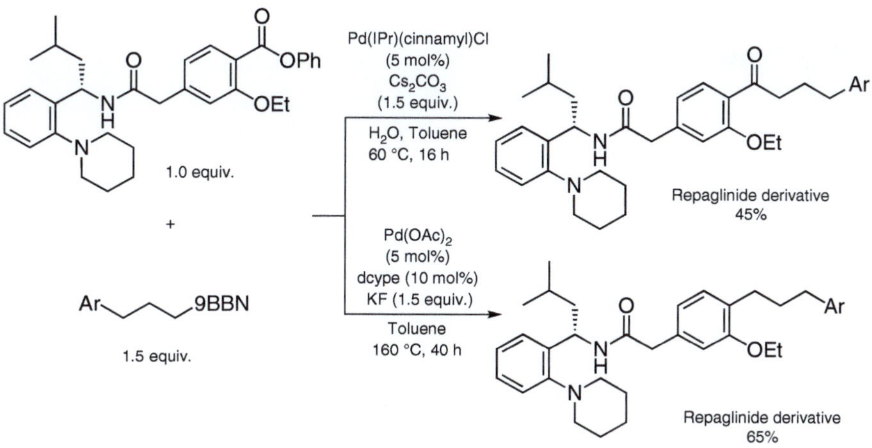

Scheme 6.34 Pd-catalyzed alkylation of esters.

6.5.2 Amidation

The direct late-stage conversion of esters to amides has been widely described. Using nickel catalysis, methyl esters have been efficiently converted to the corresponding anilides (Scheme 6.35) by reaction with anilines [137] or nitroarenes [138].

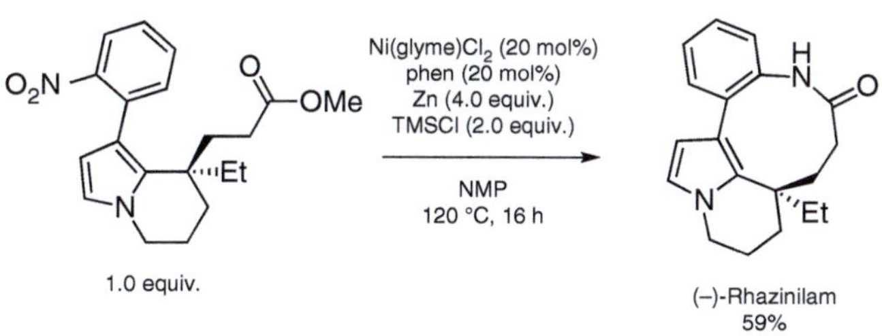

Scheme 6.35 Ni-catalyzed intramolecular amidation of esters.

The direct amidation of ester has also been reported in the absence of transition metal through a base-promoted process. Indeed, under mild conditions, LiHMDS was able to promote a series of transamidations on a broad scope of amide and to perform the direct amidation of alkyl esters [139]. If the use of a strong base limited the functional group tolerance of this approach, the synthetic utility of the method was demonstrated by the synthesis of moclobemide (Scheme 6.36).

Avoiding the use of an excess of strong base, another procedure relying on a combination of catalytic amount of Lewis acid and base, efficiently converted methyl esters to the corresponding anilides [140], including the synthesis of active compounds such as bixafen (Scheme 6.37).

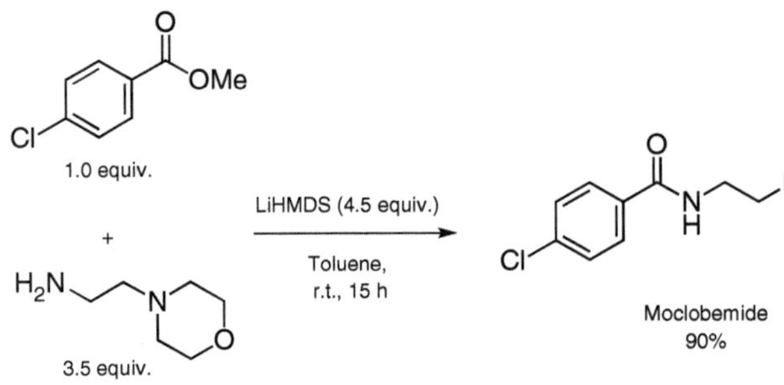

Scheme 6.36 Base-promoted amidation of esters.

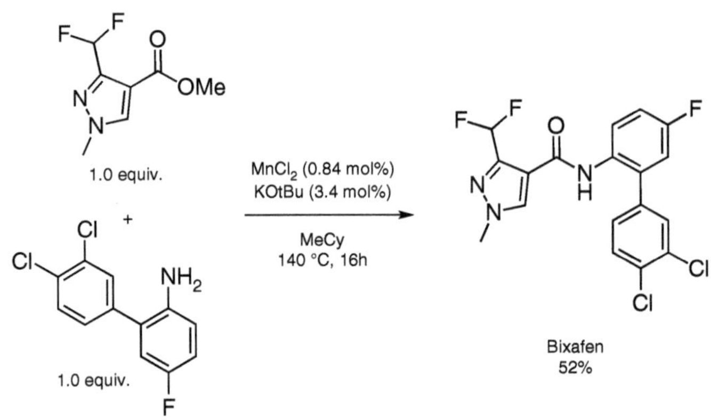

Scheme 6.37 Mn-catalyzed amidation of esters.

6.5.3 Decarbonylation

Among the possible transformations enabled by the metal-catalyzed activation of the ester C—O bond, the conversion into ethers by decarbonylation is an original strategy, competing with ether synthesis by intermolecular cross-coupling. In 2017, the possibility of using either nickel or palladium catalysis for this intramolecular transformation was described, leading to the direct formation of diaryl ethers [141], and also proved to be efficient in intermolecular coupling with an additional phenol derivative. In terms of late-stage applications, this method has been applied to the etherification of natural products, giving high yields with both nickel and palladium catalysts (Scheme 6.38).

6.6 Activation of Ethers

Ethers are among the most popular functional groups in molecules described in the field of medicinal chemistry, generally competing with amides as the most common functional groups [3]. On the other hand, the use of the reactivity of this functional group in the context of late-stage functionalization is quite limited, as ethers are inert under many conditions. However, in the last 15 years, the activation of ether

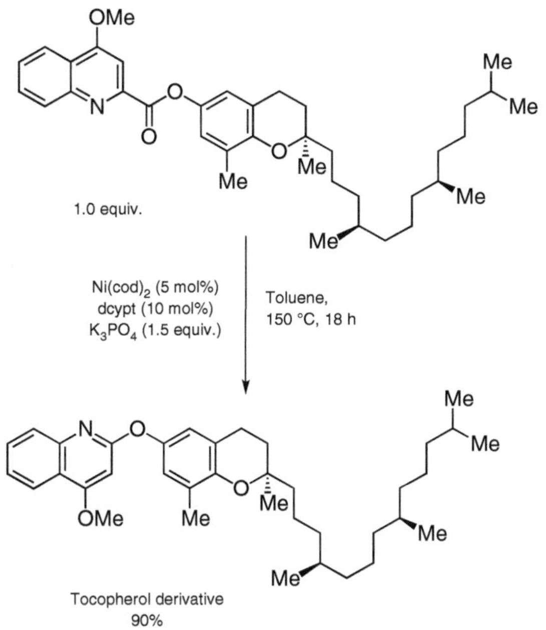

Scheme 6.38 Ni-catalyzed etherification of esters.

C—O bonds, mainly catalyzed by transition metal, has been reported for efficient cross-coupling reactions, including in late-stage applications.

6.6.1 Metal-Catalyzed Cross-Coupling

Cross-coupling involving the metal-catalyzed activation of ether C—O bonds has been reported almost exclusively using nickel catalysis [142]. This catalyst has been used for the formation of C—C bonds by activation of ethers, which have been engaged in Kumada–Tamao–Corriu, Negishi, and Suzuki–Miyaura type reactions, including those in enantioselective approaches.

To give recent examples, in the context of late-stage cross-coupling, the arylation of aromatic [143] and allylic ethers [144] has been efficiently applied to challenging structures (Scheme 6.39).

Besides arylation, nickel-catalyzed late-stage alkynylation and alkylation via ether C—O bond activation have also been efficiently performed [145], most recently for the derivatization of buflavine [146].

In addition to C—C bond formations, the metal-catalyzed late-stage strategy has been applied to C—N bond formations by nickel-catalyzed amidations [147] or borylations using nickel [148], zinc/nickel tandem [149], and iron catalysis [150] (Scheme 6.40).

6.6.2 Metal-Free Cross-Coupling

In the absence of a metal catalyst, the conversion of ethers by cleavage of C—O bonds has been carried out under basic conditions. The etherification of aryl methyl

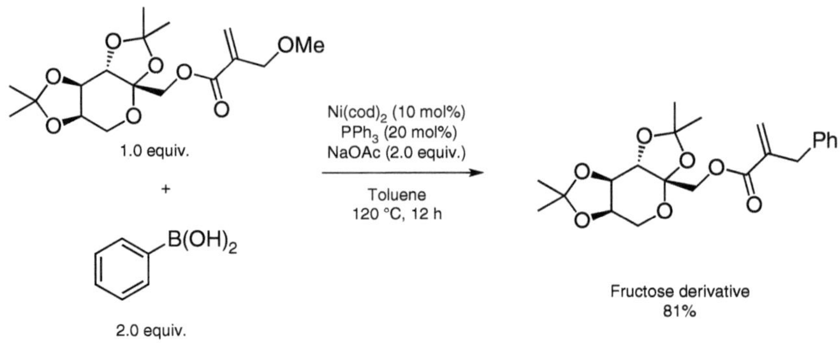

Scheme 6.39 Ni-catalyzed arylation of ethers.

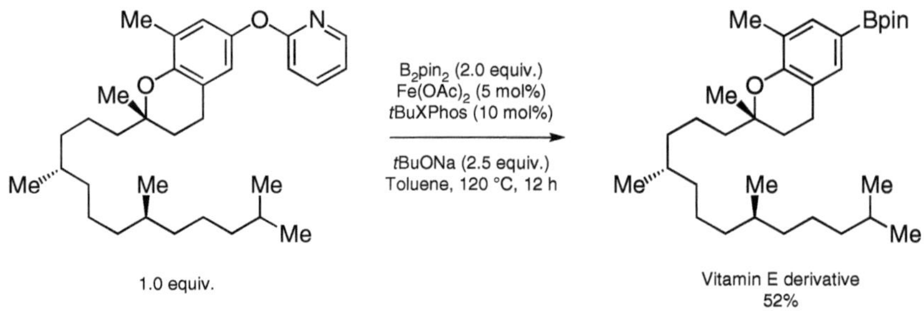

Scheme 6.40 Fe-catalyzed borylation of ethers.

ether has been performed in the presence of potassium *tert*-butoxide but required the presence of a nitrile on the aromatic core [151]. Despite this limitation, the etherification process has been applied to a wide range of primary and secondary alcohols, including in late-stage derivatization. A similar base-promoted approach has been developed for the amination [152] of ethers and applied to the preparation of bioactive compounds (Scheme 6.41).

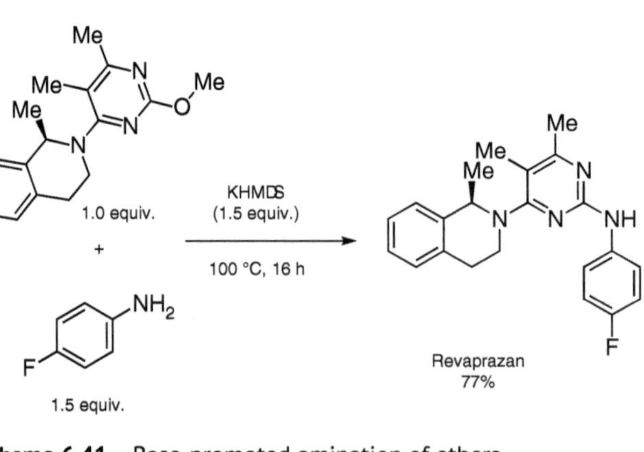

Scheme 6.41 Base-promoted amination of ethers.

Apart from the base-promoted etherification and amination of ethers, another metal-free activation of ether C—O bond is a borylation based on homolytic cleavage by the formation of an EDA complex, performed under photoactivation [153]. This reaction showed a good functional group tolerance and was used in the borylation of eugenol derivative (Scheme 6.42).

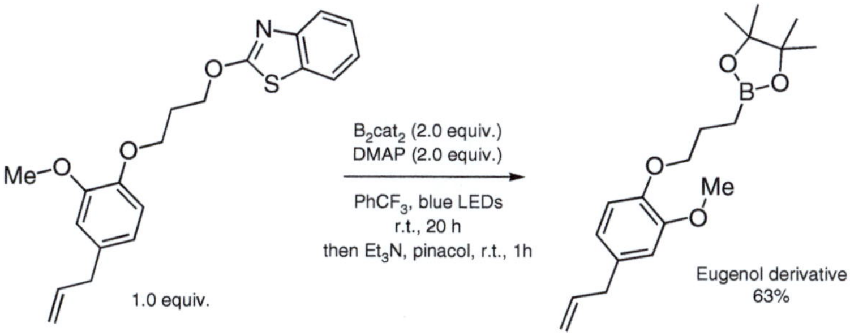

Scheme 6.42 Borylation of ether derivatives.

6.6.3 Dealkylation

Direct conversion of ethers to the corresponding alcohols by dealkylation has been widely reported but originally required harsh conditions and suffered from low selectivity, limiting late-stage application. However, since 2015, milder conditions have been developed using base [154], copper catalysis [155], or biocatalysis [156], allowing late-stage use of this strategy (Scheme 6.43).

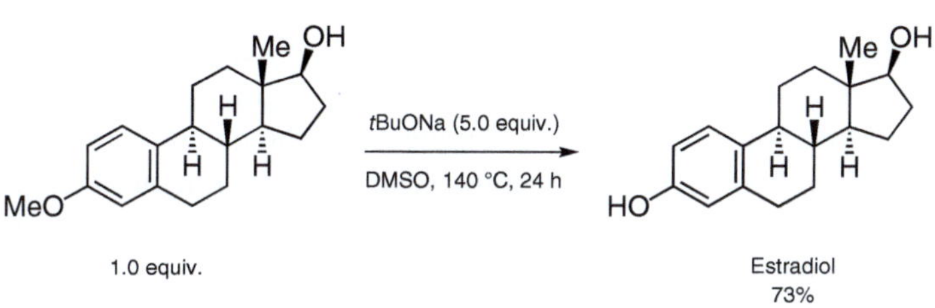

Scheme 6.43 Base-promoted dealkylation of ethers.

Beyond dealkylation, it is worth noting that several selective cleavages of ether C—O bonds, converting aryl alkyl ethers to arenes, have been reported using nickel catalysis with a broad scope of reducing agents, but late-stage applications are currently limited [157].

6.6.4 Opening of Cyclic Ethers

Among the various classes of ethers, the strained cyclic ethers such as epoxide and oxetane have also been the focus of numerous studies to investigate their reactivity, including in late-stage applications. For example, a nickel-catalyzed cross-coupling between epoxides and (hetero)aryl halides [158] has been shown to be effective even in late-stage (Scheme 6.44).

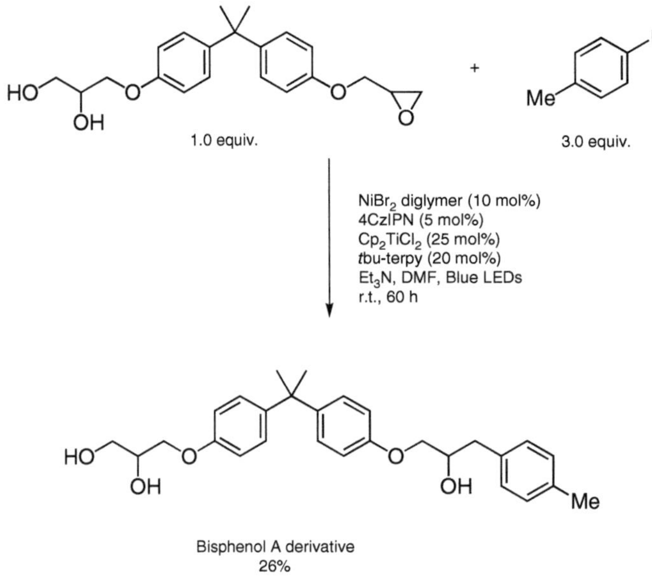

Scheme 6.44 Ni-catalyzed opening of cyclic ethers.

6.7 Summary and Conclusions

Late-stage functionalization has recently emerged as a powerful approach for the selective derivatization of compounds of interest, and in this highly active area of research, the selective activation of C—O bonds complements the much more popular C—H bond activation. Bypassing the need for prior activation or multi-step synthesis, this strategy allows the direct conversion of a wide scope of functional groups at advanced stages of synthesis. Far from being an exhaustive review of existing methods using this approach, the chapter has illustrated significant examples that highlight the diversity of functional groups that can be effectively converted using late-stage C—O bond activation.

Transition metal catalysis is one of the most widely used tools for this approach as it has the potential to increase selectivity and reduce the need for harsh reaction conditions, two crucial factors for the late-stage functionalization of complex molecules. In addition, organocatalysis and radical photoactivation are also undergoing rapid development for late-stage applications, reducing the field's dependence on sophisticated organometallic catalysts.

The rapid developments in this field offer the prospect of a wide range of applications, including increasing the efficiency of access to known compounds of interest as well as rapid exploration of the properties of vast libraries of original compounds, by making them accessible through the late-stage derivatization of advanced intermediates. Thus, we can expect the continued development of efficient catalysts and original methodologies that will further expand the application of late-stage functionalization by C—O bond activation.

References

1 (a) Castellino, N.J., Montgmery, A.P., Danon, J.J., and Kassiou, M. (2023). Late-stage functionalization for improving drug-like molecular properties. *Chemical Reviews* 123: 8127–8153. (b) Moir, M., Danon, J.J., Reekie, T.A., and Kassiou, M. (2019). An overview of late-stage functionalization in today's drug discovery. *Expert Opinion on Drug Discovery* 14: 1137–1149. (c) Börgel, J. and Ritter, T. (2020). Late-stage functionalization. *Chem* 6: 1877–1887.

2 (a) Cernak, T., Dykstra, K.D., Tyagarajan, S. et al. (2015). The medicinal chemist's toolbox for late stage functionalization of drug-like molecules. *Chemical Society Reviews* 45: 546–576. (b) Docherty, J.H., Lister, T.M., Mcarthur, G. et al. (2023). Transition-metal-catalyzed C—H bond activation for the formation of C—C bonds in complex molecules. *Chemical Reviews* 123: 7692–7760. (c) Bellotti, P., Huang, H.-M., Faber, T., and Glorius, F. (2023). Photocatalytic late-stage C—H functionalization. *Chemical Reviews* 123: 4237–4352.

3 Ertl, P., Altmann, E., and McKenna, J.M. (2020). The most common functional groups in bioactive molecules and how their popularity has evolved over time. *Journal of Medicinal Chemistry* 63: 8408–8418.

4 Goyal, V., Sarki, N., Narani, A. et al. (2023). Recent advances in the catalytic N-methylation and *N*-trideuteromethylation reactions using methanol and deuterated methanol. *Coordination Chemistry Reviews* 474: 214827.

5 Choi, G. and Hong, S.H. (2018). Selective monomethylation of amines with methanol as the C1 source. *Angewandte Chemie, International Edition* 57: 6166–6170.

6 Dang, T.T., Ramalingam, B., and Seayad, A.M. (2015). Efficient ruthenium-catalyzed N-methylation of amines using methanol. *ACS Catalysis* 5: 4082–4088.

7 Sarki, N., Goyal, V., Tyagi, N.K. et al. (2021). Simple $RuCl_3$-catalyzed N-methylation of amines and transfer hydrogenation of nitroarenes using methanol. *ChemCatChem* 13: 1722–1729.

8 (a) Vellakkaran, M., Singh, K., and Banerjee, D. (2017). An efficient and selective nickel-catalyzed direct N-alkylation of anilines with alcohols. *ACS Catalysis* 7: 8152–8158. (b) Das, J. and Banerjee, D. (2018). Nickel-catalyzed phosphine free direct N-alkylation of amides with alcohols. *The Journal of Organic Chemistry* 83: 3387–3384.

9 (a) Neumann, J., Elangovan, S., Spannenberg, A. et al. (2017). Improved and general manganese-catalyzed N-methylation of aromatic amines using methanol. *Chemistry – A European Journal* 23: 5410–5413. (b) Fertig, R., Irrgang, T., Freitag, F. et al. (2017). Manganese-catalyzed and base-switchable synthesis of amines or imines *via* borrowing hydrogen or dehydrogenative condensation. *ACS Catalysis* 8: 8525–8530.

10 Ma, Z., Zhou, B., Li, X. et al. (2022). Reusable Co-nanoparticles, for general and selective N-alkylation of amines and ammonia with alcohols. *Chemical Science* 13: 111–117.

11 Schlepphorst, C., Maji, B., and Glorius, F. (2016). Ruthenium-NHC catalyzed α-alkylation of methylene ketones provides branched products through borrowing hydrogen strategy. *ACS Catalysis* 6: 4184–4188.

12 Chakraborty, P., Gangwar, M.K., Emayavaramban, B. et al. (2019). α-alkylation of ketones with secondary alcohols catalyzed by well-defined Cp^*Co^{III}-complexes. *ChemSusChem* 12: 3463–3467.

13 Bhattacharrya, D., Sarmah, B.K., Nandi, S. et al. (2021). Selective catalytic synthesis of α-alkylated ketones and β-disubstituted ketones *via* acceptorless dehydrogenative cross-coupling of alcohols. *Organic Letters* 23: 869–875.

14 Bera, S., Kabadwal, L.M., and Banerjee, D. (2024). Harnessing alcohols as sustainable reagents for late-stage functionalization: synthesis of drugs and bio-inspired compounds. *Chemical Society Reviews* 53: 4607–4647.

15 Dong, Z. and MacMillan, D.W.C. (2021). Metallaphotoredox-enabled deoxygenative arylation of alcohols. *Nature* 598: 451–456.

16 (a) Wang, J.Z., Sakai, H.A., and MacMillan, D.W.C. (2022). Alcohols as alkylating agents: photoredox-catalyzed conjugate alkylation *via* in-situ deoxygenation. *Angewandte Chemie, International Edition* 61: e202207150. (b) Cai, Q., McWhinnie, I.M., Dow, N.W. et al. (2024). Engaging alkenes in metallaphotoredox: a triple catalytic radical sorting approach to olefin-alcohol cross-coupling. *Journal of the American Chemical Society* 146: 12300–12309.

17 Sakai, H.A. and MacMillan, D.W.C. (2022). Nontraditional fragment couplings of alcohols and carboxylic acids: $C(sp^3)$—$C(sp^3)$ cross-coupling *via* radical sorting. *Journal of the American Chemical Society* 144: 6185–6192.

18 Gould, C.A., Pace, A.L., and MacMillan, D.W.C. (2023). Rapid and modular access to quaternary carbons from tertiary alcohols *via* bimolecular homolytic substitution. *Journal of the American Chemical Society* 145: 16330–16336.

19 Carson, W.P. II, Tsymbal, A.V., Pipal, R.W. et al. (2024). Free-radical deoxygenative amination of alcohols *via* copper metallaphotoredox catalysis. *Journal of the American Chemical Society* 146: 15681–15687.

20 Chen, R., Intermaggio, N.E., Xie, J. et al. (2024). Alcohol-alcohol cross-couling enabled by S_H2 radical sorting. *Science* 383: 1350–1357.

21 Bissonnette, N.B., Bisballe, N., Tran, A.V. et al. (2024). Development of a general organophosphorus radical trap: deoxyphosphonylation of alcohols. *Journal of the American Chemical Society* 146: 7942–7949.

22 Intermaggio, N.E., Millet, A., Davis, D.L., and Macmillan, D.W.C. (2022). Deoxytrifluoromethylation of alcohols. *Journal of the American Chemical Society* 144: 111961–111968.

23 Carson, W.P. II, Sarver, P.J., Goudy, N.S., and MacMillan, D.W.C. (2023). Photoredox catalysis-enabled sulfination of alcohols and bromides. *Journal of the American Chemical Society* 145: 20767–20774.

24 Xie, H., Guo, J., Wang, Y.-Q. et al. (2020). Radical dehydroxylation alkylation of tertiary alcohols by Ti catalysis. *Journal of the American Chemical Society* 142: 16787–16794.

25 Cai, A., Yan, W., and Liu, W. (2021). Aryl radical activation of C—O bonds: copper-catalyzed deoxygenative difluoromethylation of alcohols. *Journal of the American Chemical Society* 143: 9952–9960.

26 Stache, E.E., Ertel, A., Rovis, T., and Doyle, A.G. (2018). Generation of phosphoranyl radicals *via* photoredox catalysis enables voltage-independent activation of strong C—O bonds. *ACS Catalysis* 8: 11134–11139.

27 Li, W.-D., Wu, Y., Li, S.-J. et al. (2022). Boryl radical activation of benzylic C—OH bond: cross electrophile coupling of free alcohols and CO_2 *via* photoredox catalysis. *Journal of the American Chemical Society* 144: 8551–8559.

28 (a) Aggarwal, T., Sushmita, and Verma, A.K. (2021). Achievements in fluorination using variable reagents through a deoxyfluorination reaction. *Organic Chemistry Frontiers* 8: 6452–6468. (b) Singh, A., Sindhe, H., Kamble, A. et al. (2024). Recent advances in late-stage monofluorination of natural products and their derivatives. *Organic Chemistry Frontiers* 11: 5605–5637.

29 Tang, P., Wang, W., and Ritter, T. (2011). Deoxyfluorination of phenols. *Journal of the American Chemical Society* 133: 11482–11484.

30 Fujimoto, T. and Ritter, T. (2015). PhenoFluorMix: pratical chemoselective deoxyfluorination of phenols. *Organic Letters* 17: 544–547.

31 Jelen, J. and Tavčar, G. (2023). Deoxyfluorination of electro-deficient phenols. *Organic Letters* 25: 3649–3653.

32 Sladojevich, F., Arlow, S.I., Tang, P., and Ritter, T. (2013). Late-stage deoxyfluorination of alcohols with PhenoFluor. *Journal of the American Chemical Society* 135: 2470–2473.

33 Nielsen, M.K., Ugaz, C.R., Li, W., and Doyle, A.G. (2015). PyFluor: a low-cost, stable, and selective deoxyfluorination reagent. *Journal of the American Chemical Society* 137: 9571–9574.

34 Li, L., Ni, C., Wang, F., and Hu, J. (2016). Deoxyfluorination of alcohols with 3,3-difuoro-1,2-diarylcyclopropenes. *Nature Communications* 7: 13320.

35 Goldberg, N.W., Shen, X., Li, J., and Ritter, T. (2016). AlkylFluor: deoxyfluorination of alcohols. *Organic Letters* 18: 6102–6104.

36 Guo, J., Kuang, C., Rong, J. et al. (2019). Rapid deoxyfluorination of alcohols with *N*-tosyl-4-chlorobenzenesulfonimidoyl fluoride (SulfoxFluor) at room temperature. *Chemistry – A European Journal* 25: 7259–7264.

37 Zhang, W., Gu, Y.-C., Lin, J.-H., and Xiao, J.-C. (2020). Dehydroxylative fluorination of tertiary alcohols. *Organic Letters* 22: 6642–6646.

38 Bertrand, X., Pucheault, M., Chabaud, L., and Paquin, J.-F. (2023). Synthesis of tertiary fluorides through an acid-mediated deoxyfluorination of tertiary alcohols. *The Journal of Organic Chemistry* 88: 14527–14539.

39 Sood, D.E., Champion, S., Dawson, D.M. et al. (2020). Deoxyfluorination with CuF_2: enabled by using a lewis base activating group. *Angewandte Chemie, International Edition* 59: 8460–8463.

40 (a) Longwitz, L., Jopp, S., and Werner, T. (2019). Organocatalytic chlorination of alcohols by P(III)/P(V) redox cycling. *The Journal of Organic Chemistry* 84: 7863–7870. (b) Polat, E. and Cakici, M. (2022). Deoxygenative chlorination of aldehydes and alcohols with dichloromethyl methyl ether and $TiCl_4$. *European Journal of Organic Chemistry* 2022: e202201106.

41 van Kalkeren, H.A., Leenders, S.H.A.M., Hommersom, C.R.A. et al. (2011). *In situ* phosphine oxide reduction: a catalytic Appel reaction. *Chemistry – A European Journal* 17: 11290–11295.

42 Zhao, Y. and Antonietti, M. (2018). Visible-light-drive conversion of alcohols into iodide derivatives with iodoform. *ChemPhotoChem.* 2: 720–724.

43 Yu, D.-G. and Shi, Z.-J. (2011). Mutual activation: Suzuki-Miyaura coupling through direct cleavage of the sp^2 C—O bond of naphtholate. *Angewandte Chemie, International Edition* 50: 7097–7100.

44 (a) Li, Z., Sun, W., Wang, X. et al. (2021). Electrochemically enabled, nickel-catalyzed dehydroxylative cross-coupling of alcohols with aryl halides. *Journal of the American Chemical Society* 143: 3536–3543. (b) Guo, P., Wang, K., Jin, W.-J. et al. (2021). Dynamic kinetic cross-electrophile arylation of benzyl alcohols by nickel catalysis. *Journal of the American Chemical Society* 143: 513–523. (c) Chi, B.K., Widness, J.K., Gilbert, M.M. et al. (2022). In-situ bromination enables formal cross-electrophile coupling of alcohols with aryl and alkenyl halides. *ACS Catalysis* 12: 580–586.

45 Zhang, L.-L., Gao, Y.-Z., Cai, S.-H. et al. (2024). Ni-catalyzed enantioconvergent deoxygenative reductive cross-coupling of unactivated alkyl alcohols and aryl bromides. *Nature Communications* 15: 2733.

46 Xie, P., Sun, Z., Li, S. et al. (2020). Dehydrative cross-coupling of allylic alcohols with alkynes. *Organic Letters* 22: 1599–1604.

47 Li, Z., Qiu, J., Li, B. et al. (2021). Practical allylation with unactivated allylic alcohols under mild conditions. *Organic Chemistry Frontiers* 8: 3354–3359.

48 Cai, X., Xing, H., Qiu, J. et al. (2021). Dehydrative allylation of alkenyl sp^2 C—H bonds. *Organic Letters* 23: 4368–4373.

49 Xie, H., Wang, S., Wang, Y. et al. (2022). Ti-catalyzed reductive dehydroxylative vinylation of tertiary alcohols. *ACS Catalysis* 12: 1018–1023.

50 Guo, J., Ma, H.-R., Xiong, W.-B. et al. (2022). Iridium-catalyzed enantioselective alkynylation and kinetic resolution of alkyl allylic alcohols. *Chemical Science* 13: 13914–13921.

51 (a) Schilling, W., Riemer, D., Zhang, Y. et al. (2018). Metal-free catalyst for visible-light-induced oxidation of unactivated alcohols using air/oxygen as an oxidant. *ACS Catalysis* 8: 5425–5430. (b) Nikitas, N.F., Tzaras, D.I., Triandafillidi, I., and Kokotos, C.G. (2020). Photochemical oxidation of benzylic primary and secondary alcohols utilizing air as the oxidant. *Green Chemistry* 22: 471–477.

52 Riemer, D., Mandaviya, B., Schilling, W. et al. (2018). CO_2-catalyzed oxidation of benzylic and allylic alcohols with DMSO. *ACS Catalysis* 8: 3030–3034.

53 Budweg, S., Junge, K., and Beller, M. (2019). Transfer-dehydrogenation of secondary alcohols catalyzed by manganese NNN-pincer complexes. *Chemical Communications* 55: 14143–14146.

54 Fuse, H., Mistunuma, H., and Kanai, M. (2020). Catalytic acceptorless dehydrogenation of aliphatic alcohols. *Journal of the American Chemical Society* 142: 4493–4499.

55 (a) Nippa, D.F., Atz, K., Hohler, R. et al. (2024). Enabling late-stage drug diversification by high-throughput experimentation with geometric deep learning. *Nature Chemistry* 16: 239–248. (b) Bisht, R., Haidar, C., Hassan, M.M.M. et al. (2022). Metal-catalyzed C—H bond activation and borylation. *Chemical Society Reviews* 51: 5042–5100.

56 Chen, Z., Liu, Y., Zeng, C. et al. (2023). Borylation of phenols using sulfuryl fluoride activation. *Green Chemistry* 25: 7998–8006.

57 Guan, W., Chang, Y., and Lin, S. (2023). Electrochemically driven deoxygenative borylation of alcohols and carbonyl compounds. *Journal of the American Chemical Society* 145: 16966–16972.

58 (a) Mizuno, M. and Shioiri, T. (1997). Efficient method for the one-pot azidation of alcohols using bis(*p*-nitrophenyl)phosphorazidate. *Chemical Communications* 1997: 2165–2166. (b) He, L., Wanunu, M., Byun, H.-S., and Bittman, R. (1999). Regioselective and stereospecific azidation of 1,2- and 1,3-diols by azidomethylsilane *via* a Mitsunobu reaction. *The Journal of Organic Chemistry* 64: 6049–6055.

59 (a) Sampath Kumar, H.M., Subba Reddy, B.V., Anjaneyulu, S., and Yadav, J.S. (1998). An expedient and highly selective conversion of alcohols to azides using a $NaN_3/BF_3.Et_2O$ system. *Tetrahedron Letters* 39: 7385–7388. (b) Soltani Rad, M.N., Behrouz, S., and Khalafi-Nezhad, A. (2007). A simple one-pot procedure for the direct conversion of alcohols into azides using TsIm. *Tetrahedron Letters* 48: 3445–3449.

60 Yang, X.-G., Du, F.-H., Li, J.-J., and Zhang, C. (2022). Late-stage dehydroxyazidation of alcohols promoted by trifunctional hypervalent azido-iodine(III) reagents. *Chemistry – A European Journal* 28: e202200272.

61 Guo, J., Wang, X., Ni, C. et al. (2022). SulfoxFluor-enabled deoxyazidation of alcohols with NaN_3. *Nature Communications* 13: 2752.

62 Lee, A.-L., Mooner, D.T., and McKee, H. (2024). Direct C—H functionalization of azoles *via* Minisci reactions. *Organic & Biomolecular Chemistry* 22: 9145–9164.

63 (a) Zhao, W.-M., Chen, X.-L., Yuan, J.-W. et al. (2014). Silver catalyzed decarboxylative direct C2-alkylation of benzothiazoles with carboxylic acids. *Chemical Communications* 50: 2018–2020. (b) Kan, J., Huang, S., Lin, J. et al. (2015). Silver-catalyzed arylation of (hetero)arenes by oxidative decarboxylation of aromatic carboxylic acids. *Angewandte Chemie, International Edition* 54: 2199–2209.

64 Zhang, X., Li, S., Qiu, F. et al. (2024). Photocatalyzed Minisci-type reactions for late-stage functionalization of pharmaceutically relevant compounds. *Green Chemistry* 26: 3595–3626.

65 Garza-Sanchez, R.A., Tlahuext-Aca, A., Tavakoli, G., and Glorius, F. (2017). Visible light-mediated direct decarboxylative C—H functionalization of heteroarenes. *ACS Catalysis* 7: 4057–4061.

66 Lai, X.-L., Shu, X.-M., Song, J., and Xu, H.-C. (2020). Eletrophotocatalytic decarboxylative C—H functionalization of heteroarenes. *Angewandte Chemie, International Edition* 59: 10626–10632.

67 Sherwood, T.C., Li, N., Yazdani, A.N., and Murali Dhar, T.G. (2018). Organocatalyzed, visible-light photoredox-mediated, one-pot minisci reaction using carboxylic acids *via* *N*-(acyloxy)phthalimides. *The Journal of Organic Chemistry* 83: 3000–3012.

68 (a) Genovino, J., Lian, Y., Zhang, Y. et al. (2018). Metal-free-visible light C—H alkylation of heteroaromatics *via* hypervalent iodine-promoted decarboxylation. *Organic Letters* 20: 3229–3232. (b) Jouffroy, M. and Kong, J. (2019). Direct C—H carbamoylation of nitrogen-containing heterocycles. *Chemistry – A European Journal* 25: 2217–2221. (c) Kim, C., Jeong, J., Vellakkaran, M., and Hong, S. (2022). Photocatalytic decarboxylative pyridylation of carboxylic acids using *in situ*-generated amidyl radicals as oxidants. *ACS Catalysis* 12: 13225–13233.

69 Zhang, X.-Y., Weng, W.-Z., Liang, H. et al. (2018). Visible-light-initiated, photocatalyst-free decarboxylative coupling of carboxylic acids with N-heterocycles. *Organic Letters* 20: 4686–4690.

70 Sutherland, D.R., Veguillas, M., Oates, C.L., and Lee, A.-L. (2018). Metal-, photocatalyst-, and light-free, late-stage C—H alkylation of heteroarenes and 1,4-quinones using carboxylic acids. *Organic Letters* 20: 6863–6867.

71 Zuo, Z., Ahneman, D.T., Chu, L. et al. (2014). Merging photoredox with nickel catalysis: Coupling of α-carboxyl sp^3-carbons with aryl halides. *Science* 345: 437–440.

72 Shea, M.D., Mansoor, U.F., and Hopkins, B.A. (2020). A metallaphotoredox method for the expansion of benzyl SAR on electron-deficient amines. *Organic Letters* 22: 1052–1055.

73 Wang, H., Liu, C.-F., Song, Z. et al. (2020). Engaging α-fluorocarboxylic acids directly in decarboxylative C—C bond formation. *ACS Catalysis* 10: 4451–4459.

74 Guo, J., Norris, D., Ramirez, A. et al. (2023). Unlocking tertiary acids for metallaphotoredox $C(sp^2)$—$C(sp^3)$ decarboxylative cross-couplings. *ACS Catalysis* 13: 11910–11918.

75 Zhou, Q.-Q., Guo, W., Ding, W. et al. (2015). Decarboxylative alkynylation and carbonylative alkynylation of carboxylic acids enabled by visible-light photoredox catalysis. *Angewandte Chemie, International Edition* 54: 11196–11199.

76 Noble, A., McCarver, S.J., and MacMillan, D.W.C. (2015). Merging photoredox and nickel catalysis: decarboxylative cross-coupling of carboxylic acids with vinyl halides. *Journal of the American Chemical Society* 137: 624–627.

77 Cao, H., Jiang, H., Feng, H. et al. (2018). Photo-induced decarboxylative Heck-type coupling of unactivated aliphatic acids and terminal alkenes in the absence of sacrificial hydrogen acceptors. *Journal of the American Chemical Society* 140: 16360–16367.

78 Lu, J., Yao, Y., Li, L., and Fu, N. (2023). Dual transition metal electrocatalysis: direct decarboxylative alkenylation of aliphatic carboxylic acids. *Journal of the American Chemical Society* 145: 26774–26782.

79 Noble, A., Mega, R.S., Pflästerer, D. et al. (2018). Visible-light-mediated decarboxylative radical additions to vinyl boronic esters: rapid access to γ-amino boronic esters. *Angewandte Chemie, International Edition* 57: 2155–2159.

80 Shu, C., Mega, R.S., Andreassen, B.J. et al. (2018). Synthesis of functionalized cyclopropanes from carboxylic acids by a radical addition-polar cyclization cascade. *Angewandte Chemie, International Edition* 57: 15430–15434.

81 Kautzky, J.A., Wang, T., Evans, R.W., and MacMillan, D.W.C. (2018). Decarboxylative trifluoromethylation of aliphatic carboxylic acids. *Journal of the American Chemical Society* 140: 6522–6526.

82 Ernouf, G., Chirkin, E., Rhyman, L. et al. (2019). Photochemical strain-release-driven cyclobutylation of $C(sp^3)$-centered radicals. *Angewandte Chemie, International Edition* 59: 2618–2622.

83 Wang, D., Zhu, N., Chen, P. et al. (2017). Enantioselective decarboxylative cyanation employing cooperative photoredox catalysis and copper catalysis. *Journal of the American Chemical Society* 139: 15632–15635.

84 (a) Lai, X.-L., Chen, M., Wang, Y. et al. (2022). Photoelectrochemical asymmetric catalysis enables direct and enantioselective decarboxylative cyanation. *Journal of the American Chemical Society* 144: 20201–20206. (b) Yang, K., Wang, Y., Luo, S., and Fu, N. (2022). Electrophotochemical metal-catalyzed enantioselective decarboxylative cyanation. *Chemistry – A European Journal* 29: e202203962. (c) Yuan, Y., Yang, J., and Zhang, J. (2023). Cu-catalyzed enantioselective decarboxylative cyanation *via* the synergistic merger of photocatalysis and electrochemistry. *Chemical Science* 14: 705–710.

85 Li, Z., Zhang, G., Song, Y. et al. (2023). Copper-catalyzed enantioselective decarboxylative cyanation of benzylic acids promoted by hypervalent iodine(III) reagents. *Organic Letters* 25: 3023–3028.

86 Song, H.-T., Ding, W., Zhou, Q.-Q. et al. (2016). Photocatalytic decarboxylative hydroxylation of carboxylic acids drive by visible light and using molecular oxygen. *The Journal of Organic Chemistry* 81: 7250–7255.

87 Li, P., Zbieg, J.A., and Terrett, J.A. (2021). A platform for decarboxylative couplings *via* photoredox catalysis: direct access to carbocations from carboxylic acids for carbon–oxygen bond formation. *ACS Catalysis* 11: 10997–11004.

88 Mondal, S., Mondal, S., Halder, D. et al. (2024). Iridium/palladium dual photocatalysis for oxidative decarboxylative esterification of alcohols using α-keto acids. *Organic Chemistry Frontiers* 11: 6367–6379.

89 Denkler, L.M., Shekar, M.A., Ngan, T.S.J. et al. (2024). A general iron-catalyzed decarboxylative oxygenation of aliphatic carboxylic acids. *Angewandte Chemie, International Edition* 63: e202403292.

90 Yu, Q., Zhou, D., Yu, P. et al. (2024). Silver-catalyzed decarboxylative nitrooxylation of aliphatic carboxylic acids. *Organic Letters* 26: 5856–5861.

91 (a) Liang, Y., Zhang, X., and MacMillan, D.W.C. (2018). Decarboxylative sp^3 C—N coupling *via* dual copper and photoredox catalysis. *Nature* 559: 83–88.

(b) Nguyen, V.T., Nguyen, V.D., Haug, G.C. et al. (2020). Visible-light-enabled direct decarboxylative N-alkylation. *Angewandte Chemie, International Edition* 59: 7921–7927.

92 Yatham, V.R., Bellotti, P., and König, B. (2019). Decarboxylative hydrazination of unactivated carboxylic acids by cerium photocatalysis. *Chemical Communications* 55: 3489–3492.

93 (a) Kweon, J., Park, B., Kim, D., and Chang, S. (2024). Decarboxylative stereoretentive C—N coupling by harnessing aminating reagent. *Nature Communications* 15: 3788. (b) Kweon, J., Lee, M., Kim, D., and Chang, S. (2024). Enhancing the sustainability and scalability of transition-metal-free stereoretentive decarboxylative amidation with dioxazolones. *Organic Process Research and Development* 28: 3691–3697.

94 (a) Wen, Q., Chen, S., Shi, C. et al. (2022). Iron-catalyzed decarbonylative borylation enables the one-pot diversification of (hetero)aryl and alkyl carboxylic acids. *Cell Reports Physical Science* 3: 100995. (b) Wei, Q., Lee, Y., Liang, W. et al. (2022). Photocatalytic direct borylation of carboxylic acids. *Nature Communications* 13: 7112. (c) Yin, C., Liao, Y., He, B. et al. (2023). Rh-catalyzed direct decarbonylative borylation of carboxylic acids. *Chemistry – A European Journal* 29: e202300184.

95 Na, C.G., Ravelli, D., and Alexanian, E. (2020). Direct decarboxylative functionalization of carboxylic acids *via* O—H hydrogen atom transfer. *Journal of the American Chemical Society* 142: 44–49.

96 Pedersen, P.S., Blakemore, D.C., Chinigo, G.M. et al. (2023). One-pot synthesis of sulfonamides from unactivated acids and amines *via* aromatic decarboxylative halosulfonylation. *Journal of the American Chemical Society* 145: 21189–21196.

97 (a) Yin, F., Wang, Z., Li, Z., and Li, C. (2012). Silver-catalyzed decarboxylative fluorinative of aliphatic carboxylic acids in aqueous solution. *Journal of the American Chemical Society* 134: 10401–10404. (b) Yu, Q., Zhou, D., Ma, J., and Song, C. (2024). Decarboxylative nucleophilic fluorination of aliphatic carboxylic acids. *Organic Letters* 26: 4257–4261.

98 Amani, J. and Molander, G.A. (2017). Direct conversion of carboxylic acids to alkyl ketones. *Organic Letters* 19: 3612–3615.

99 Badir, S.O., Dumoulin, A., Matsui, J.K., and Molander, G.A. (2018). Synthesis of reversed C-acyl glycosides through Ni/photoredox dual catalysis. *Angewandte Chemie, International Edition* 57: 6610–6613.

100 Yang, B. and Tang, R.-Y. (2024). Direct synthesis of dialkyl ketones from deoxygenative cross-coupling of carboxylic acids and alcohols. *Chemical Science* 15: 18405–18410.

101 (a) Zhang, M., Xie, J., and Zhu, C. (2018). A general deoxygenation approach for synthesis of ketones from aromatic carboxylic acids and alkenes. *Nature Communications* 9: 3517. (b) Guo, Y.-Q., Wang, R., Song, H. et al. (2020). Visible-light-induced deoxygenation/defluorination protocol for synthesis of γ,γ-difluoroallylic ketones. *Organic Letters* 22: 709–713.

102 Bay, A.V., Fitzpatrick, K.P., Betori, R.C., and Scheidt, K.A. (2020). Combined photoredox and carbene catalysis for the synthesis of ketones from carboxylic acids. *Angewandte Chemie, International Edition* 59: 9143–9148.

103 Ning, Y., Wang, S., Li, M. et al. (2021). Site-specific Umpolung amidation of carboxylic acids *via* triplet synergistic catalysis. *Nature Communications* 12: 4637.

104 (a) Miao, Y.-Q., Kang, J.-X., Ma, Y.-N., and Chen, X. (2021). Visible light-mediated synthesis of amides from carboxylic acids and amine-boranes. *Green Chemistry* 23: 3595–3599. (b) Gu, C., Wang, S., Zhang, Q., and Xie, J. (2022). Visible-light-mediated amidation from carboxylic acids and tertiary amines *via* C—N cleavage. *Chemical Communications* 58: 5873–5876.

105 Li, L.-J., Zhou, Z.-Q., Liu, Z.-K. et al. (2023). Organo-cyanamides: convenient reagents for catalytic amidation of carboxylic acids. *Chemical Communications* 59: 438–441.

106 Yan, Q., Yuan, Q.-J., Shatskiy, A. et al. (2024). General approach to amides through decarboxylative radical cross-coupling of carboxylic acids and isocyanides. *Organic Letters* 26: 3380–3385.

107 Qiu, X.-F., Liu, D.-Y., Zhang, W.-F., and Hu, X.-G. (2021). Late-stage transformation of carboxylic acids to *N*-trifluoroethyl-imides with trifluoromethyl diazomethane. *Synthesis* 53: 691–670.

108 Zheng, Y., Zhao, Y., Tao, S. et al. (2021). Green esterification of carboxylic acids promoted by *tert*-butyl nitrite. *European Journal of Organic Chemistry* 2021: 2713–2718.

109 Pieber, B., Malik, J.A., Cavedon, C. et al. (2019). Semi-heterogeneous dual nickel/photocatalysis using carbon nitrides: esterification of carboxylic acids with aryl halides. *Angewandte Chemie, International Edition* 58: 9575–9580.

110 Zheng, T., Xu, Y., Xiao, X. et al. (2024). Remote selective cross-dehydrogenative coupling reactions of aryl C—H bonds with carboxylic acids at room temperature. *European Journal of Organic Chemistry* 27: e202400939.

111 Pradhan, T.R., Lee, H.E., Gonzalez-Montiel, G.A. et al. (2020). Highly chemoselective esterification from *O*-aminoallylation of carboxylic acids: metal- and reagent-free hydrocarboxylation of allenamides. *Chemistry – A European Journal* 26: 13826–13831.

112 Scattolin, T., Deckers, K., and Schoenebeck, F. (2017). Direct synthesis of acyl fluorides from carboxylic acids with the bench-stable solid reagent $(Me_4N)SCF_3$. *Organic Letters* 19: 5740–5743.

113 Song, H.-X., Tian, Z.-Y., Xiao, J.-C., and Zhang, C.-P. (2020). Tertiary-amine-initiated synthesis of acyl fluorides from carboxylic acids and $CF_3SO_2OCF_3$. *Chemistry – A European Journal* 26: 16261–16265.

114 Liang, Y., Zhao, Z., Taya, A., and Shibata, N. (2021). Acyl fluorides from carboxylic acids, aldehydes, or alcohols under oxidative fluorination. *Organic Letters* 23: 847–852.

115 Mao, R., Bera, S., Cheseaux, A., and Hu, X. (2019). Deoxygenative trifluoromethylthiolation of carboxylic acids. *Chemical Science* 10: 9555–9559.

116 Griffin, J.D., Zeller, M.A., and Nicewicz, D.A. (2015). Hydrodecarboxylation of carboxylic and malonic acid derivatives *via* organic photoredox catalysis: substrate scope and mechanistic insight. *Journal of the American Chemical Society* 137: 11340–11348.

117 Nguyen, V.T., Nguyen, V.D., Haug, G.C. et al. (2019). Alkene synthesis by photocatalytic chemoenzymatically compatible dehydrodecarboxylation of carboxylic acids and biomass. *ACS Catalysis* 9: 9485–9498.

118 Liu, C., Qin, Z.-X., Ji, C.-L. et al. (2019). Highly-chemoselective step-down reduction of carboxylic acids to aromatic hydrocarbons *via* palladium catalysis. *Chemical Science* 10: 5736–5742.

119 Lunic, D., Sanosa, N., Funes-Ardoiz, I., and Teskey, C.J. (2022). Mild and chemoselective carboxylic acid reduction promoted by borane catalysis. *Angewandte Chemie, International Edition* 61: e202207647.

120 (a) Hatano, M. and Ishihara, K. (2008). Recent progress in the catalytic synthesis of tertiary alcohols from ketones with organometallic reagents. *Synthesis* 2008: 1647–1675. (b) Mushtaq, A., Zahoor, A.F., Ahmad, M.N. et al. (2024). Accessing the synthesis of natural products and their analogues enabled by the Barbier reaction: a review. *RSC Advances* 14: 33536–33567.

121 Yang, Y., Perry, I.B., Lu, G. et al. (2016). Copper-catalyzed asymmetric addition of olefin-derived nucleophiles to ketones. *Science* 353: 144–150.

122 Cai, Y., Ruan, L.-X., Rahman, A., and Shi, S.-L. (2021). Fast enantio- and chemoselective arylation of ketones with organoboronic esters enabled by nickel/*N*-heterocyclic carbene catalysis. *Angewandte Chemie, International Edition* 60: 5262–5267.

123 (a) Vu, M.D., Das, M., Guo, A. et al. (2019). Visible-light photoredox enables ketone carbonyl alkylation for easy access to tertiaryl alcohols. *ACS Catalysis* 9: 9009–9014. (b) Peng, X., Hirao, Y., Yabu, S. et al. (2023). A catalytic alkylation of ketones *via* sp^3 C—H bond activation. *The Journal of Organic Chemistry* 88: 6333–6346. (c) Wang, Q.-L., Huang, H., Zhu, M. et al. (2023). Visible-light-induced stoichiometric coupling of alkylarenes and trifluoromethyl ketones. *Organic Letters* 25: 3800–3805. (d) Mayerhofer, V.J., Lippolis, M., and Teskey, C.J. (2024). Dual-catalysed intermolecular reductive coupling of dienes and ketones. *Angewandte Chemie, International Edition* 63: e202314870. (e) Tan, J., Yang, L., Su, H. et al. (2024). Visible-light-driven asymmetric aldol reaction of ketones and glycinates *via* synergistic Lewis acid/photoredox catalysis. *Chemical Science* 15: 16050–16058.

124 Trowbridge, A., Reich, D., and Gaunt, M.J. (2018). Multicomponent synthesis of tertiary alkylamines by photocatalytic olefin-hydroaminoalkylation. *Nature* 561: 522–527.

125 Dong, J., Wang, Z., Wang, X. et al. (2019). Ketones and aldehydes as alkyl radical equivalents for C—H functionalization of heteroarenes. *Science Advances* 5: eaax9955.

126 Li, J., Wang, H., Qiu, Z. et al. (2020). Metal-free direct deoxygenative borylation of aldehydes and ketones. *Journal of the American Chemical Society* 142: 13011–13020.

127 Zhang, L. and Nagib, D.A. (2023). Carbonyl cross-metathesis *via* deoxygenative gem-di-metal catalysis. *Nature Chemistry* 16: 107–113.

128 Sun, Y., Wang, Y., and Zuo, Z. (2024). Carbonate-assisted selectively deoxygenative cross-coupling reaction between aryl and aliphatic aldehydes: a straightforward route to access α-alkylated aryl ketones. *Organic Chemistry Frontiers* 11: 6510–6517.

129 Renz, M. and Meunier, B. (1999). 100 years of Baeyer–Villiger oxidations. *European Journal of Organic Chemistry* 1999: 737–750.

130 Liu, J., Zhang, C., Zhang, Z. et al. (2019). Nitromethane as a nitrogen donor in Schmidt-type formation of amides and nitriles. *Science* 367: 281–285.

131 Sietmann, J., Ong, M., Mück-Lichtenfeld, C. et al. (2021). Desymmetrization of prochiral cyclobutanones *via* nitrogen insertion: a concise route to chiral γ-lactams. *Angewandte Chemie, International Edition* 60: 9719–9723.

132 Ong, M., Arnold, M., Walz, A.W., and Wahl, J.M. (2022). Stereospecific nitrogen insertion using amino diphenylphosphinates: an aza-Baeyer–Villiger rearrangement. *Organic Letters* 24: 6171–6175.

133 Brägger, Y., Green, O., Bhawal, B.N., and Morandi, B. (2023). Late-stage molecular editing enabled by ketone chain-walking isomerization. *Journal of the American Chemical Society* 145: 19496–19502.

134 Guo, L., Hsiao, C.-C., Yue, H. et al. (2016). Nickel-catalyzed C_{sp2}—C_{sp3} cross-coupling *via* C—O bond activation. *ACS Catalysis* 6: 4438–4442.

135 Chatupheeraphat, A., Liao, H.-H., Srimontree, W. et al. (2018). Ligand-controlled chemoselective C(acyl)—O bond vs C(aryl)—C bond activation of aromatic esters in nickel catalyzed C(sp^2)—C(sp^3) cross-couplings. *Journal of the American Chemical Society* 140: 3724–3735.

136 Masson-Makdissi, J., Vandavasi, J.K., and Newman, S.G. (2018). Switchable selectivity in the Pd-catalyzed alkylative cross-coupling of esters. *Organic Letters* 20: 4094–4098.

137 Hie, L., Nathel, N.F.F., Hong, X. et al. (2016). Nickel-catalyzed activation of acyl C—O bonds of methyl esters. *Angewandte Chemie, International Edition* 55: 2810–2814.

138 Cheung, C.W., Ploeger, M.L., and Hu, X. (2017). Direct amidation of esters with nitroarenes. *Nature Communications* 10: 14878.

139 Li, C., Ji, C.-L., Hong, X., and Szostak, M. (2019). Highly chemoselective, transition-metal-free transamidation of unactivated amides and direct amidation of alkyl esters by N—C/O—C cleavage. *Journal of the American Chemical Society* 141: 11161–11172.

140 Niggli, N.E., Vollgraff, T., Winter, C. et al. (2023). Synthesis of anilides by aminolysis of unactivated esters using $MnCl_2$ in combination with strong bases as catalyst. *Advanced Synthesis and Catalysis* 365: 1794–1800.

141 Takise, R., Isshiki, R., Muto, K. et al. (2017). Decarbonylative diaryl ether synthesis by Pd and Ni catalysis. *Journal of the American Chemical Society* 139: 3340–3343.

142 (a) Cornella, J., Zarate, C., and Martin, R. (2014). Metal-catalyzed activation of ethers *via* C—O bond cleavage: a new strategy for molecular diversity.

Chemical Society Reviews 43: 8081–8097. (b) Tobisu, M. and Chatani, N. (2015). Cross-couplings using aryl ethers *via* C—O bond activation enabled by nickel catalysts. *Accounts of Chemical Research* 48: 1717–1726. (c) Tollefson, E.J., Hanna, L.E., and Jarvo, E.R. (2015). Stereospecific nickel-catalyzed cross-coupling reactions of benzylic ethers and esters. *Accounts of Chemical Research* 48: 2344–2353.

143 Ambre, R., Yang, H., Chen, W.-C. et al. (2019). Nickel carbodicarbene catalyzes Kumada cross-coupling of aryl ethers with grignard reagents through C—O bond activation. *European Journal of Inorganic Chemistry* 2019: 3511–3517.

144 Li, X., Li, Y., Zhang, Z. et al. (2021). Nickel-catalyzed arylation of $C(sp^3)$—O bonds in allylic alkyl ethers with organoboron compounds. *Organic Letters* 23: 6612–6616.

145 (a) Taylor, B.L.H., Swift, E.C., Waetzig, J.D., and Jarvo, E.R. (2011). Stereospecific nickel-catalyzed cross-coupling reactions of alkyl ethers: enantioselective synthesis of diarylethanes. *Journal of the American Chemical Society* 133: 389–391. (b) Wisniewska, H.M., Swift, E.C., and Jarvo, E.R. (2013). Functional-group-tolerant, nickel-catalyzed cross-coupling reaction for enantioselective construction of tertiary methyl-bearing stereocenters. *Journal of the American Chemical Society* 135: 9083–9090.

146 Shimazumi, R., Morita, K., Yoshida, T. et al. (2021). Late-stage derivatization of buflavine by nickel-catalyzed direct substitution of a methoxy group *via* C—O bond activation. *Synthesis* 53: 3037–3044.

147 Heravi, M.M., Panahi, F., and Iranpoor, N. (2020). Nickel-catalyzed reductive amidation of aryl-triazine ethers. *Chemical Communications* 56: 1992–1995.

148 Zarate, C., Manzano, R., and Martin, R. (2015). *Ipso*-borylation of aryl ethers *via* Ni-catalyzed C-OMe cleavage. *Journal of the American Chemical Society* 137 (21): 6754–6757.

149 Lyu, H., Kevlishvili, I., Yu, X. et al. (2021). Boron insertion into alkyl ether bonds *via* zinc/nickel tandem catalysis. *Science* 372 (6538): 175–182.

150 Zeng, X., Zhang, Y., Liu, Z. et al. (2020). Iron-catalyzed borylation of aryl ethers *via* cleavage of C—O bonds. *Organic Letters* 22 (8): 2950–2955.

151 Wang, X., Li, C., Wang, X. et al. (2018). Metal-free etherification of aryl methyl ether derivatives by C—OMe bond cleavage. *Organic Letters* 20: 4267–4272.

152 Wang, X., Yang, Q.-X., Long, C.-Y. et al. (2019). Anticancer-active *N*-heteroaryl amines synthesis: nucleophilic amination of *N*-heteroaryl alkyl ethers with amines. *Organic Letters* 21: 5111–5115.

153 Wang, H., Zhao, J.-F., Zhu, X.-L. et al. (2023). Photoinduced borylation of the inert $C(sp^3)$—O bond of alkyl heteroaryl ethers. *Organic Letters* 25: 6485–6489.

154 Zhang, M.-X., Hu, X.-H., Xu, Y.-H., and Loh, T.-P. (2015). Selective dealkylation of alkyl aryl ethers. *Asian Journal of Organic Chemistry* 4: 1047–1049.

155 Liu, L., Li, Z., Chen, C. et al. (2018). Efficient dealkylation of aryl alkyl ethers catalyzed by Cu_2O. *Tetrahedron* 74: 2447–2453.

156 Grimm, C., Lazzarotto, M., Pompei, S. et al. (2020). Oxygen-free regioselective biocatalytic demethylation of methyl-phenyl ethers *via* methyltransfer employing veratrol-*O*-demethylase. *ACS Catalysis* 10: 10375–10380.

157 (a) Álvarez-Bercedo, P. and Martin, R. (2010). Ni-catalyzed reduction of inert C—O bonds: a new strategy for using aryl ethers as easily removable directing groups. *Journal of the American Chemical Society* 132: 17352–17353. (b) Sergeev, A.G. and Hartwig, J.F. (2011). Selective, nickel-catalyzed hydrogenolysis of aryl ethers. *Science* 332: 439–443. (c) Wiensch, E.M., Todd, D.P., and Montgomery, J. (2017). Silyloxyarenes as versatile coupling substrates enabled by nickel-catalyzed C—O bond cleavage. *ACS Catalysis* 7: 5568–5571. (d) Igarashi, T., Haito, A., Chatani, N., and Tobisu, M. (2018). Nickel-catalyzed reductive cleavage of carbon–oxygen bonds in anisole derivatives using diisopropylaminoborane. *ACS Catalysis* 8: 7475–7483.

158 Parasram, M., Shields, B.J., Ahmad, O. et al. (2020). Regioselective cross-electrophile coupling of epoxide and (hetero)aryl iodides *via* Ni/Ti/photoredox catalysis. *ACS Catalysis* 10: 5821–5827.

7

C—N Bond Activation Under Redox-Neutral Conditions

Xu Chen[1], *Qun Zhao*[1], *and Michal Szostak*[2]

[1] *Jiangnan University, School of Biotechnology and Key Laboratory of Industrial Biotechnology of Ministry of Education, No. 1800, Lihu Avenue, Wuxi 214122, China*
[2] *Rutgers University, Department of Chemistry, 73 Warren Street, Newark, NJ 07102, USA*

7.1 Introduction

The C—N bond is extensively prevalent in organic synthesis, existing abundantly in numerous organic molecules, pharmaceuticals, natural products, synthetic polymers, and biomacromolecules. As a result, the construction of C—N bonds has become a central topic of research in organic synthesis. Several approaches, including transition-metal-catalyzed C—N cross-coupling [1], hydroamination [2], and C—H bond amination [3], have been developed for achieving this goal. From a synthetic perspective, the reverse disconnection, that converts C—N bonds to other chemical bonds, is also highly significant due to the wide availability of nitrogen-containing compounds, their distinct synthetic properties, and the well-established methods for the construction of C—N bonds. In principle, the ubiquitous presence of C—N bonds, together with their orthogonal means of activation, makes them ideal for late-stage functionalization of complex molecules.

In the most general context, employing transition metals for bond activation has emerged as the foremost strategy for the synthesis of complex molecules. However, the cleavage of inert C—N bonds catalyzed by transition metals still presents a significant challenge due to the high bond dissociation energy [4] and the stability of unactivated nitrogen-containing compounds. Despite these challenges, the past two decades have witnessed a rapidly growing interest in this field, resulting in the development of novel methods for C—N bond activation [5]. Numerous approaches, including nitrogen quaternization, oxidative addition of neutral C—N bonds, and the employment of directing groups, have been utilized for C—N bond activation.

This chapter focuses on the significant advancements achieved in transition-metal-catalyzed C—N bond activation under the most synthetically useful redox-neutral conditions. The content is organized according to the class of nitrogen-containing compounds undergoing C—N bond cleavage, such as aryl

Late-Stage Functionalization and Diversification in Organic Synthesis: Methods and Applications,
First Edition. Edited by Tatiana Besset.

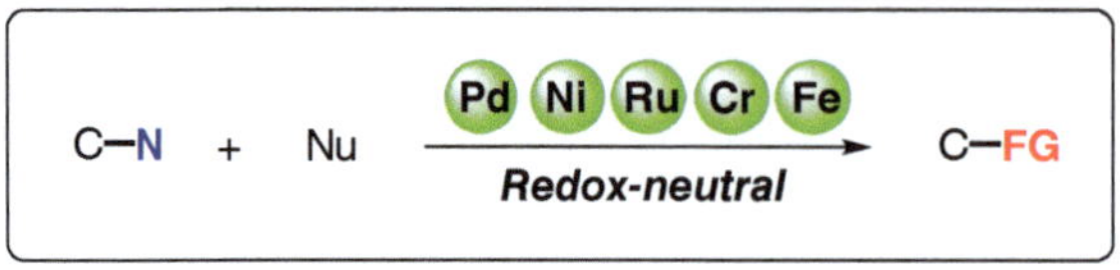

Scheme 7.1 Transition-metal-catalyzed C–N bond activation under redox-neutral conditions.

quaternary ammonium salts, arylamines, and nitroarenes, and the specific transition-metal catalyst employed for C—N bond activation (Scheme 7.1).

While not the central focus of this chapter, it should be noted that several other methods have been developed that utilize transition-metal-catalyzed C—N bond activation. These processes include activation of 1° amines using Katritzky salts [6], activation of amide C(acyl)—N bonds by transition metals by both acyl and decarbonylative pathways [7], and the C—N cleavage of unactivated aziridines through strain release [8]. While these significant methodologies are beyond the scope of this chapter, readers interested in further details are encouraged to refer to the referenced reviews for more information.

7.2 C–N Activation of Ammonium Salts

A major challenge in the direct activation of amines lies in the typically high bond dissociation energy of the C—N bond. One promising approach to address this issue is to quaternize the nitrogen center, thereby reducing the strength of the C—N bond. Furthermore, such arylammonium salts are readily derived from commercially available aromatic amines and easily prepared at a reasonable cost. Over the past two decades, there has been a considerable interest in the transition-metal-catalyzed activation of C—N bonds in ammonium salts [5, 9].

7.2.1 Ni-Catalyzed C–N Activation of Ammonium Salts

In 1988, Wenkert and coworkers reported the seminal study on the first cross-coupling of aryltrimethylammonium iodides with aryl and alkyl Grignard reagents (Scheme 7.2) [10]. This pioneering work showcased the feasibility of oxidative addition of a Ni(0) species into a C—N bond. However, the Wnekert method required demanding reaction conditions with high temperature and prolonged reaction time, resulting in low to moderate yields (18–63%), which left much room for improvement.

In 2003, the MacMillan group achieved a breakthrough in the nickel catalyzed C—N activation of aryltrimethylammonium salts, developing the first Suzuki–Miyaura cross-coupling of aryltrimethylammonium salts (Scheme 7.3) [11]. Compared to Wenkert's earlier work, this reaction uses a novel Ni(0)/NHC catalytic system instead of Ni(dppp)Cl_2, which is based on a strongly σ-donating NHC ligand, thus offering the advantage of broad scope and superior yield (up to 98%) for a range of aniline and boronic acid substrates.

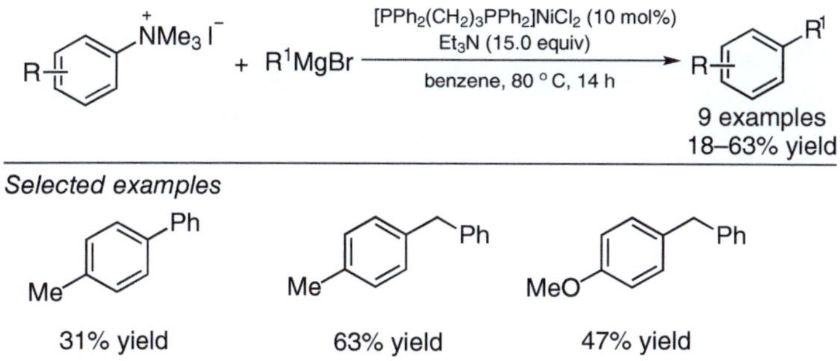

Scheme 7.2 The first Kumada cross-coupling of aryltrimethylammonium salts.

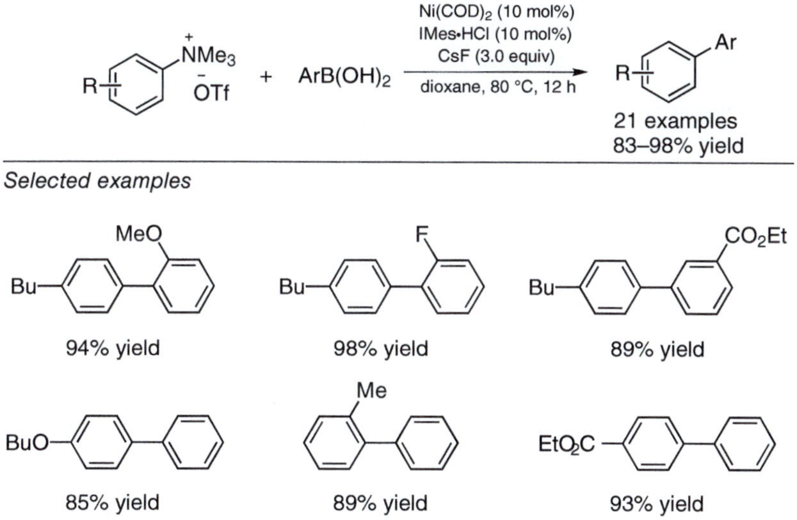

Scheme 7.3 The first Suzuki–Miyaura cross-coupling of aryltrimethylammonium salts.

Based on the studies by Wenkert and MacMillan that nickel complexes can activate the aryl–ammonium bond in both Kumada and Suzuki–Miyaura cross-coupling, Wang group developed a highly efficient process for nickel-catalyzed coupling of aryltrimethylammonium iodide salts with aryl- and alkylzinc reagents (Scheme 7.4) [12]. The authors found that readily available complex, $Ni(PCy_3)_2Cl_2$, gave the best results compared with the PPh_3, PEt_3, and dppe-ligated complexes and afforded to the desired products in up to 99% yield. The protocol tolerates a broad range of functional groups, such as OMe, NMe_2, COOEt, PhC(O), CN, and CF_3.

In 2014, the Wang group also developed the first Buchwald–Hartwig type amination of aryltrimethylammonium salts (Scheme 7.5) [13]. They found that aryltrimethylammonium triflates and amines in the presence of $Ni(COD)_2$ (10 mol%), IPr·HCl (10 mol%) and *t*BuONa (2.0 equiv.) could react smoothly to generate aromatic amines in moderate to excellent yields. Various aliphatic and aromatic amines as well as activated and unactivated aryltrimethylammonium

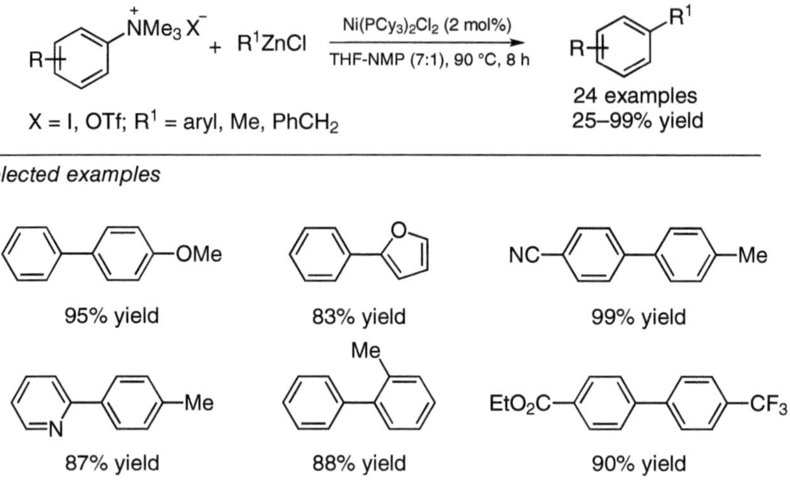

Scheme 7.4 The first Negishi cross-coupling of aryltrimethylammonium salts.

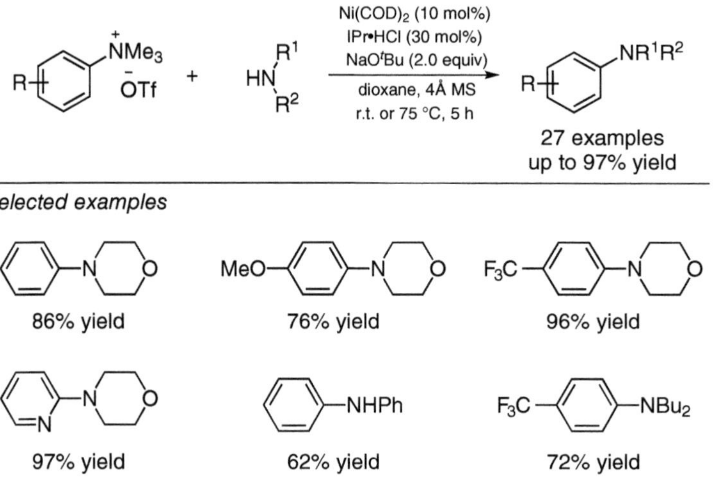

Scheme 7.5 Nickel-catalyzed cross-coupling of aryltrimethylammonium triflates with amines.

triflates can participate in this reaction to afford the corresponding aminated product, demonstrating the broad substrate scope of this transformation.

In addition to these reactions, Ni/NHC combination has also been shown to be an efficient catalytic system in the Stille cross-coupling of quaternary ammonium salts [14].

Organoboron compounds are important building blocks in organic synthesis as the carbon–boron bond could be easily transformed into a carbon–carbon or carbon–heteroatom bond [15]. Consequently, the direct conversion of aryl ammonium salts into organoboronates is a reaction of significant importance. In this context, Itami and coworkers reported a nickel-catalyzed C—N borylations of aryl ammonium salts with B_2pin_2 using the combination of $Ni(COD)_2$ (10 mol%) and PBu_3 (20 mol%) as a catalyst system, which delivered the corresponding borylated products in up to 89% yield (Scheme 7.6) [16]. Interestingly, aryl ammonium salts bearing an electron-donating group displayed higher reactivity than those containing an electron-withdrawing group as aryl ammonium salts with electron-withdrawing substituents, such as CF_3, afforded the corresponding products in diminished yield (37%) compared to their electron-donating counterparts.

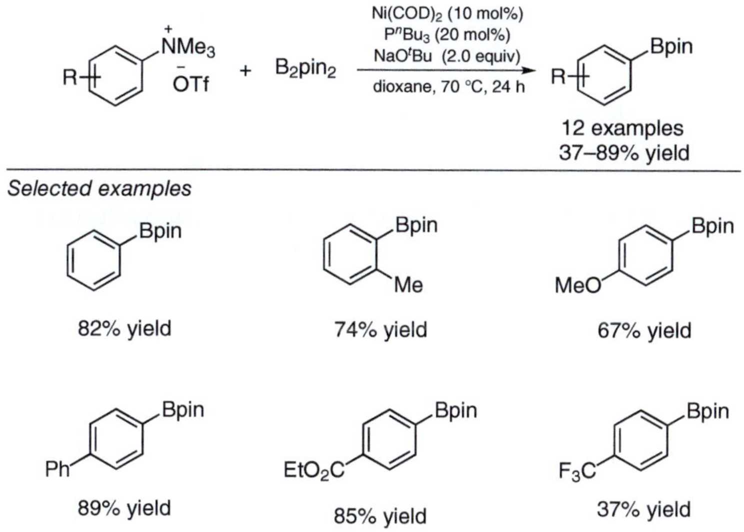

Scheme 7.6 Nickel-catalyzed C—N borylation of aryltrimethylammonium salts.

7.2.2 Pd-Catalyzed C—N Activation of Ammonium Salts

After Wenkert's seminal discovery of the Kumada cross-coupling of aryltrimethylammonium salts, it was not until 2010 that Reeves reported the first palladium-catalyzed cross-coupling of aryltrimethylammonium triflates with Grignard reagents (Scheme 7.7) [17]. Palladium was found to be more efficient than nickel for catalyzing this transformation. The reaction proceeded under mild ambient temperature reaction conditions and uses only a slight excess of Grignard reagents, leading to a wide range of functional groups tolerance.

The Sonogashira cross-coupling reaction plays an important role as a unique method for C(sp)—C(sp^2) bond formation for the preparation of aryl alkynes. In 2019, Cao and Dang group disclosed the first example of aryltrimethylammonium salts in Sonogashira cross-coupling reactions catalyzed by a stable Pd(II)–NHC complex (Scheme 7.8) [18]. The reactions proceeded smoothly under mild conditions with a stoichiometric ratio of alkynes, providing the coupled product in up to 93% yield. A diverse array of alkyl, aryl, and heteroaryl alkyne substrates can be used with the added advantage of accommodating diverse functional groups. The authors proposed that the mechanism involves the following steps according to DFT

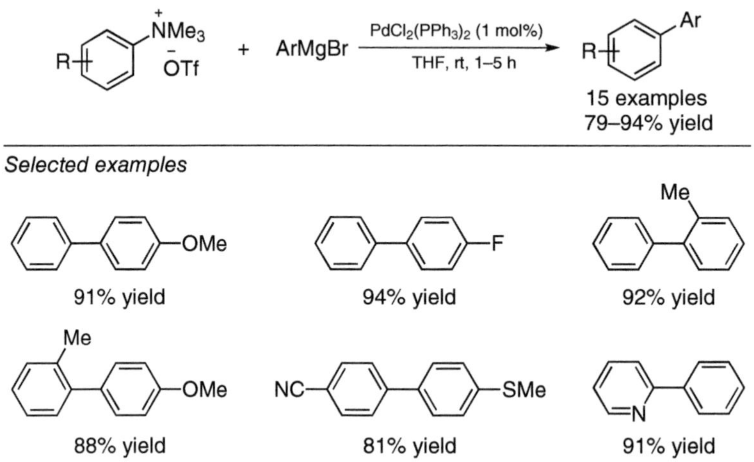

Scheme 7.7 Palladium-catalyzed Kumada cross-coupling of aryltrimethylammonium salts.

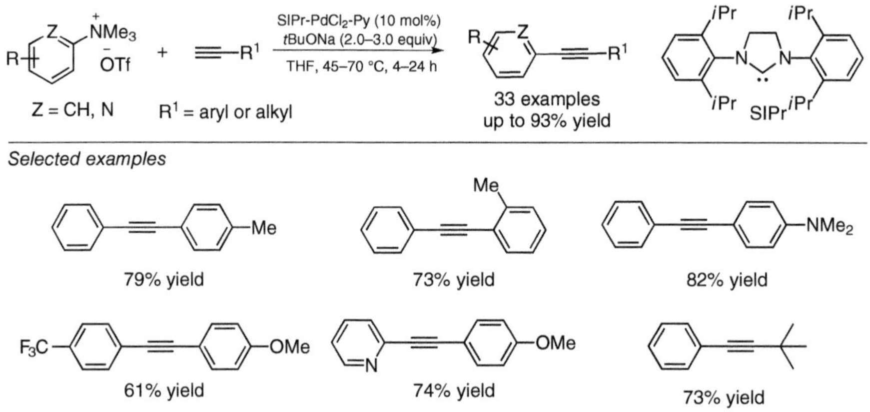

Scheme 7.8 Palladium-catalyzed Sonogashira cross-coupling of aryltrimethylammonium salts.

calculations: C—N bond oxidative addition to Pd(0), coordination of alkyne to Pd(II) center, transmetalation between alkyne and Pd(II) complex, and C—C reductive elimination, resulting in the formation of the alkyne product and regeneration of the catalyst (Scheme 7.9). DFT calculations also show that the C—N oxidative addition is the turnover-limiting step.

Subsequently, a method for the Sonogashira cross-coupling of aryltrimethylammonium salts using bench-stable Pd(II)–NHC precatalysts based on aniline stabilization, such as [Pd(NHC)(3-CF_3-An)Cl_2], has been published [18]. These Pd(II)–NHCs with aniline or cinnamyl coordination are highly involved in the activation of inert N—C bonds, including aryl and B-alkyl Suzuki–Miyaura cross-coupling of 2-pyridyl trimethylammonium salts [18].

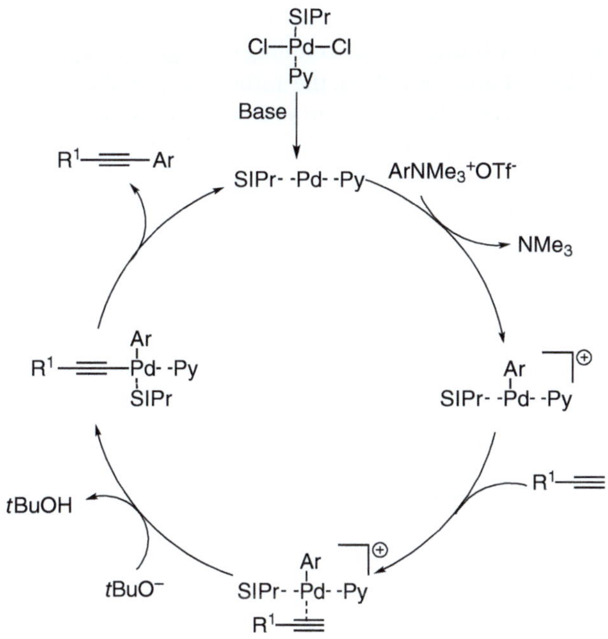

Scheme 7.9 Proposed catalytic cycle for the Sonogashira cross-coupling of aryltrimethylammonium salts.

7.2.3 Fe-Catalyzed C–N Activation of Ammonium Salts

Iron, as an earth-abundant and environmentally benign transition metal, has been used in the Kumada reaction of aryltrimethylammonium triflates with alkylmagnesium reagents by the Wang group [19]. The reaction proceeds at room temperature in a mixed solvent of THF and NMP using 10 mol% of $Fe(acac)_3$ as catalyst, delivering the cross-coupling product in up to 99% yield (Scheme 7.10). It should be noted

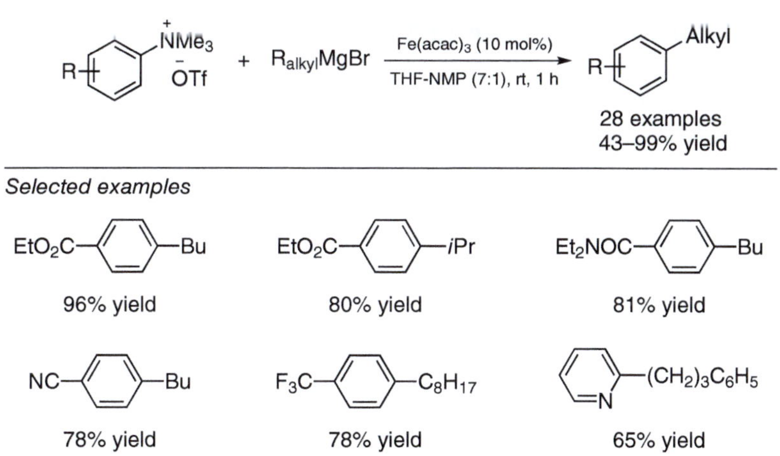

Scheme 7.10 Iron-catalyzed Kumada cross-coupling of aryltrimethylammonium salts.

that challenging 2° alkyl Grignard reagents containing β-hydrogen could also be tolerated. Based on the mechanism of Fe-catalyzed cross-coupling of Grignard reagents with aryl halides proposed by Fürstner and coworkers, the authors proposed the following mechanism (Scheme 7.11): (1) the alkyl Grignard reagent reduces $Fe(acac)_3$ to form the active catalyst $[Fe(MgX)_2]$; (2) the reaction of aryltrimethylammonium triflate with $[Fe(MgX)_2]$ generates $ArFe(OTf)(MgX)_2$ and releases Mg(X)(OTf) to yield ArFe(MgX); and (3) transmetalation with RMgX produces $ArFe(R)(MgX)_2$, which undergoes reductive elimination to form the final product and regenerate the active catalyst $[Fe(MgX)_2]$.

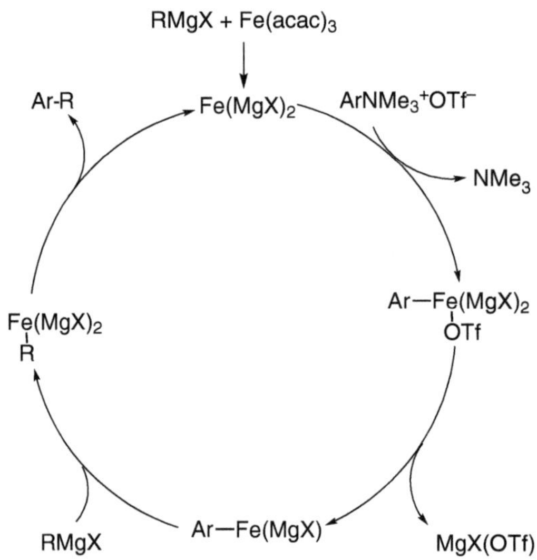

Scheme 7.11 Proposed mechanism for the iron-catalyzed Kumada cross-coupling of aryltrimethylammonium salts.

7.3 C—N Activation of Arylamines

Due to the prevalence of aryl C—N bonds and a wide range of well-established methods for their construction, including Ullman coupling [1a], Buchwald–Hartwig amination [1b, 1c] and aryl C—H amination [3], the direct activation of neutral aryl carbon–nitrogen bonds in aniline derivatives holds a major significance in organic synthesis. However, the thermodynamic stability of aryl amines and the strong coordinating ability of their N center present major challenges in this area. In the past decade, the directing group strategy has offered a solution to the activation of neutral aryl C—N bonds via direct oxidative addition to transition metals.

7.3.1 Ru-Catalyzed C—N Activation of Arylamines

Based on their earlier finding that $RuH_2(CO)(PPh_3)_3$ could function as an excellent catalyst for the functionalization of C—H [20] and C—O bonds [21], in 2007,

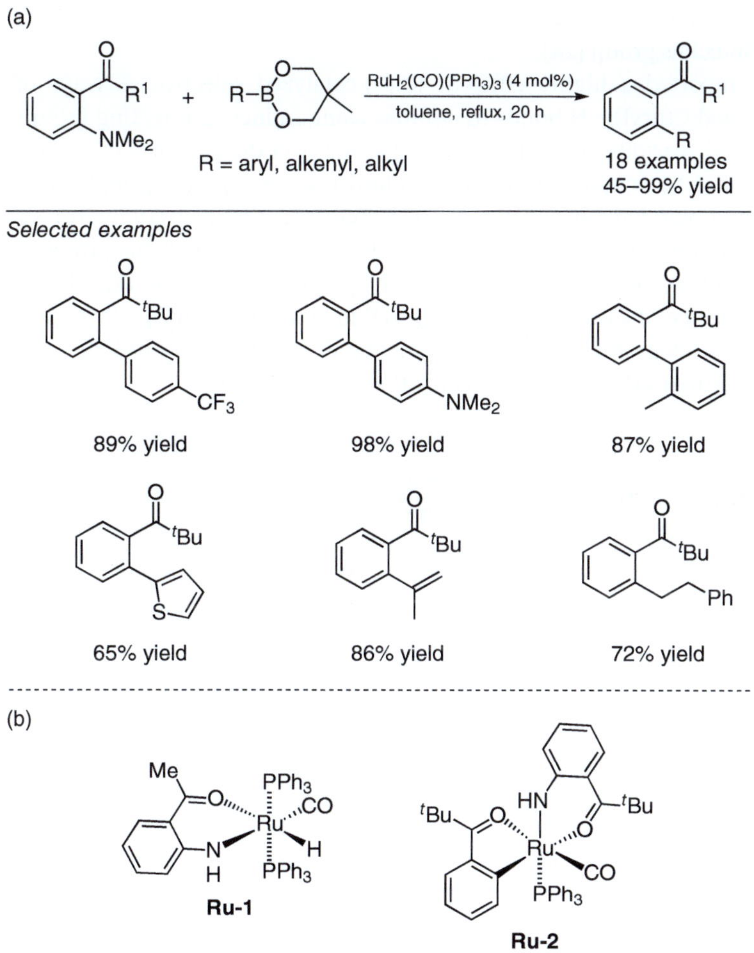

Scheme 7.12 First Ru-catalyzed C(sp^2)–N activation/C–C cross-coupling of anilines. (a) Ru-catalyzed C–N activation. (b) X-ray structures of ruthenium intermediates.

Kakiuchi group reported the first example of a catalytic C—C bond formation via the cleavage of aryl C—N bonds in anilines catalyzed by $RuH_2(CO)(PPh_3)_3$ using carbonyl directing groups (Scheme 7.12a) [22]. Various organoboronates encompassing aryl, heteroaryl, alkenyl, and alkylboronates can be used in this reaction, delivering the corresponding products in excellent up to 99% yield. Key mechanistic steps include (1) oxidative addition of an unactivated aryl C—N bond to the Ru complex; (2) transmetalation between the Ru–NR_2 species and organoboronate; and (3) reductive elimination.

It should be noted that the same group later succeeded in determining X-ray structures of aryl ruthenium intermediates **Ru-1** and **Ru-2** (Scheme 7.12b), showing the initial coordination of the *ortho*-carbonyl group to the ruthenium catalyst and the cleavage of the C—N bond, respectively [23]. These findings support the proposed catalytic cycle for Ru(0)/(II)-catalyzed C—N bond activation.

This methodology was later extended to the use of amides as *ortho*-chelating groups by the Snieckus group [24].

In 2019, we reported a highly efficient Ru(0)-catalyzed selective cleavage of neutral unactivated C(aryl)—N bonds in anilines using imines as directing groups (Scheme 7.13) [25]. Various organoboranes and anilines in the presence of 5 mol% of $Ru_3(CO)_{12}$ could be coupled smoothly to afford the corresponding high-value biaryl ketones and biaryl aldehydes in up to 95% yield after facile *in situ* hydrolysis demonstrating the high efficiency and excellent functional group of this protocol. Notably, this reaction showed excellent selectivity for activation of C(aryl)—N bonds in the presence of C(aryl)—H, C(aryl)—OMe and even C(aryl)—Br bonds, offering a novel strategy for constructing functionalized biaryls by utilizing the orthogonal properties of the C—N bond activation by Ru catalysis.

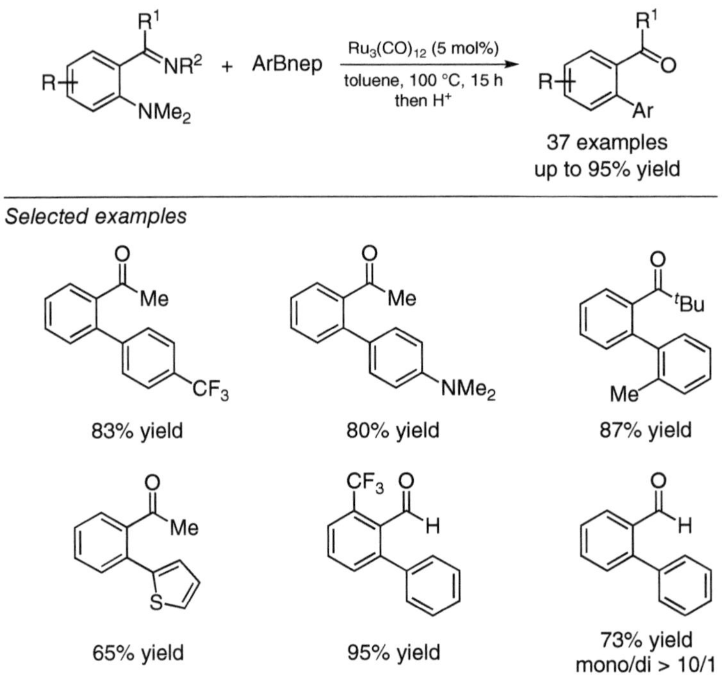

Scheme 7.13 Ru-catalyzed C(sp^2)–N activation/C–C cross-coupling of anilines.

7.3.2 Ni-Catalyzed C–N Activation of Arylamines

In 2014, Chatani and Tobisu group demonstrated an impressive example of catalytic C(aryl)—N bond cleavage reactions of *N*-aryl amides and *N*-aryl carbamates by a nickel-catalyzed reduction and borylation of in the absence of *ortho*-directing groups (Scheme 7.14) [26]. In this study, HBPin was used for the reduction of *N*-aryl amides to synthesize arenes (Scheme 7.14a), while $B_2(nep)_2$ was found to be most effective for the borylative cleavage of *N*-aryl amides for the synthesis of aryl-boronates (Scheme 7.14b). While this method is limited to highly conjugated *N*-aryl

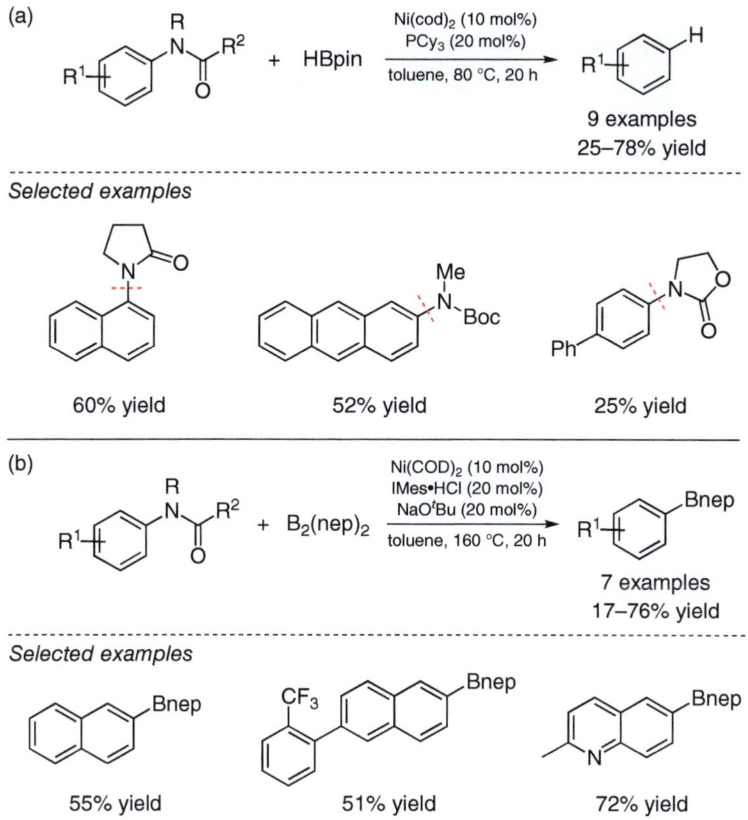

Scheme 7.14 (a) Ni-catalyzed reduction of *N*-aryl amides using HBPin and (b) Ni-catalyzed borylation of *N*-aryl amides using $B_2(nep)_2$.

amides as a significant decrease in yield was observed for benzene derivatives, it is noteworthy that this transformation represents a rare example of transition metal-catalyzed cleavage of amide C(alkyl/aryl)—N bonds instead of the more reactive amide C(acyl)—N bonds.

In 2018, the Shi group reported the first Suzuki–Miyaura cross-coupling of *N*,*N*-dialkyl aryl amines without directing groups (Scheme 7.15) [27]. This reaction employs $NiBr_2$ as the catalyst, strongly σ-donating $IMes^{Me}$ as the ligand, and magnesium as the additive. Magnesium was shown to play a dual role as a reductant to reduce Ni(II) species to active Ni(I) catalyst and as a promoter to facilitate the Ni(I)/Ni(III) catalytic cycle. One limitation of this protocol is that the substrate scope is limited to conjugated arenes. The authors proposed a mechanism based on Ni(I)/Ni(III) catalytic cycle (Scheme 7.16). The active catalyst, $NiBr(NHC)_2$, is generated from the reduction of Ni(II) by Mg and then undergoes oxidative addition, transmetalation, and reductive elimination to yield the final cross-coupling product.

In 2018, the Shi group also extended this methodology to the direct borylation of *N*,*N*-dialkyl aryl amines via C—N bond activation [28].

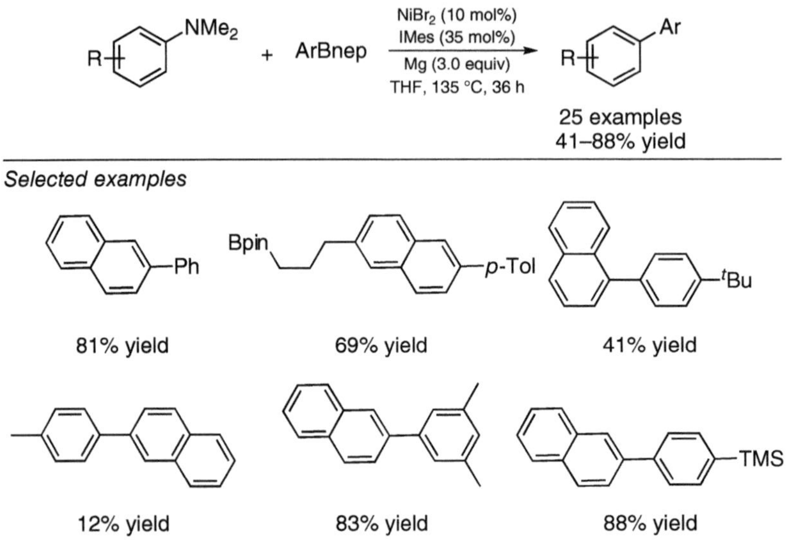

Scheme 7.15 Ni-catalyzed Suzuki–Miyaura cross-coupling of aryl amines.

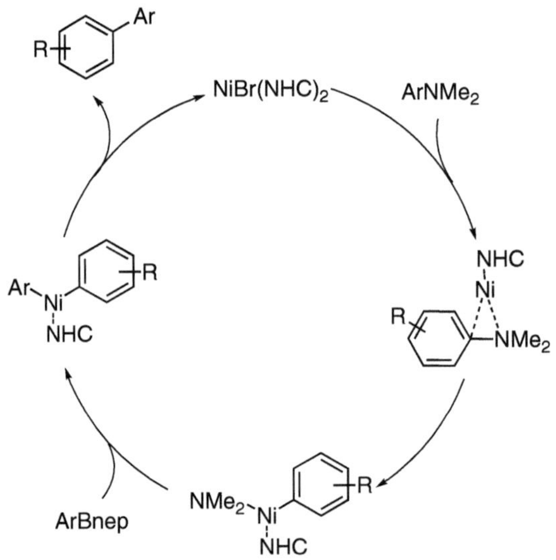

Scheme 7.16 Proposed mechanism for the Suzuki–Miyaura cross-coupling of aryl amines.

In 2020, Xia and coworkers developed a Ni-catalyzed aryl C—N bond cleavage of challenging mono-protected anilines in the absence of *ortho*-directing groups (Scheme 7.17) [29]. The reaction uses *N*-arylsulfonamides as substrates, $Ni(PCy_3)_2Cl_2$ (10 mol%) as a catalyst, and a new *N*-heterocyclic carbene ligand derived from benzoimidazole (15 mol%), affording various biaryls in good to excellent yields (up to 95%). In contrast to the previous methods reported by Chatani [26]

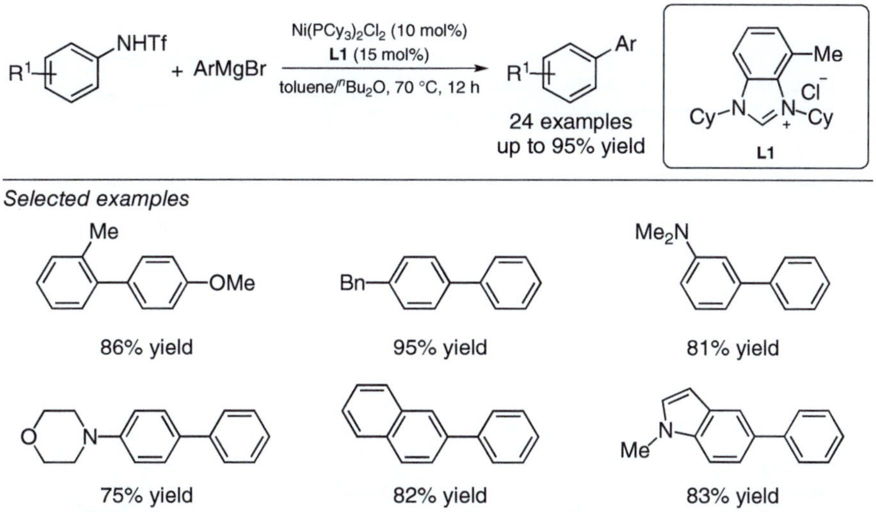

Scheme 7.17 Ni-catalyzed Kumada cross-coupling reaction of *N*-arylsulfonamides.

and Shi [27], which are broadly restricted to π-extended anilines, this reaction works well with both simple mono-protected and π-extended anilines, thus representing a significant step forward in the transition-metal-catalyzed nondirected aryl C—N bond cleavage of anilines.

In 2004, the Robins group reported an interesting example of nickel–NHC-catalyzed Suzuki–Miyaura cross-coupling of azole-substituted purine derivatives with arylboronic acids (Scheme 7.18) [30]. The reaction proceeds smoothly under mild conditions with 10 mol% of $Ni(COD)_2$/SIPr or $Ni(COD)_2$/IPr, converting imidazolyl, benzimidazolyl, and 1,2,4-triazolylpurine (2′-deoxy)nucleoside derivatives to the corresponding 6-arylpurine (2′-deoxy)nucleosides in up to 85% yield. This methodology provides a new avenue for modifying C6-position of purine nucleosides from readily accessible 6-azolyl precursors vs. the less synthetically accessible 6-halopurine 2′-deoxynucleosides. However, it should be noted that this reaction is limited to the specific purine scaffold.

7.3.3 Cr-Catalyzed C–N Activation of Arylamines

In addition to ruthenium and nickel, directing group-assisted activation of aryl C—N bonds in anilines has also been demonstrated using chromium. In 2017, Zeng and coworkers reported the first example of chromium-catalyzed selective cleavage of C(aryl)—N bonds in neutral anilines with Grignard reagents (Scheme 7.19) [31]. The reaction proceeds efficiently at mild room temperature conditions using simple and inexpensive chromium(II) chloride salt as a precatalyst and imines as directing groups, furnishing *ortho*-arylated and *ortho*-alkylated aromatic aldehydes in good to excellent yields of up to 96%. Mechanistic studies demonstrated that the reactive catalyst is a low-valent chromium species generated *in situ* by the reduction of $CrCl_2$ with a Grignard reagent. DFT calculations suggested that the catalytic cycle begins

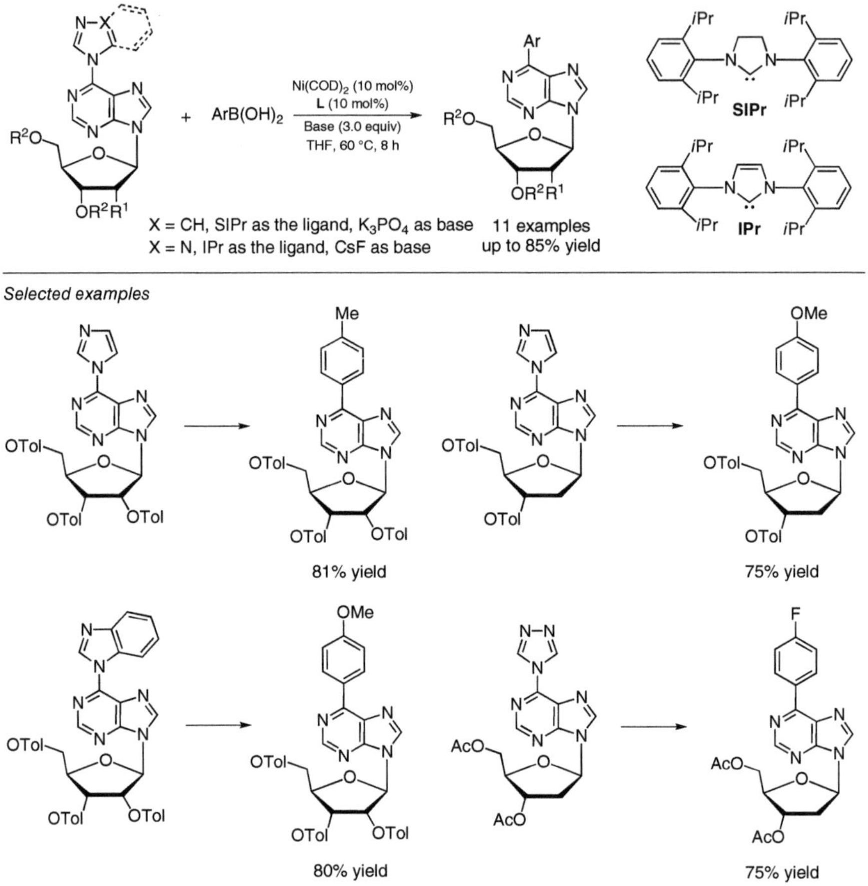

Scheme 7.18 Ni-catalyzed Suzuki–Miyaura cross-coupling of azole-substituted purine derivatives.

with a low-valent chromium, which coordinates with substrate in the presence of an imine directing group (Scheme 7.20). Subsequently, a facile insertion of the C(aryl)—N bond to chromium(0) occurs, which is followed by transmetalation and reductive elimination to produce the final product and regenerate the active catalyst.

7.3.4 Miscellaneous Examples

It should also be noted that palladium has been extensively used in the functionalization of allylamines via π-allyl complexes generated from the oxidative addition of C—N bond to Pd(0). Allylic amines are versatile building blocks in synthetic organic chemistry. A common strategy for activating C—N bond of allylamines is through quaternization of the nitrogen atom. The reader is encouraged to consult specific reviews on the activation of C—N bond of allylic amines [5b, 9a].

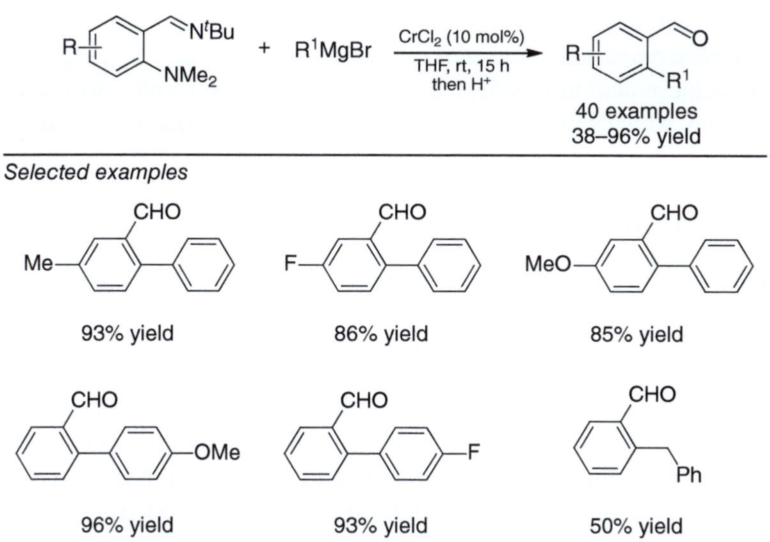

Scheme 7.19 Cr-catalyzed Kumada cross-coupling of anilines.

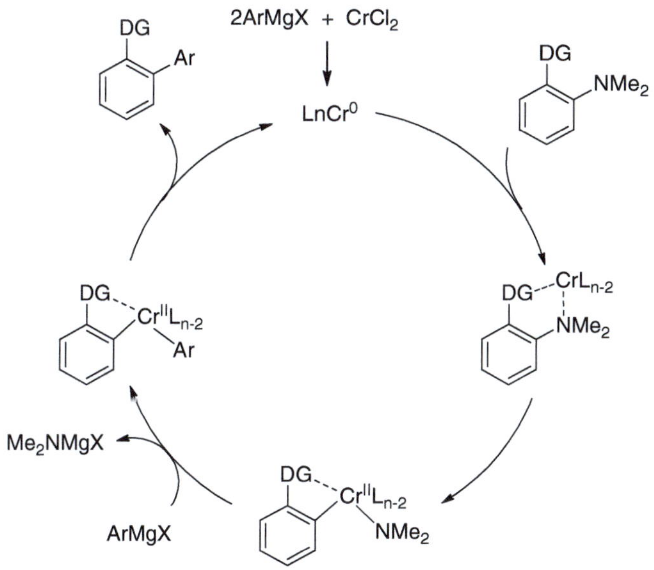

Scheme 7.20 Proposed mechanism for the Cr-catalyzed Kumada cross-coupling of anilines.

7.4 C–N Bond Activation in Nitroarenes

Nitroarenes are abundant chemical feedstocks, which can be synthesized by the classic electrophilic aromatic nitration of arenes. The direct utilization of nitroarenes as electrophilic partners has attracted an increased attention from the chemical community due to its advantages in orthogonal selectivity, atom economy, and step economy. While several metals, including rhodium, palladium, and copper, have been employed in the denitrative coupling of nitroarenes [32], the most significant achievement in this field is the development of a well-controlled Pd/BrettPhos system, which will be discussed in detail in the subsequent sections.

7.4.1 Pd-Catalyzed C–N Activation of Nitroarenes

In an early example, in 2013, Chang, Wu, Yu, and coworkers reported a Pd-catalyzed coupling of arylboronic acids with nitroarenes in the presence of water for the synthesis of diaryl ethers [33]. However, the generality of nitroarenes was limited to electron-deficient substrates. In 2017, Nakao reported a seminal study on a palladium-catalyzed denitrative Suzuki–Miyaura cross-coupling of nitroarenes (Scheme 7.21) [34]. Various nitroarenes could react with arylboronic acids irrespective of the electronic and steric effects, generating various biaryls in good yield of up to 84%. Mechanistic study demonstrated that (cod)Pd(CH_2SiMe_3)$_2$, a Pd(0) precursor, could react with nitroarenes in the presence of BrettPhos to form BrettPhosPd(Ph)(NO_2) via the oxidative addition of the Ph—NO_2 bond, representing the first example of an oxidative addition of an unactivated Ar—NO_2 bond onto a Pd(0) center. Based on DFT calculations, a plausible catalytic cycle was proposed, involving oxidative addition of an Ar—NO_2 bond to a Pd(0) center and forming Ar—Pd(II)—NO_2 (Scheme 7.22). This intermediate then undergoes

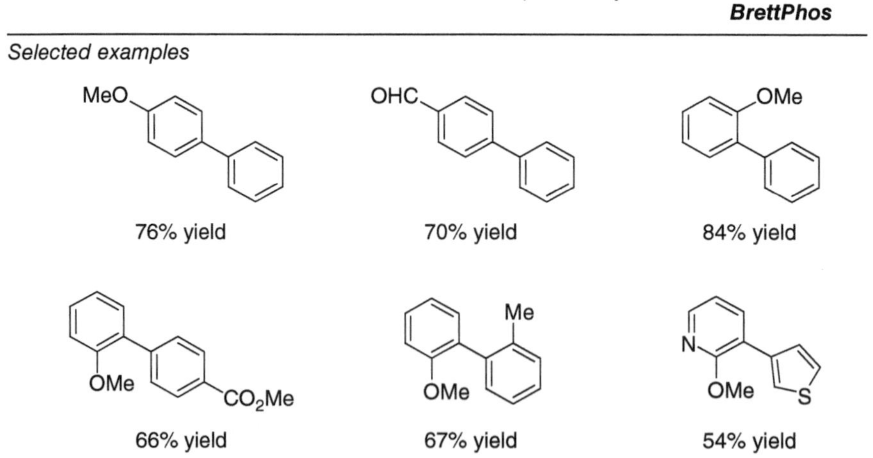

Scheme 7.21 Pd-catalyzed Suzuki–Miyaura cross-coupling of nitroarenes.

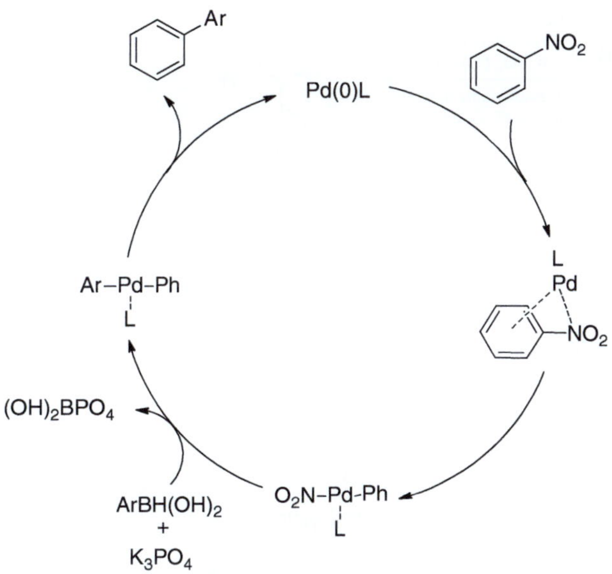

Scheme 7.22 Proposed mechanism of Pd-catalyzed Suzuki–Miyaura cross-coupling of nitroarenes.

transmetalation with arylboronic acid. The following reductive elimination generates the biaryl product. The direct observation of oxidative addition of the Ar—NO_2 bond to Pd(0) in this impressive transformation opened the door for an array of developments in C—NO_2 bond functionalization.

In 2020, the You group extended this methodology to the denitrative methylation of nitroarenes by using trimethylboroxine as a nucleophile [35].

In the same year, the Nakao group also reported the first palladium-catalyzed Buchwald–Hartwig amination of nitroarenes using $Pd(acac)_2$/BrettPhos as a catalyst and K_3PO_4 as a base (Scheme 7.23) [36]. A range of nitroarenes with electron-withdrawing or electron-donating groups could couple with diarylamines, arylamines, and alkylamines, affording substituted arylamines in moderate to good yields of up to 81%. Mechanistically, the reaction was proposed to involve the following steps (Scheme 7.24): (1) coordination of nitroarenes to Pd(0)L; (2) oxidative addition of the Ar—NO_2 bond to Pd(0); (3) transmetalation of amine with Ar—Pd—NO_2 complex; (4) reductive elimination of arylpalladium amide to form the arylamine product and regenerate Pd(0) catalyst. The facile generation of arylamine product from a stoichiometric reaction of the oxidative addition adduct, Ar—Pd(L)—NO_2, with benzylic amine strongly supporting the proposed catalytic cycle.

Transition-metal-catalyzed cross-coupling of an aryl electrophile with terminal alkynes (Sonogashira reaction) is one of the most powerful methods for the construction of C(sp^2)—C(sp) bonds. In 2020, the You group reported a palladium-catalyzed denitrative Sonogashira cross-coupling of nitroarenes with terminal alkynes, affording a wide range of aryl alkynes in moderate to excellent yields of up to 96% (Scheme 7.25) [37]. The utility of this denitrative alkynylation

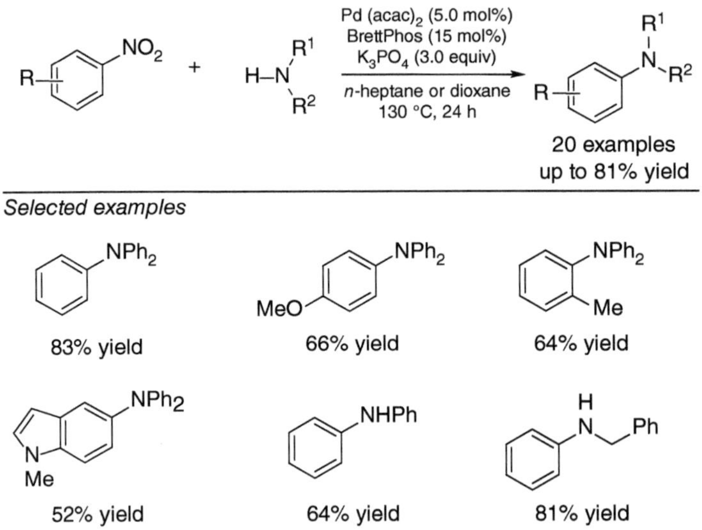

Scheme 7.23 Pd-catalyzed Buchwald–Hartwig amination of nitroarenes.

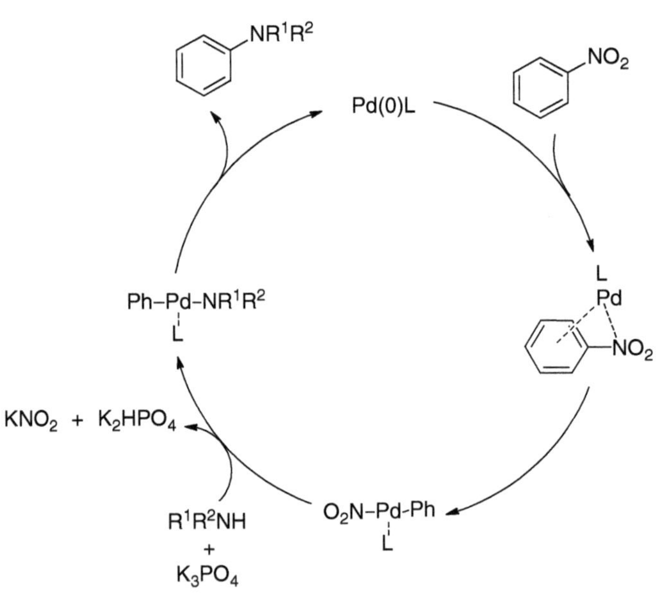

Scheme 7.24 Proposed mechanism for the Buchwald–Hartwig amination of nitroarenes.

has been demonstrated in the orthogonal cross-coupling and the synthesis of phenanthrene, 9,9′-bifluorenylidene, and spirobifluorene derivatives.

Aryl nitriles have played a key role in organic chemistry as versatile intermediates in the synthesis of carboxylic acids, aldehydes, amides, and amines [38]. In 2024, Yamaguchi reported a direct synthesis of aryl nitriles from nitroarenes

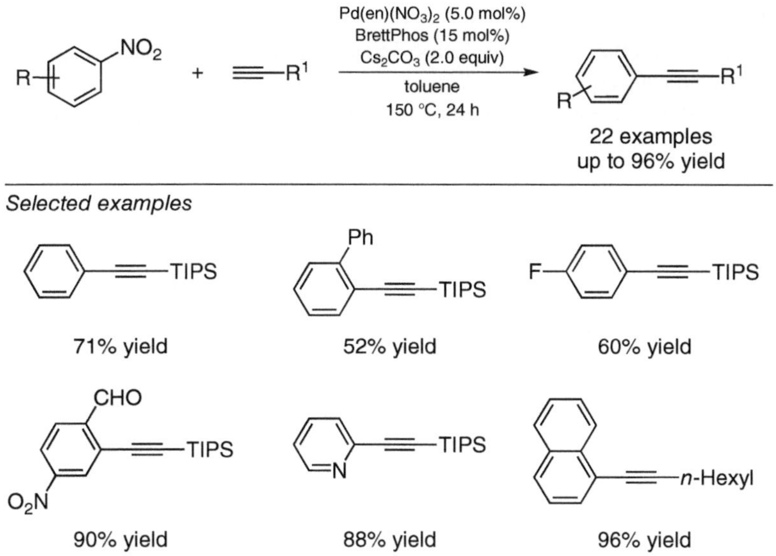

Scheme 7.25 Pd-catalyzed Sonogashira cross-coupling of nitroarenes.

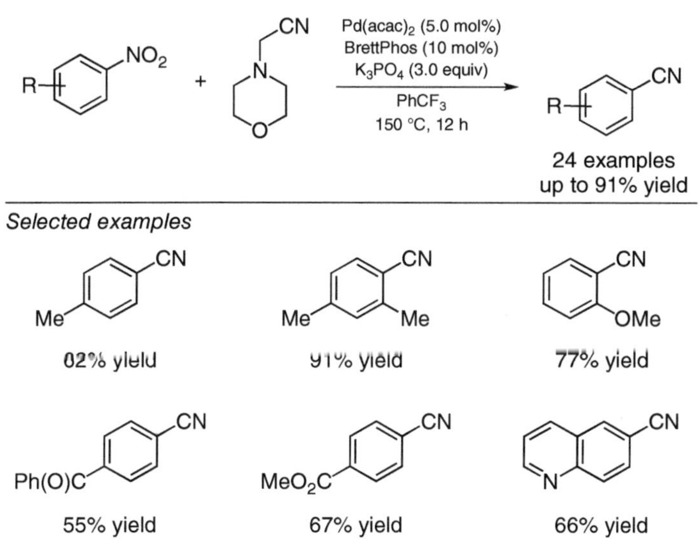

Scheme 7.26 Pd-catalyzed cyanation of nitroarenes.

and aminoacetonitrile using $Pd(acac)_2$ (5 mol%)/BrettPhos (10 mol%) as a catalyst system and K_3PO_4 (3.0 equiv.) as a base, furnishing a wide range of aryl nitriles in moderate to excellent yields of up to 91% (Scheme 7.26) [39]. Notably, this methodology achieved a halogen- and waste-free catalytic cyanation using aminoacetonitrile as the cyano source.

In addition to these reactions, Pd/BrettPhos catalytic system has also been used in the denitrative intramolecular C—H arylation [40], Mizoroki–Heck reaction [41], and reductive denitration [42].

A related Pd(II)–NHC system based on well-defined air- and moisture-stable palladium–imidazo[1,5-*a*]pyridin-3-ylidene complexes has been reported [42].

7.5 Summary

The examples discussed in this chapter clearly demonstrate that transition-metal-catalyzed C—N bond activation under redox-neutral conditions has become a robust strategy for the application of nitrogen-containing compounds, including aryl quaternary ammonium salts, arylamines, and nitroarenes. To overcome the inherent challenges associated with activating inert C—N bonds, several distinct strategies have been introduced. The main strategies include quaternization of the nitrogen center, introduction of directing groups, and direct oxidative addition of neutral C—N bonds using strongly nucleophilic catalyst systems. Despite the very significant progress in this important field, major challenges remain, including harsh reaction conditions, limited studies on the reaction mechanism in many cases, and limited scope of electrophiles, thus providing ample opportunities for future development. Given the prevalence of C—N bonds in numerous organic compounds, the development of highly efficient methods for activating C—N bonds holds a very significant promise for exploring new avenues in design and late-stage scaffold diversification of complex molecules.

Acknowledgments

The authors acknowledge the generous support from Jiangnan University (Q.Z.), the Wuxi Science and Technology Development Fund Project (K20241034, X.C.), the NIH (R35GM133326, M.S.), the NSF (CAREER CHE-1650766, M.S.), and Rutgers University (M.S.).

References

1 (a) Monnier, F. and Taillefer, M. (2009). Catalytic C—C, C—N, and C—O Ullmann-type coupling reactions. *Angewandte Chemie, International Edition* 48: 6954–6971. (b) Guram, A.S. and Buchwald, S.L. (1994). Palladium-catalyzed aromatic aminations with in situ generated aminostannanes. *Journal of the American Chemical Society* 116: 7901–7902. (c) Paul, F., Patt, J., and Hartwig, J.F. (1994). Palladium-catalyzed formation of carbon-nitrogen bonds. Reaction intermediates and catalyst improvements in the hetero cross-coupling of aryl halides and tin amides. *Journal of the American Chemical Society* 116: 5969–5970.

2 Müller, T.E., Hultzsch, K.C., Yus, M. et al. (2008). Hydroamination: direct addition of amines to alkenes and alkynes. *Chemical Reviews* 108: 3795–3892.

3 Park, Y., Kim, Y., and Chang, S. (2017). Transition metal-catalyzed C—H amination: scope, mechanism, and applications. *Chemical Reviews* 117: 9247–9301.

4 Blanksby, S.J. and Ellison, G.B. (2003). Bond dissociation energies of organic molecules. *Accounts of Chemical Research* 36: 255–263.

5 (a) Ouyang, K., Hao, W., Zhang, W.-X. et al. (2015). Transition-metal-catalyzed cleavage of C—N single bonds. *Chemical Reviews* 115: 12045–12090. (b) Wang, Q., Su, Y., Li, L. et al. (2016). Transition-metal catalysed C—N bond activation. *Chemical Society Reviews* 45: 1257–1272. (c) Boit, T.B., Bulger, A.S., Dander, J.E. et al. (2020). Activation of C—O and C—N bonds using non-precious-metal catalysis. *ACS Catalysis* 10: 12109–12126. (d) García-Cárceles, J., Bahou, K.A., and Bower, J.F. (2020). Recent methodologies that exploit oxidative addition of C—N bonds to transition metals. *ACS Catalysis* 10: 12738–12759. (e) Liu, J., Yang, Y., Ouyang, K. et al. (2021). Transition-metal-catalyzed transformations of C—N single bonds: advances in the last five years, challenges and prospects. *Green Synthesis and Catalysis* 2: 87–122.

6 (a) Basch, C.H., Liao, J., Xu, J. et al. (2017). Harnessing alkyl amines as electrophiles for nickel-catalyzed cross couplings via C—N bond activation. *Journal of the American Chemical Society* 139: 5313–5316. (b) He, F.-S., Ye, S., and Wu, J. (2019). Recent advances in pyridinium salts as radical reservoirs in organic synthesis. *ACS Catalysis* 9: 8943–8960. (c) Rössler, S.L., Jelier, B.J., Magnier, E. et al. (2020). Pyridinium salts as redox-active functional group transfer reagents. *Angewandte Chemie, International Edition* 59: 9264–9280. (d) Yousif, A.M., Colarusso, S., and Bianchi, E. (2023). Katritzky salts for the synthesis of unnatural amino acids and late-stage functionalization of peptides. *European Journal of Organic Chemistry* 26: e202201274.

7 (a) Meng, G., Shi, S., and Szostak, M. (2016). Cross-coupling of amides by N—C bond activation. *Synlett* 27: 2530–2540. (b) Feng, M., Zhang, H., and Maulide, N. (2022). Challenges and breakthroughs in selective amide activation. *Angewandte Chemie, International Edition* 61: e202212213.

8 Huang, C.-Y. and Doyle, A.G. (2014). The chemistry of transition metals with three-membered ring heterocycles. *Chemical Reviews* 114: 8153–8198.

9 (a) Zhao, B., Rogge, T., Ackermann, L. et al. (2021). Metal-catalysed C—Het (F, O, S, N) and C—C bond arylation. *Chemical Society Reviews* 50: 8903–8953. (b) Tu, Y. and Zhao, J. (2021). Recent advances in the Pd-catalyzed coupling of arylhydrazines and ammonium salts via C—N bond cleavage. *Chemical Record* 21: 3442–3457.

10 Wenkert, E., Han, A.-L., and Jenny, C.-J. (1988). Nickel-induced conversion of carbon–nitrogen into carbon–carbon bonds. One-step transformations of aryl, quaternary ammonium salts into alkylarenes and biaryls. *Journal of the Chemical Society, Chemical Communications* 975–976.

11 Blakey, S.B. and MacMillan, D.W.C. (2003). The first Suzuki cross-couplings of aryltrimethylammonium salts. *Journal of the American Chemical Society* 125: 6046–6047.

12 Xie, L.-G. and Wang, Z.-X. (2011). Nickel-catalyzed cross-coupling of aryltrimethylammonium iodides with organozinc reagents. *Angewandte Chemie, International Edition* 50: 4901–4904.

13 Zhang, X.-Q. and Wang, Z.-X. (2014). Nickel-catalyzed cross-coupling of aryltrimethylammonium triflates and amines. *Organic & Biomolecular Chemistry* 12: 1448–1453.

14 Wang, D.-Y., Kawahata, M., Yang, Z.-K. et al. (2016). Stille coupling via C—N bond cleavage. *Nature Communications* 7: 12937.

15 Sandford, C. and Aggarwal, V.K. (2017). Stereospecific functionalizations and transformations of secondary and tertiary boronic esters. *Chemical Communications* 53: 5481–5494.

16 Zhang, H., Hagihara, S., and Itami, K. (2015). Making dimethylamino a transformable directing group by nickel-catalyzed C—N borylation. *Chemistry – A European Journal* 21: 16796–16800.

17 Reeves, J.T., Fandrick, D.R., Tan, Z. et al. (2010). Room temperature palladium-catalyzed cross coupling of aryltrimethylammonium triflates with aryl grignard reagents. *Organic Letters* 12: 4388–4391.

18 Chen, Q., Gao, F., Tang, H. et al. (2019). Sonogashira cross-coupling of aryltrimethylammonium salts. *ACS Catalysis* 9: 3730–3736.

19 Guo, W.-J. and Wang, Z.-X. (2013). Iron-catalyzed cross-coupling of aryltrimethylammonium triflates and alkyl Grignard reagents. *Tetrahedron* 69: 9580–9585.

20 Kakiuchi, F. and Murai, S. (2002). Catalytic C—H/Olefin coupling. *Accounts of Chemical Research* 35: 826–834.

21 Kakiuchi, F., Usui, M., Ueno, S. et al. (2004). Ruthenium-catalyzed functionalization of aryl carbon–oxygen bonds in aromatic ethers with organoboron compounds. *Journal of the American Chemical Society* 126: 2706–2707.

22 Ueno, S., Chatani, N., and Kakiuchi, F. (2007). Ruthenium-catalyzed carbon–carbon bond formation via the cleavage of an unreactive aryl carbon–nitrogen bond in aniline derivatives with organoboronates. *Journal of the American Chemical Society* 129: 6098–6099.

23 Koreeda, T., Kochi, T., and Kakiuchi, F. (2009). Cleavage of C–N bonds in aniline derivatives on a ruthenium center and its relevance to catalytic C—C bond formation. *Journal of the American Chemical Society* 131: 7238–7239.

24 Zhao, Y. and Snieckus, V. (2014). Beyond directed ortho metalation: ruthenium-catalyzed amide-directed CAr—N activation/C—C coupling reaction of anthranilamides with organoboronates. *Organic Letters* 16: 3200–3203.

25 Zhao, Q., Zhang, J., and Szostak, M. (2019). Ruthenium(0)-catalyzed cross-coupling of anilines with organoboranes by selective carbon–nitrogen cleavage. *ACS Catalysis* 9: 8171–8177.

26 Tobisu, M., Nakamura, K., and Chatani, N. (2014). Nickel-catalyzed reductive and borylative cleavage of aromatic carbon–nitrogen bonds in *N*-aryl amides and carbamates. *Journal of the American Chemical Society* 136: 5587–5590.

27 Cao, Z.-C., Xie, S.-J., Fang, H. et al. (2018). Ni-catalyzed cross-coupling of dimethyl aryl amines with arylboronic esters under reductive conditions. *Journal of the American Chemical Society* 140: 13575–13579.

28 Cao, Z.-C., Li, X.-L., Luo, Q.-Y. et al. (2018). Direct borylation of tertiary anilines via C—N bond activation. *Organic Letters* 20: 1995–1998.

29 Zhang, Z.-B. and Xia, J.-B. (2020). NHC-nickel catalyzed C—N bond cleavage of mono-protected anilines for C—C cross-coupling. *Organic Letters* 22: 9609–9613.

30 Liu, J. and Robins, M.J. (2004). Azoles as Suzuki cross-coupling leaving groups: syntheses of 6-arylpurine 2′-deoxynucleosides and nucleosides from 6-(imidazol-1-yl)- and 6-(1,2,4-triazol-4-yl)purine derivatives. *Organic Letters* 6: 3421–3423.

31 Cong, X., Fan, F., Ma, P. et al. (2017). Low-valent, high-spin chromium-catalyzed cleavage of aromatic carbon–nitrogen bonds at room temperature: a combined experimental and theoretical study. *Journal of the American Chemical Society* 139: 15182–15190.

32 (a) Muto, K., Okita, T., and Yamaguchi, J. (2020). Transition-metal-catalyzed denitrative coupling of nitroarenes. *ACS Catalysis* 10: 9856–9871. (b) Kashihara, M. and Nakao, Y. (2021). Cross-coupling reactions of nitroarenes. *Accounts of Chemical Research* 54: 2928–2935.

33 Peng, D., Yu, A., Wang, H. et al. (2013). A 2,2′-bipyridine-palladacycle catalyzed the coupling of arylboronic acids with nitroarenes. *Tetrahedron* 69: 6884–6889.

34 Yadav, M.R., Nagaoka, M., Kashihara, M. et al. (2017). The Suzuki–Miyaura coupling of nitroarenes. *Journal of the American Chemical Society* 139: 9423–9426.

35 Feng, B., Yang, Y., and You, J. (2020). A methylation platform of unconventional inert aryl electrophiles: trimethylboroxine as a universal methylating reagent. *Chemical Science* 11: 6031–6035.

36 Inoue, F., Kashihara, M., Yadav, M.R. et al. (2017). Buchwald–Hartwig amination of nitroarenes. *Angewandte Chemie, International Edition* 56: 13307–13309.

37 Feng, B., Yang, Y., and You, J. (2020). Palladium-catalyzed denitrative Sonogashira-type cross-coupling of nitrobenzenes with terminal alkynes. *Chemical Communications* 56: 790–793.

38 Nakao, Y. (2021). Metal-mediated C—CN bond activation in organic synthesis. *Chemical Reviews* 121: 327–344.

39 Iizumi, K., Tanaka, H., Muto, K. et al. (2024). Palladium-catalyzed denitrative synthesis of aryl nitriles from nitroarenes and organocyanides. *Organic Letters* 26: 3977–3981.

40 Asahara, K.K., Okita, T., Saito, A.N. et al. (2019). Pd-catalyzed denitrative intramolecular C—H arylation. *Organic Letters* 21: 4721–4724.

41 Okita, T., Asahara, K.K., Muto, K. et al. (2020). Palladium-catalyzed Mizoroki–Heck reaction of nitroarenes and styrene derivatives. *Organic Letters* 22: 3205–3208.

42 Kashihara, M., Yadav, M.R., and Nakao, Y. (2018). Reductive denitration of nitroarenes. *Organic Letters* 20: 1655–1658.

8

Late-Stage Functionalization by C—S Bond Activation

Jun-Jie Chen and Huan-Ming Huang

School of Physical Science and Technology, ShanghaiTech University, 393 Middle Huaxia Road, Pudong, Shanghai, 201210, China

8.1 Introduction

Late-stage functionalization refers to a type of chemical modification that is performed on a molecule at a relatively late stage of its synthesis. This approach allows chemists to add specific functional groups or other moieties to a complex molecule, without a need to make significant changes to the overall structure of the molecule during earlier stages of its synthesis [1].

This method can be particularly useful in the synthesis of complex, biologically active molecules such as drugs, natural products, and other bioactive compounds. By adding functional groups at a late stage, chemists can modify the properties of the molecule in a targeted and controlled manner, without disrupting the overall architecture of the molecule [2].

One of the main advantages of late-stage functionalization is that it allows the synthesis of a wide range of structurally diverse molecules from a single intermediate or scaffold. This approach can significantly streamline the synthesis process, reducing the number of steps required to produce a desired molecule and potentially improving the overall yield and efficiency of the synthesis. Another advantage of late-stage functionalization is that it can be used to install functional groups that are difficult or impossible to incorporate during earlier stages of synthesis. For example, some functional groups may be unstable or reactive under the conditions required for earlier steps in the synthesis, making it challenging to incorporate them at an earlier stage. By performing the functionalization at a later stage, chemists can avoid these issues and successfully add the desired functional groups to the molecule.

Overall, late-stage functionalization is a powerful tool in the synthesis of complex molecules, and it has become an increasingly important approach in the field of organic chemistry [3]. As techniques for late-stage functionalization continue to improve and evolve, it is likely that this approach will play an even greater role in the

Late-Stage Functionalization and Diversification in Organic Synthesis: Methods and Applications,
First Edition. Edited by Tatiana Besset.

synthesis of new and useful molecules in the future. In this chapter, we focused on the recent advances in late-stage functionalization by C—S bond activation enabled by ionic and radical synthetic strategies.

8.2 Transition Metal-Catalyzed Late-Stage Functionalization by C—S Bond Activation

The Suzuki–Miyaura cross-coupling reaction has become one of the most valuable synthetic tools in synthetic and medicinal chemistry through carbon–carbon bond formation. Different sulfonates are common coupling partners in this Nobel-winning synthetic transformation. Recently, Niu and coworkers utilized 1-sulfonyl glycals **1.1** as coupling partner to couple with aryl boronic acids **1.2** with a catalytic amount of $Ni(COD)_2$ and $Cy_3P{\cdot}HBF_4$ to produce a library of aryl glycals in good to excellent yields (Scheme 8.1) [4]. Interestingly, this nickel-catalyzed

Niu [4]

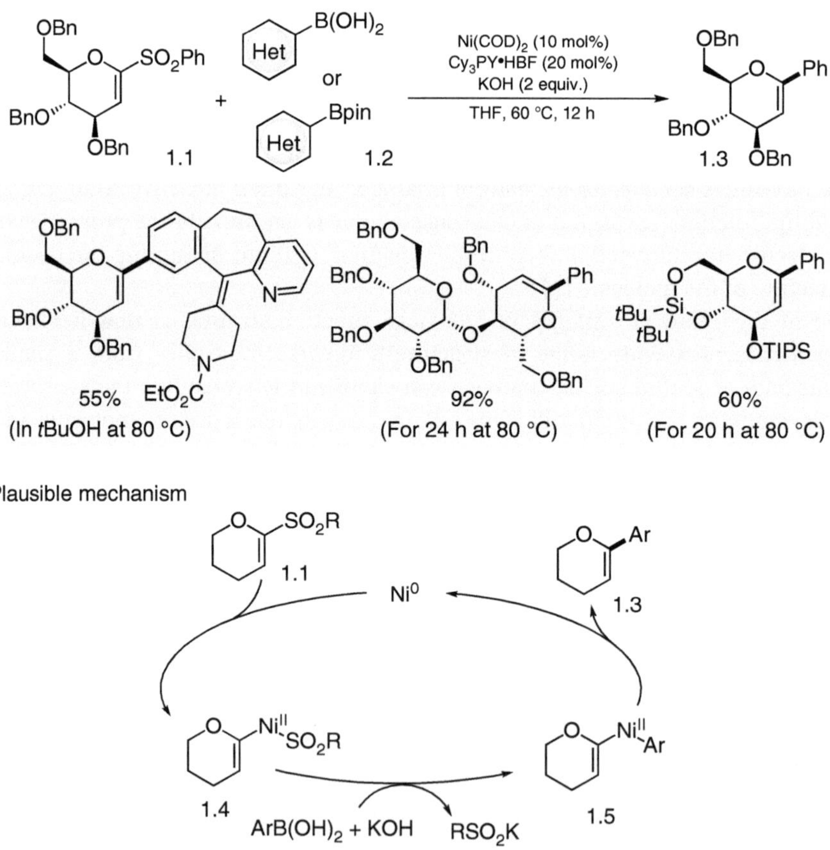

Scheme 8.1 Ni-catalyzed Suzuki–Miyaura cross-coupling of α-oxo-vinylsulfones.

approach has been applied in the synthesis of complex architectures. As they proposed, the Ni(0) catalytic species undergoes oxidative addition with suitable 1-sulfonyl glycal **1.1** through C—S bond activation. After transmetalation with suitable aryl boronic acid with the help of KOH, the Ni(II) intermediate **1.5** could undergo reductive elimination to produce the final coupling product **1.3**. Meanwhile, the Ni(0) species could be released to undergo another catalytic cycle. Remarkably, this mild approach was proceeded under mild conditions which tolerated a broad scope of heterocycles and functional groups.

Site-selective functionalization of complex aromatic rings remains an unsolved challenge in synthetic chemistry. Recently, the Ritter group has demonstrated a series of late-stage functionalization of arenes via the isolated aryl thianthrenium salts using thianthrene *S*-oxide (TTSO) [5]. In 2021, Ritter and coworkers presented an undirected *para*-selective two-step C—H alkylation of complex arenes useful for late-stage functionalization (Scheme 8.2) [6]. The robustness of this synthetic transformation has been applied in the late-stage functionalization of complex arenes to achieve the site-selective alkylation using a simple palladium catalyst. As they suggested, Pd(0) catalytic species undergo oxidative addition with aryl thianthrenium salt to form Pd(II) species **2.4**, which could undergo transmetalation with alkyl zinc reagents to generate Pd(II) intermediates **2.5**. After reductive elimination, the final carbon–carbon bond formation product **2.3** could be formed smoothly.

Tritium labeling is a critical tool for investigating the pharmacokinetic and pharmacodynamic properties of drugs, autoradiography, receptor binding, and receptor occupancy studies. Ritter and his team developed two elegant palladium-catalyzed coupling reactions to achieve the late-stage C—H tritiation of complex architectures successfully (Scheme 8.3) [7].

Meanwile, Wang and Peng also developed a transient mediator (TTSO) approach for *para*-selective functionalization of monosubstituted simple arenes [8]. Recently, Wang and Zhou reported a simplified procedure for site-selective borylation of simple arenes using this mediator approach with a Pd/phosphine catalyst (Scheme 8.4) [9]. This catalytic site-selective borylation has been applied in the late-stage functionalization of bioactive scaffolds.

In the same year, Wang and his team successfully achieved an efficient copper-catalyzed carboxylation of aryl thianthrenium salts with carbon dioxide (CO_2) at room temperature (Scheme 8.5) [10]. The substrate scope is very broad, and complex arenes could also be tolerated to deliver the corresponding carboxylic acids in good to excellent yields with only 1 atm of CO_2. This mild and practical synthetic approach is initiated by Cu(0) species, which are formed by CuCl with $ZnEt_2$ via releasing H_2 and C_2H_4. Next, Cu(0) reduces the aryl thianthrenium salts **5.1** to form aryl Cu(I) intermediate, which reacts with CO_2 to generate carboxylate Cu(I) intermediate **5.3**. With the help of $ZnEt_2$, the copper catalyst is released, and the corresponding zinc carboxylate **5.4** could be further protonated to afford the final carboxylic acid **5.2**.

In 2023, Feng, Xue, and Liu developed an elegant example regarding site-selective silylation via C—S bond cleavage (Scheme 8.6) [11]. Interestingly, selective cleavage of C(aryl)—S bonds can occur in the presence of iron catalyst and

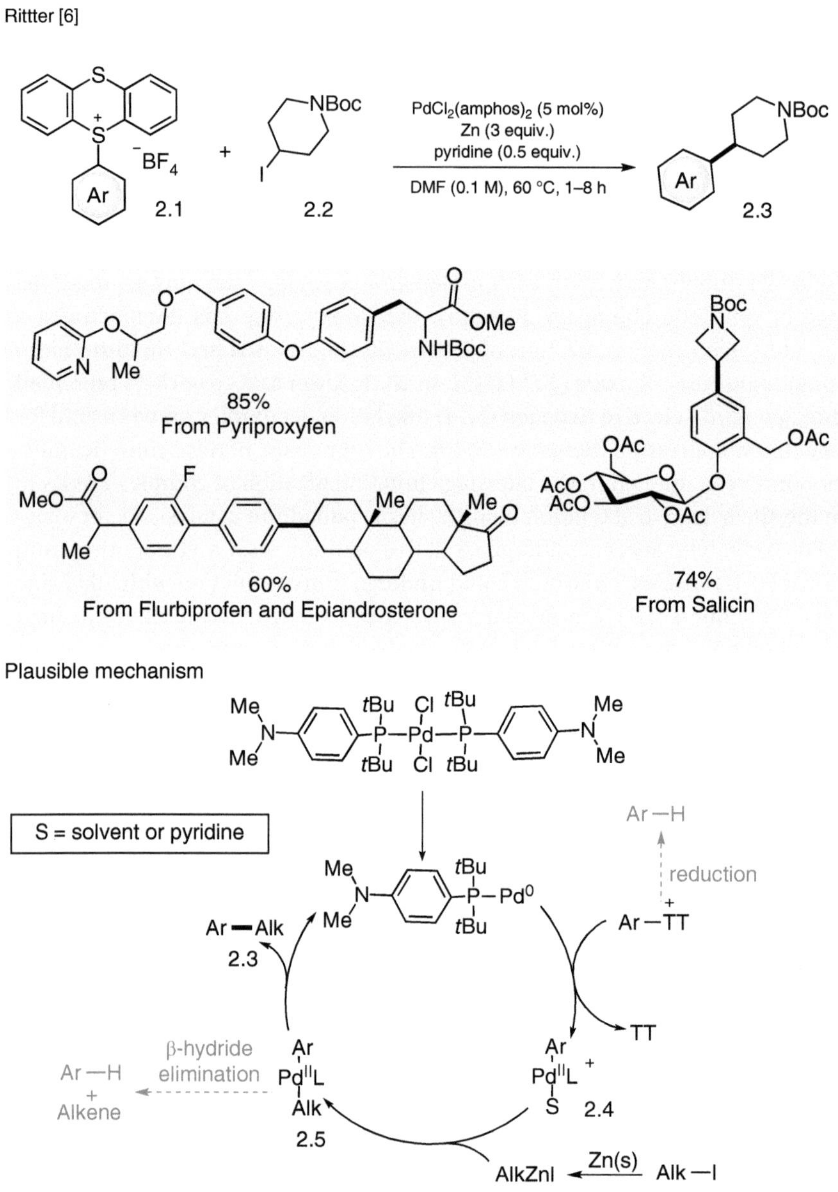

Scheme 8.2 Site-selective C–H alkylation of complex arenes.

β-diketimine ligand. Meanwhile, the cleavage of C(alkyl)—S bonds can be achieved by *t*-BuONa without the use of transition metals. The corresponding silylated products could be formed efficiently in good to excellent yields with high site-selectivity. These two elegant synthetic approaches have been applied in the late-stage functionalization of complex architectures with moderate to good yields to produce the corresponding complex silylated products. As they proposed, iron-silyl species **6.5**

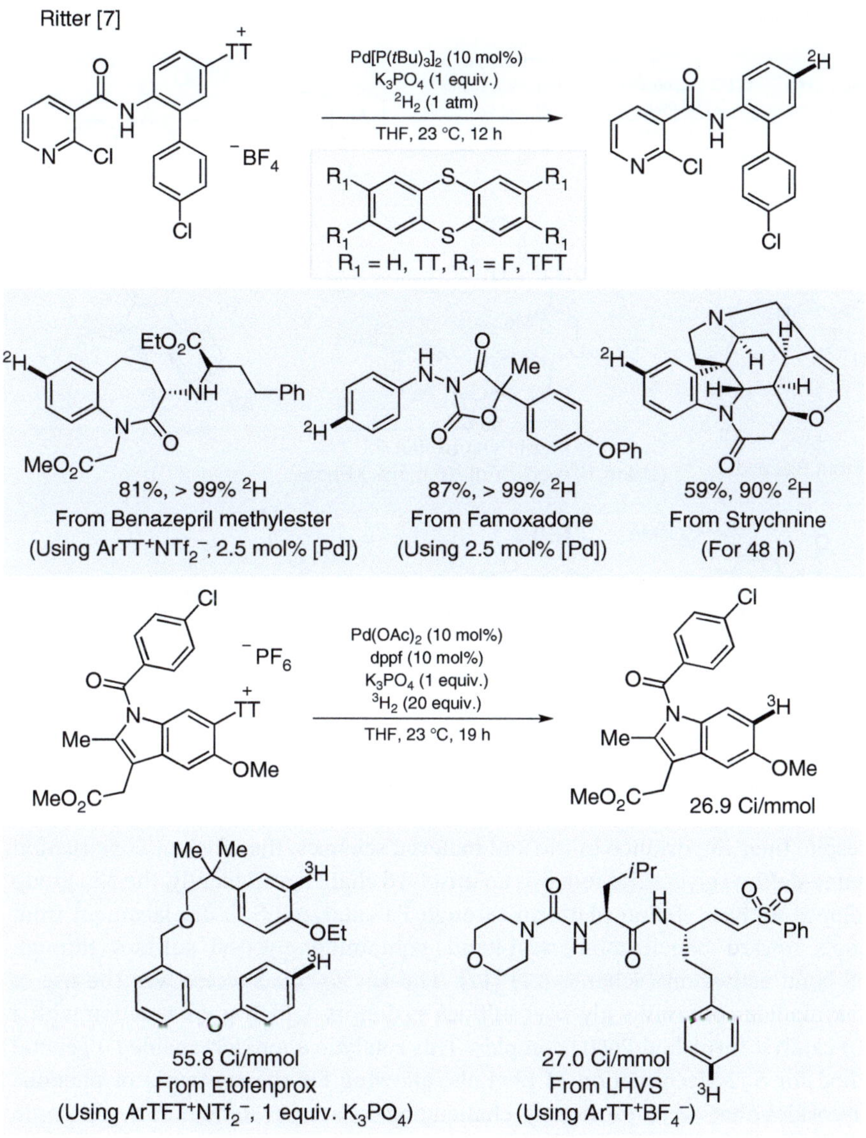

Scheme 8.3 Site-selective C–H tritiation of aryl thianthrenium salts.

is *in situ* generated, which then react with the aryl alkyl sulfide to afford the desired product **6.3** through a metathesis silylation process. Meanwhile, the iron complex **6.7** could react with Bpin-$SiEt_3$ and *t*-BuONa to regenerate iron catalytic species **6.8**. In the C(alkyl)—S bond silylation reaction, silylborane reagent could be activated by using *t*-BuONa to yield the complex **6.9**, which could be then triggered by the additional *t*-BuONa to generate the silyl radical through a radical mechanism. The newly formed silyl radical could react with aryl alkyl sulfide to produce the alkylsilicane products **6.4**.

Wang and Zhou [9]

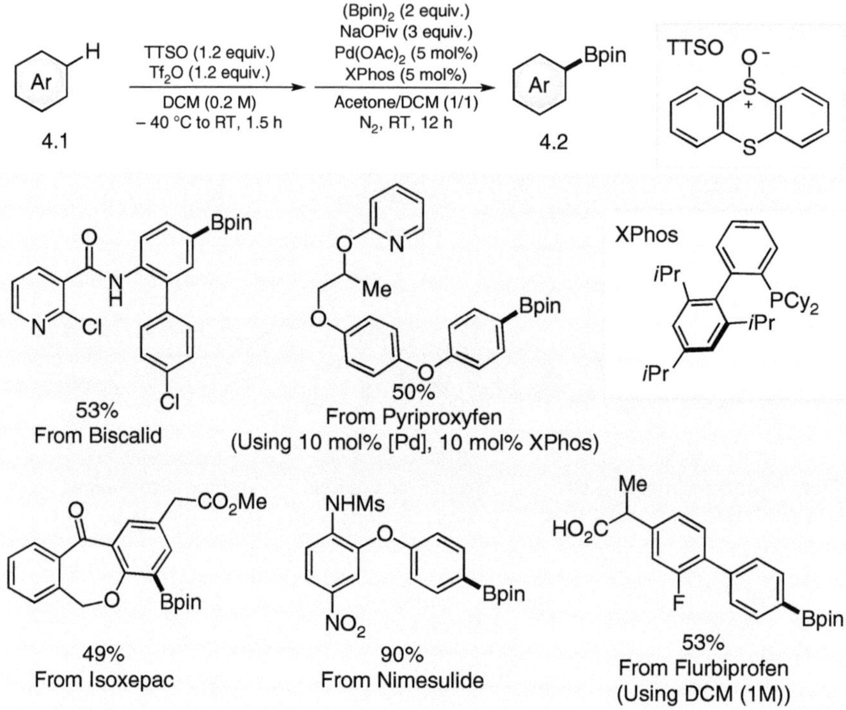

Scheme 8.4 Pd-catalyzed site-selective borylation of simple arenes via thianthrenation.

Despite their importance in life and material sciences, the efficient construction of stereo-defined glycosides remains an unsolved challenge. Recently, the Niu group developed a glycosylation platform through Pd-catalyzed S_N2 displacement from phenols toward bench-stable, aryl-iodide-containing glycosyl sulfides through C—S bond activation (Scheme 8.7) [12]. The key to this success was the use of the aryl iodide-containing glycosyl sulfides as donors, which upon reaction with a Pd(0)-catalyst furnished Pd(II) complex. This catalytic approach enabled a general method for S_N2-glycosylation of phenols, allowing for the synthesis of phenolic *O*-glycosides that were previously challenging to access. Remarkably, this mild method has been applied in the late-stage modification of complex architectures with good to excellent yields and excellent regioselectivity.

8.3 Late-Stage Functionalization by C—S Bond Activation under Electrochemical Condition

Electrochemistry has appeared as an emerging synthetic approach to construct carbon–carbon and carbon–heteroatom bond formation. Malins and Connal successfully developed two programmable modification of peptides under tunable electrochemical oxidation (Scheme 8.8) [13]. Interestingly, the most electron-rich

Wang [10]

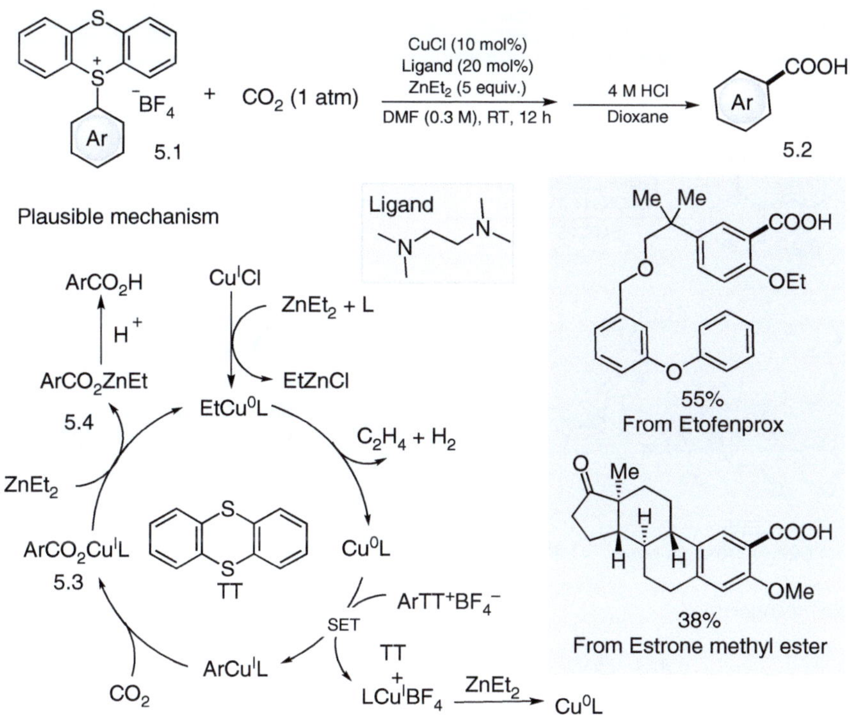

Scheme 8.5 Copper-catalyzed carboxylation of aryl thianthrenium salts with CO_2.

auxiliary, bearing a low oxidation potential, is selectively oxidized, while the more electron-deficient group, displaying a higher oxidation potential, remains intact. Thus, the authors could achieve the site-selective and tunable electrochemical modification of a peptide bearing two discretely oxidizable sites as shown in Scheme 8.8.

In 2023, Cao, Kong, and their team developed a practical fluorosulfonylation of both aryl and alkyl thianthrenium salts under electrochemical conditions (Scheme 8.9) [14]. This simple approach does not need external redox reagents or transition metal catalysts. Additionally, this electrochemical method has also been applied in the late-stage modification of complex drugs in good to excellent yields with complete site-selectivity.

8.4 Late-Stage Functionalization by C—S Bond Activation through Photoredox Chemistry

Independently, Procter group also discovered site-selective formal C—H/C—H coupling of arenes by interrupted Pummerer activation using commercial dibenzothiophene *S*-oxide (DBTSO) as the process mediator (Scheme 8.10) [15]. Based on

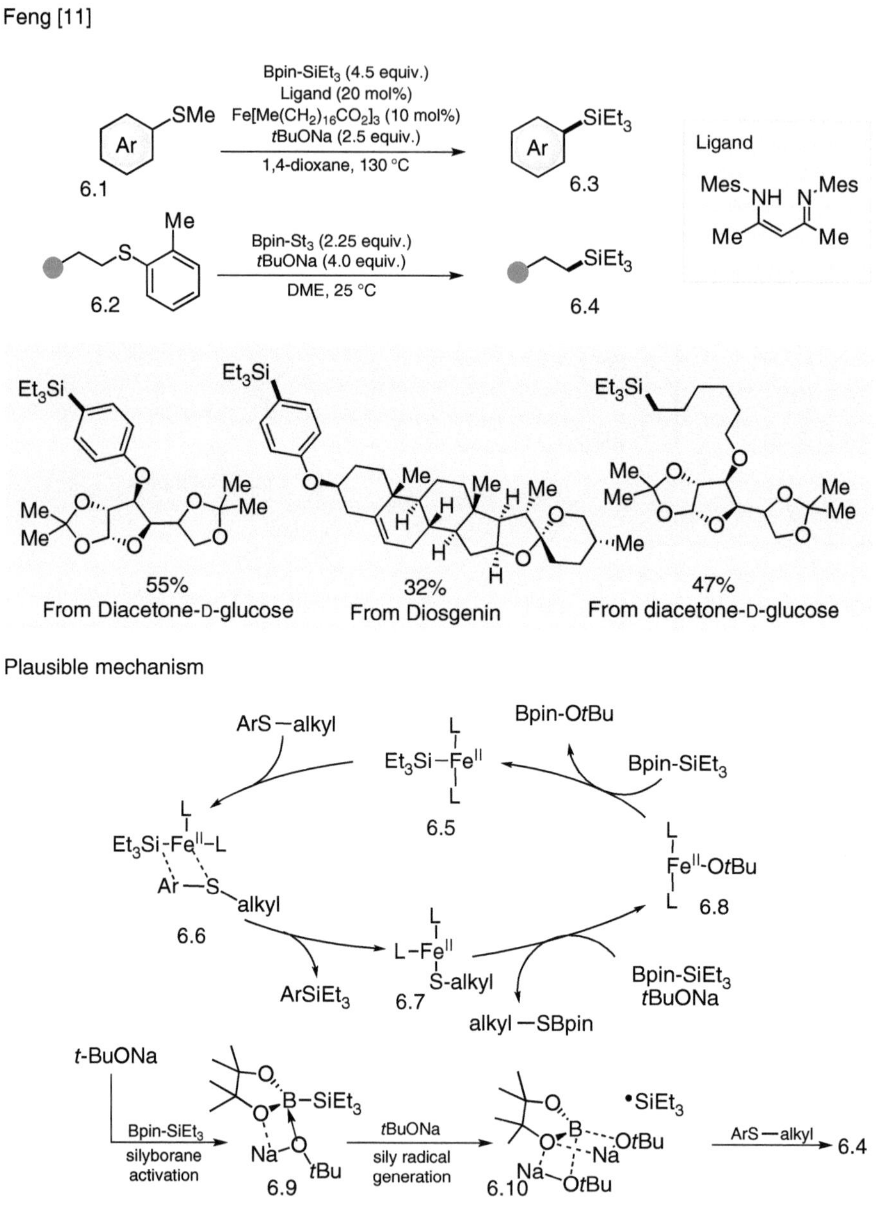

Scheme 8.6 Site-selective silylation via C—S bond cleavage.

their long interest in interrupted Pummerer reaction [16], they designed an innovative conceptual design to achieve the formal C—H/C—H coupling of arenes and successfully utilized this elegant synthetic approach in the synthesis of complex drugs and natural product-based architectures. As they proposed, the arenes could be activated to form the sulfonium intermediate **10.1**, which could be reduced to form the aryl radical **10.3** with suitable photo-excited organophotocatalyst. The aryl

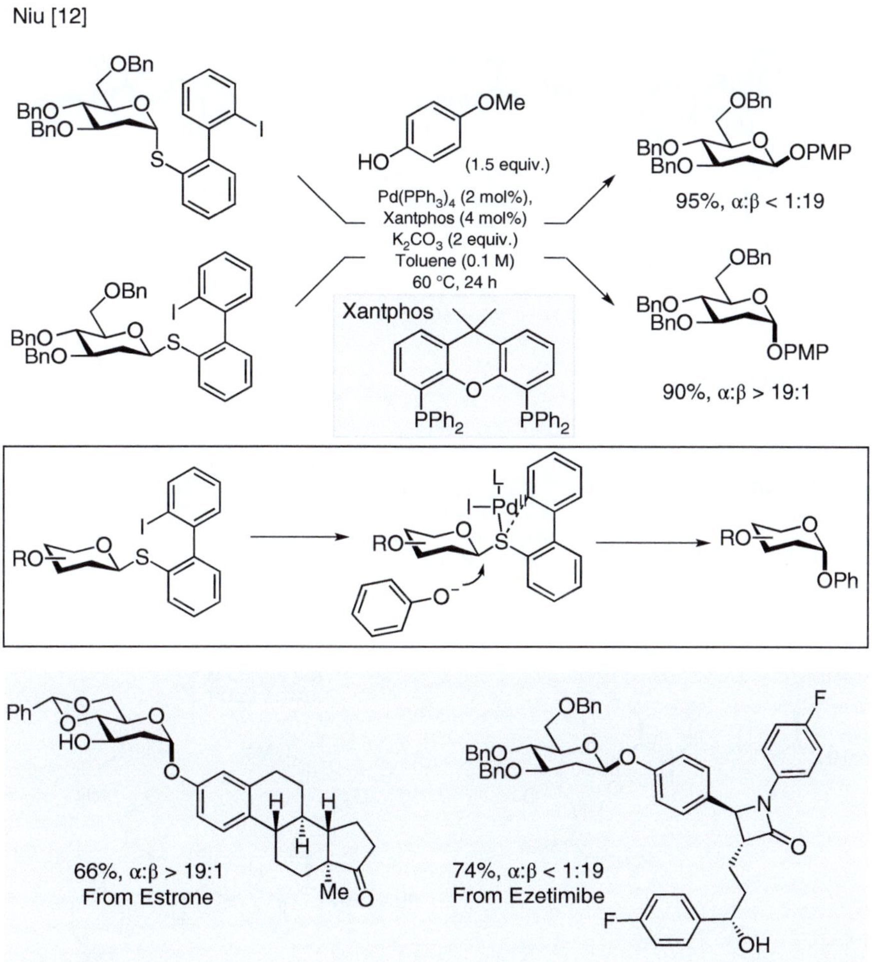

Scheme 8.7 Palladium catalysis enables cross-coupling-like S_N2-glycosylation of phenols.

radical **10.3** could be trapped by excess pyrrole to form the intermediate **10.4**, which could be oxidized to form the carbon cation intermediate **10.5** using the reduced organophotocatalyst. The final coupling product **10.2** could be formed efficiently after deprotonation.

Meanwhile, Wang and Peng developed a transient mediator approach that enabled *para*-selective functionalization of monosubstituted benzene derivatives through *in situ* generation of versatile sulfonium salts, which could be simply converted to versatile *para*-borylation of various monosubstituted simple aromatic rings under photoredox conditions (Scheme 8.11, upper) [8]. In 2021, Shi, Lu, and their team overcame the challenge regarding the generation of non-stabilized alkyl radicals from sulfonium salts and developed a mild synthetic approach to generate non-stabilized alkyl radicals under mild photoredox conditions to convert to alkylboronates in good to excellent yields (Scheme 8.11, down) [17].

Malins and Connal [13]

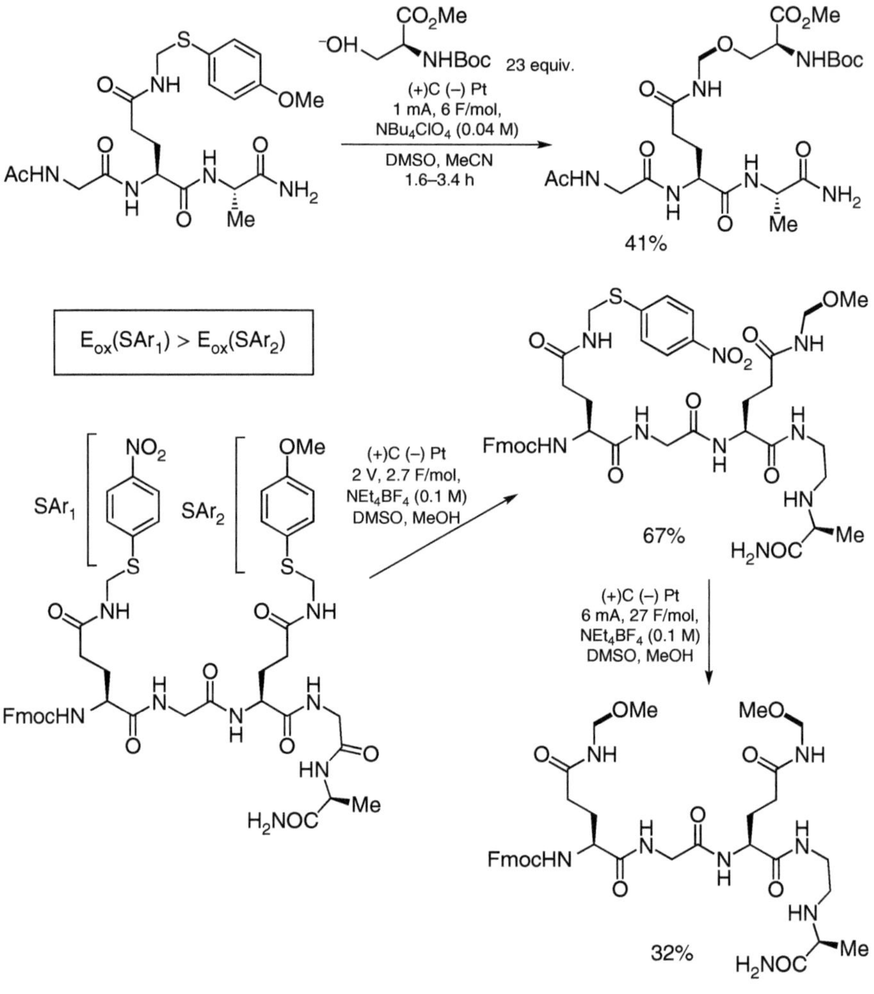

Scheme 8.8 Tunable electrochemical peptide modifications via C–S bond cleavage.

In 2022, Molander and his team developed a mild approach to build C—S bonds between thianthrenium salts and thiols through electron donor–acceptor (EDA) complex photoactivation strategy (Scheme 8.12) [18]. Interestingly, the complex sulfides could be produced in good to excellent yields. Additionally, this mild photocatalytic approach does not require additional photocatalysts, and the desired EDA complex could be formed with simple potassium carbonates.

In the same year, Niu and Houk developed an elegant example regarding halogen-bond-assisted radical activation of glycosyl donors to achieve the stereoconvergent 1,2-*cis*-glycosylation (Scheme 8.13, top) [19]. As they proposed, sulfone **13.1** could be converted to glycosyl radical **13.2** with C_4F_9I under photochemical condition, which could be further converted to glycosyl iodide **13.3**. With the

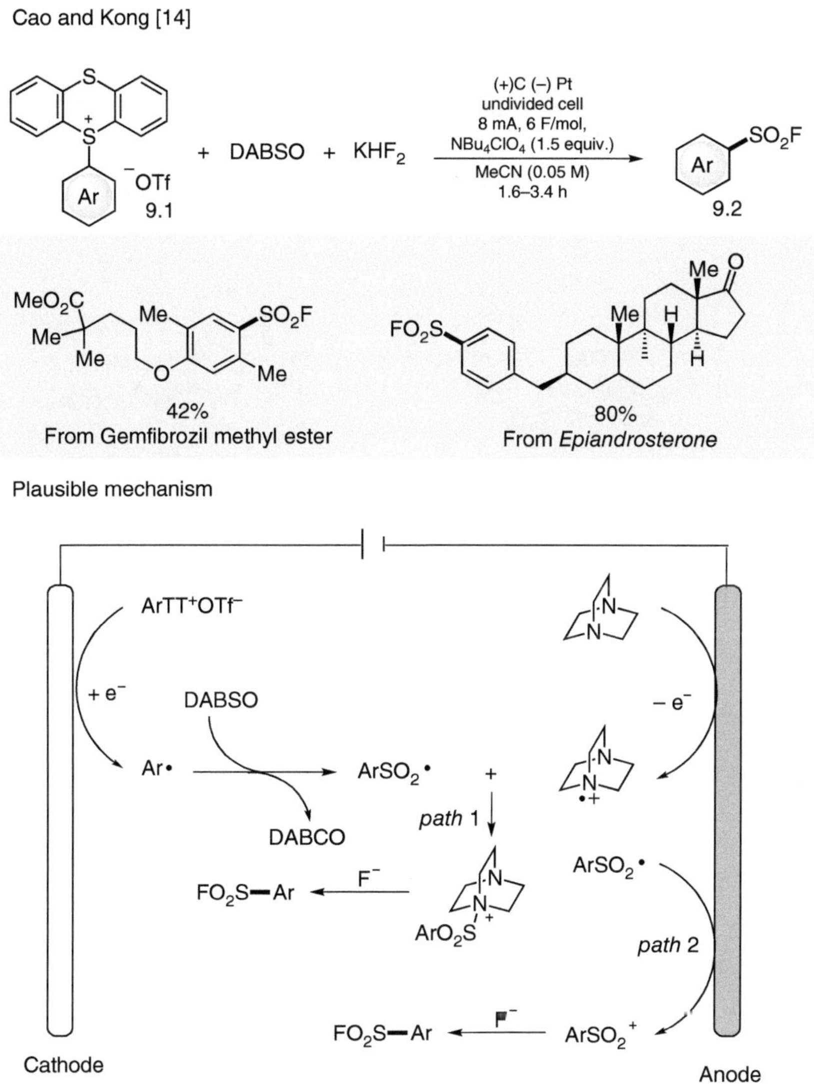

Scheme 8.9 Selective fluorosulfonylation of thianthrenium salts enabled by electrochemistry.

formation of hydrogen-bond complex, **13.4** renders the attack of the nucleophile pseudo-intramolecular, which preferentially afforded the 1,2-*cis* product **13.5**. Soon after, Niu and Wu achieved site-selective glycosylation between unprotected or minimally protected donor and acceptor sugars, producing 1,2-*cis*-*O*-glycosides (Scheme 8.13, bottom) [20].

In 2023, Procter and his team developed a general platform to selectively functionalize arene C—H bond via photo-excited EDA complex (Scheme 8.14) [21]. They successfully used a catalytic amount of triphenyl amine donors to activate the triarylsulfonium salts under visible light conditions. The corresponding aryl radical

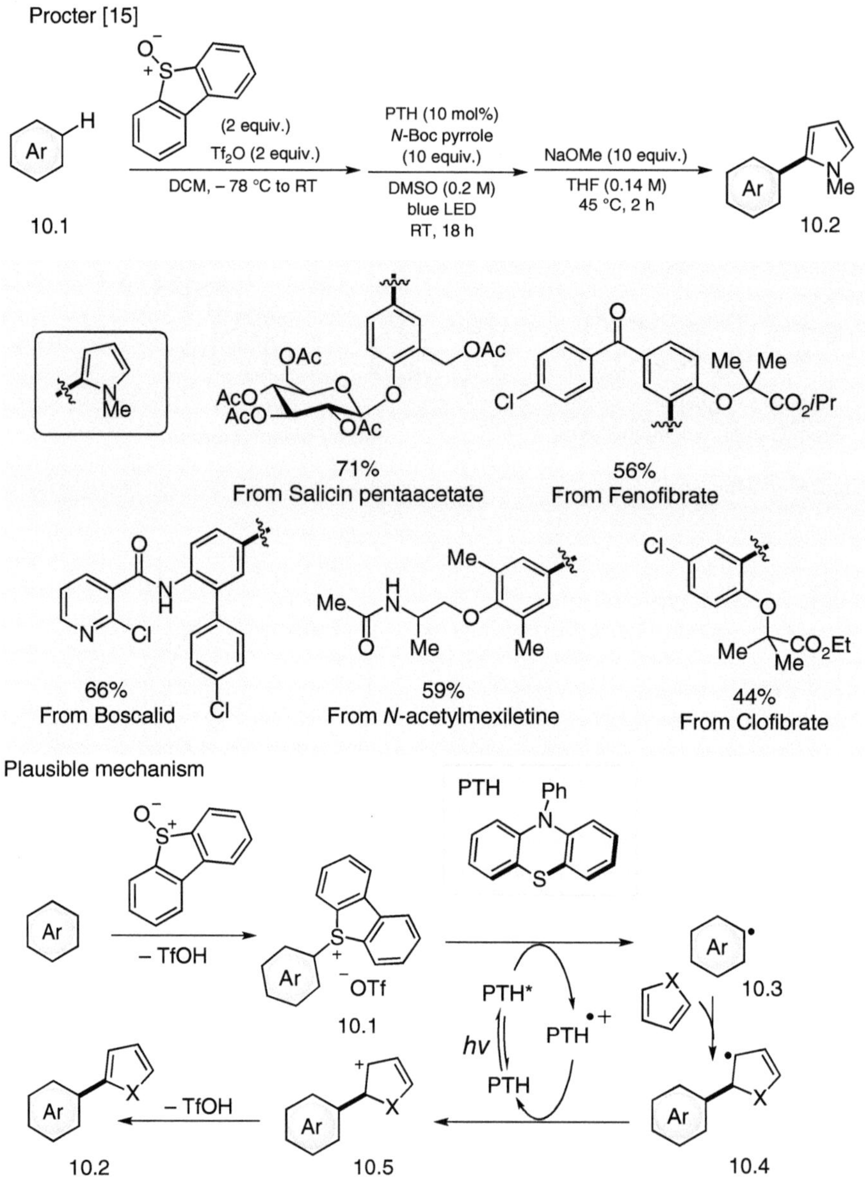

Scheme 8.10 Site-selective formal C–H/C–H coupling of arenes by interrupted Pummerer activation.

could then be trapped by 1-phenyl-1-trimethylsiloxyethylene or *tert*-butyl isocyanide to produce the corresponding ketones and aryl nitriles, respectively. This approach is the first to enable the generation of aryl radicals by means of EDA-complex photoactivation, independent of the electronics of the aromatic substrates, which has been applied in C—H alkylation and C—H cyanation successfully.

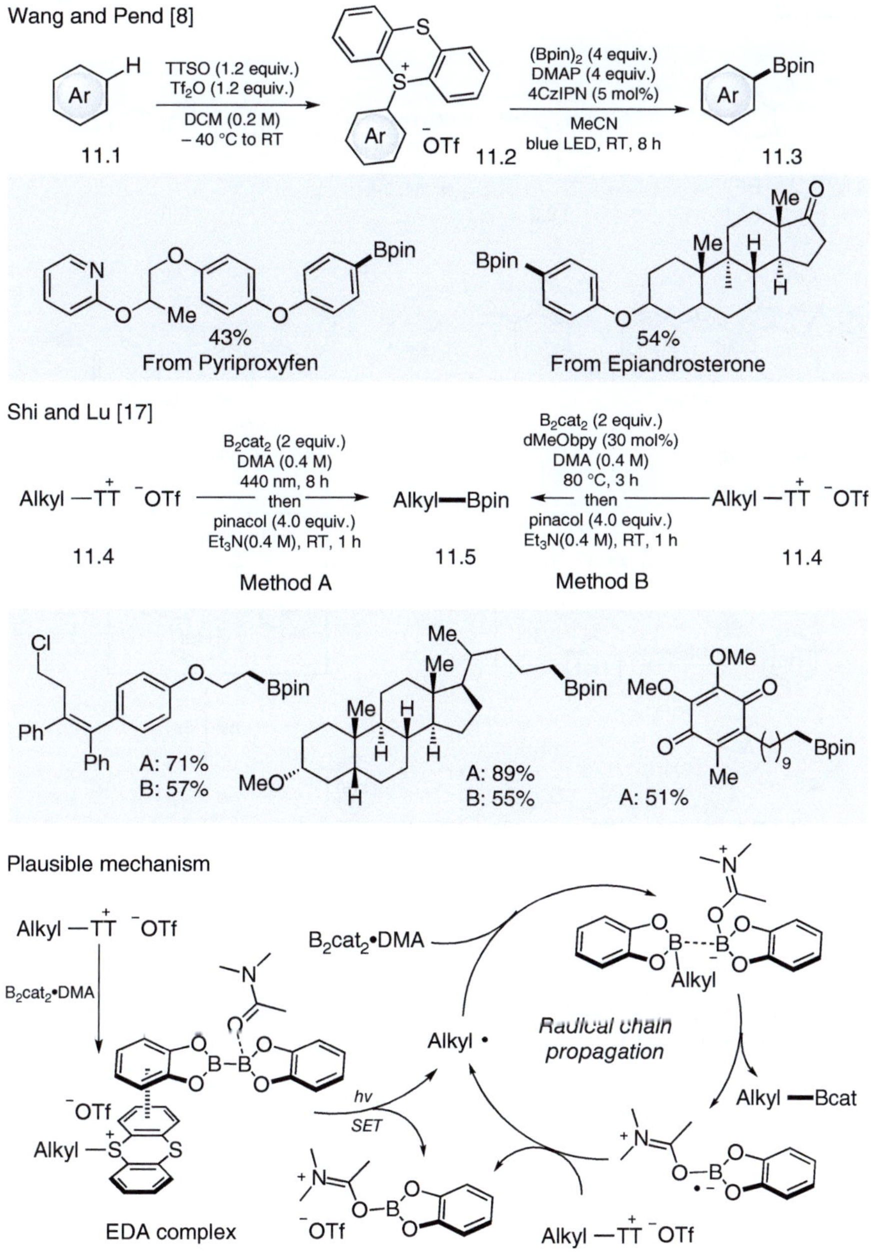

Scheme 8.11 *Para*-selective borylation of monosubstituted benzenes and C–B and C–C bond formation from *S*-(alkyl) thianthrenium salts under photoredox condition.

Molander [18]

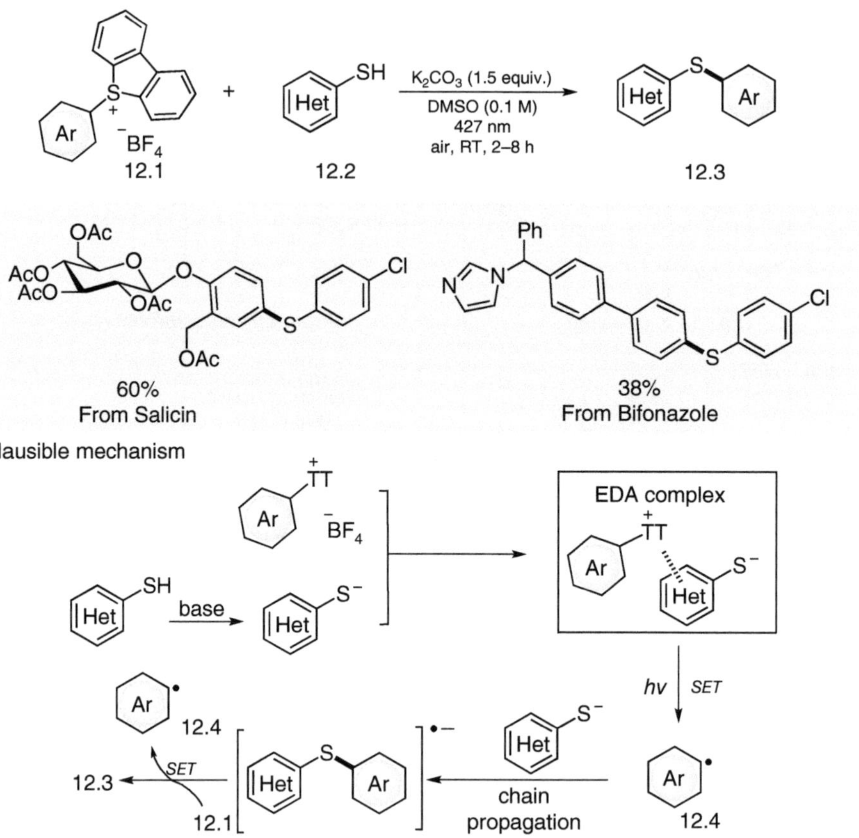

Scheme 8.12 C–S bond formation through (EDA) complex photoactivation strategy.

8.5 Late-Stage Functionalization by C–S Bond Activation Enabled by Metallaphotoredox Chemistry

The merger of photoredox and transition metal catalysis has become an attractive synthetic approach in the formation of carbon–carbon and carbon–heteroatom bonds [22]. Recently, Ritter and his team have employed their aryl sulfonium salts to achieve site-selective oxygenation [23], trifluoromethylation [24], and fluorination [25] enabled by dual photoredox and copper catalysis (Scheme 8.15). As they proposed, aryl radical **15.1** could be formed through SET process, which was then trapped by different nucleophiles and copper(II) species. After reductive elimination from the copper(III) intermediate, the final coupling product could be produced and meanwhile, the copper(I) catalysts could be regenerated.

Shi, Lu, and coworkers focused on the synthetic transformation of alkyl thianthrenium salts. They successfully utilized copper catalysis and applied it in the coupling reaction between alkyl thianthrenium salts and alkylation reagents successfully

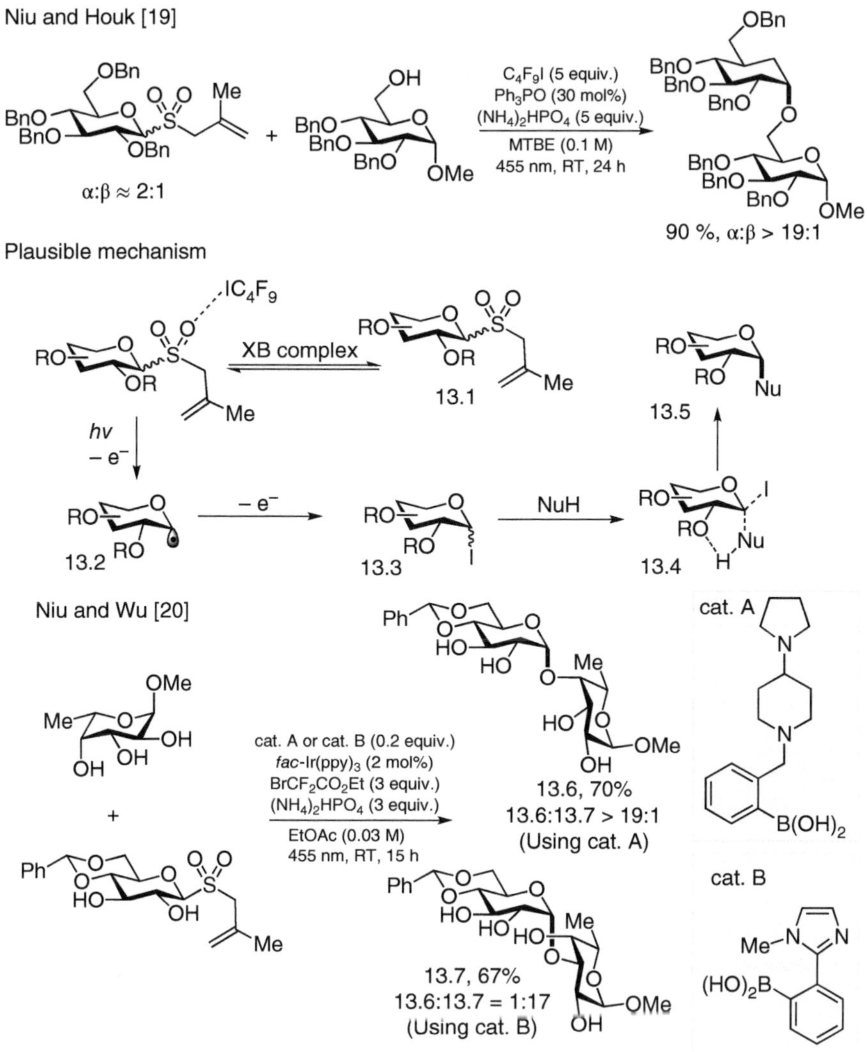

Scheme 8.13 Stereoconvergent 1,2-*cis*-glycosylation.

(Scheme 8.16, top) [26]. Alternatively, Wang, Zhou, and Wu successfully achieved photoinduced thiocyanation and selenocyanation of arylthianthrenium (ArTT) salts (Scheme 8.16, bottom) [27].

Very recently, Ritter and his team successfully achieved azidoarylation of alkenes *via* photoinduced copper catalysis (Scheme 8.17) [28]. The substrate scope is relatively broad, and various natural product-based complex architectures could be tolerated in good to excellent yields. As they proposed, the copper catalysis could be photo-excited to more reducing copper catalysis, which could reduce the ArTT salts to generate aryl radical intermediates. After trapping by activated alkene motifs, the electrophilic radical **17.4** could react with copper azide intermediates to produce the three-component coupling products smoothly.

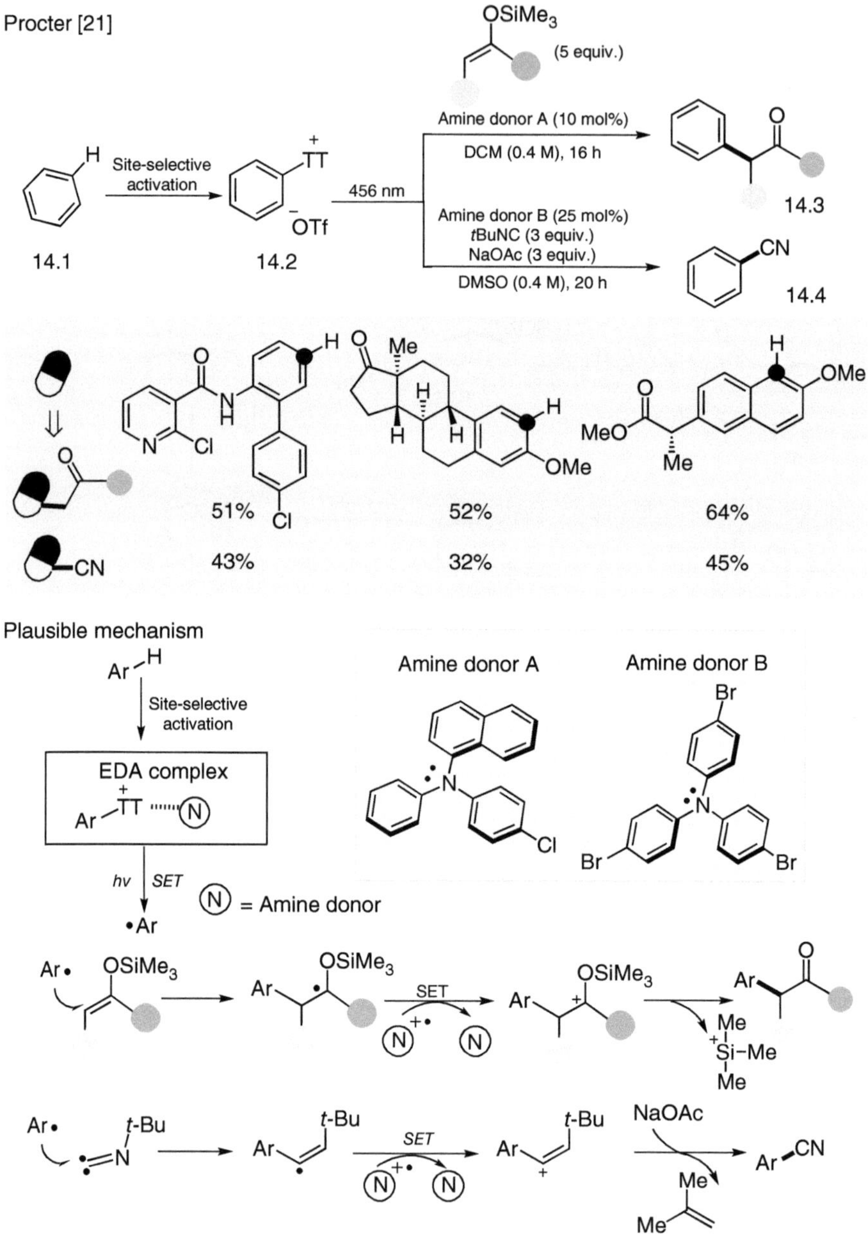

Scheme 8.14 A general arene C–H functionalization strategy via photo-excited EDA complexes.

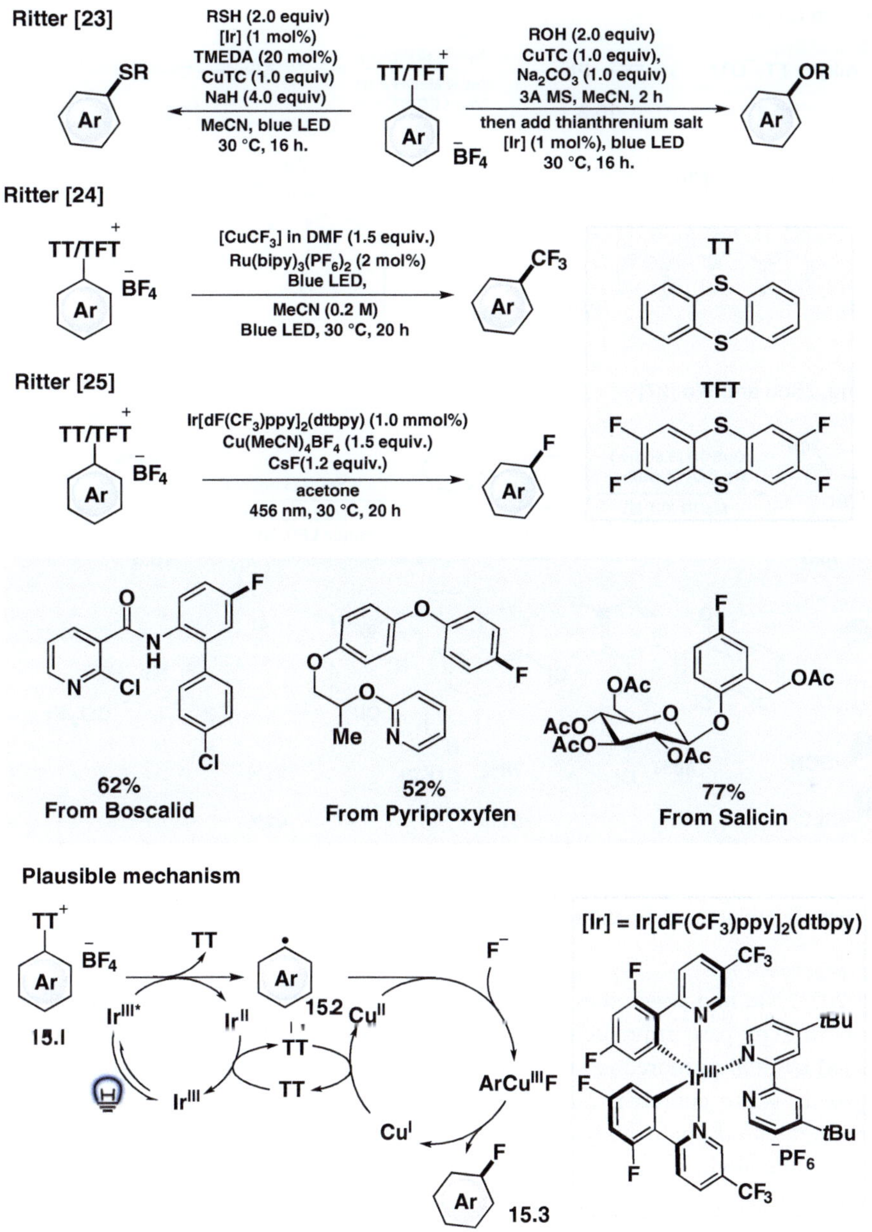

Scheme 8.15 Site-selective late-stage oxygenation, trifluoromethylation, and fluorination enabled by dual photoredox and copper catalysis.

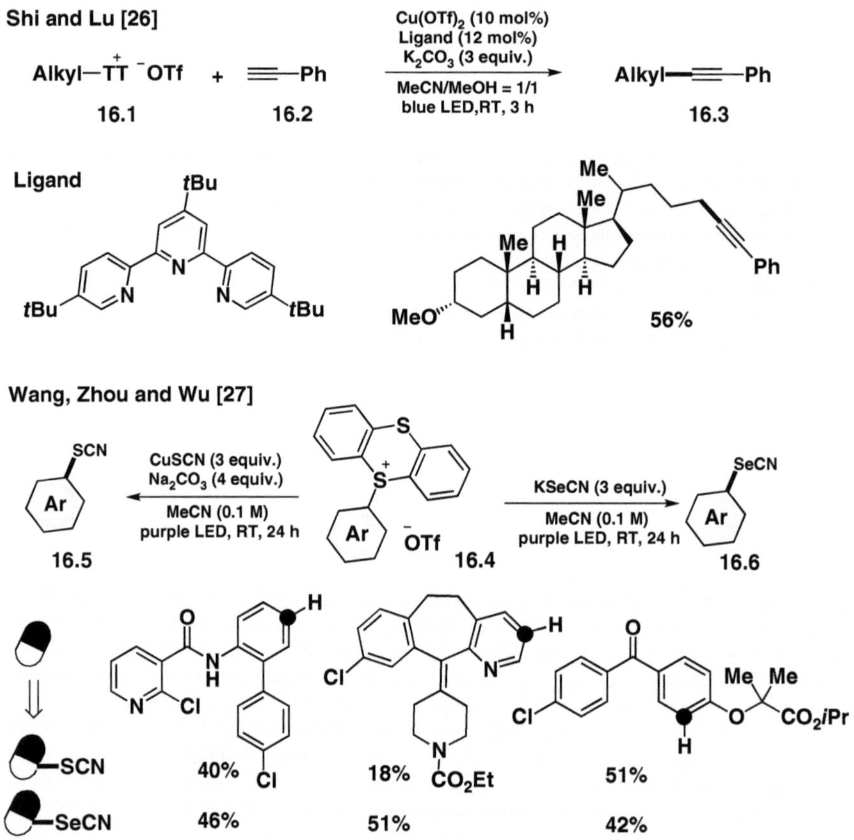

Scheme 8.16 Copper-catalyzed alkyl thianthrenium salts with alkylation reagents and photoinduced thiocyanation and selenocyanation of arylthianthrenium salts.

In 2023, Niu and coworkers reported a direct and stereoselective cross-coupling approach to prepare unprotected aryl C-glycosides from readily available materials enabled by dual photoredox and nickel catalysis (Scheme 8.18, top) [29]. As they proposed, sulfur-centered radical **18.2** could be formed with photo-excited ruthenium-based photocatalyst, which then added to alkene substrate **18.1**. After releasing the allylic motif and SO_2, the glycosyl radical **18.5** could react with aryl halide with the aid of photocatalysis to produce the coupling product **18.7**. In the same year, Ritter and coworkers successfully achieved the carbon–heteroatom bond coupling with electron-rich aryls with nickel catalysis under visible light condition (Scheme 8.18, bottom) [30]. They successfully used ArTT salts to overcome the limitation regarding electron-rich aryl electrophiles and successfully demonstrated nickel-catalyzed C–heteroatom bond-forming reaction, including amination, oxygenation, sulfuration, and halogenation.

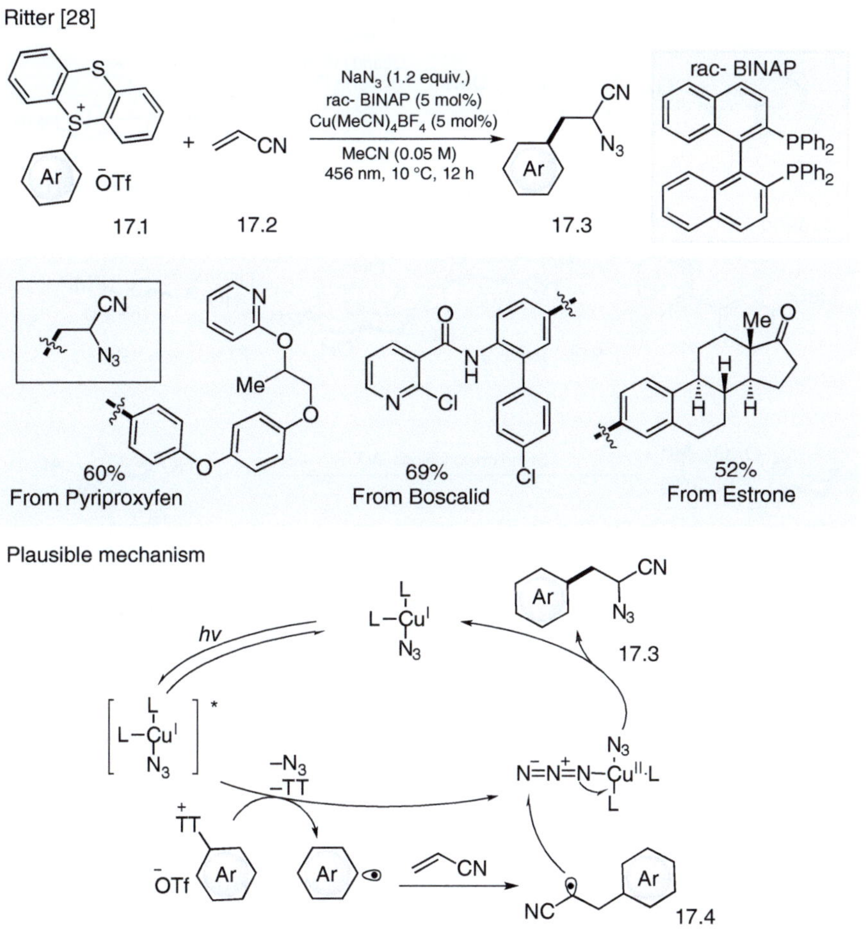

Scheme 8.17 Photoinduced copper-catalyzed late-stage azidoarylation of alkenes via arylthianthrenium salts.

8.6 Late-Stage Functionalization by C—S Bond Activation Enabled by Traditional Radical Approaches

In 2021, Ritter and coworkers demonstrated an excellent example regarding late-stage heteroarylation of hetero(aryl)sulfonium salts enabled by α-amino alkyl radicals under visible light conditions (Scheme 8.19) [31]. This mild synthetic approach did not require photoredox catalysis and visible light. The α-amino alkyl radicals could be formed through sequential single-electron oxidation and deprotonation of an amine, which could then reduce the thianthrenium salts to produce the aryl radical. After trapped by heteroaryl, the two different aromatic rings could be coupled efficiently.

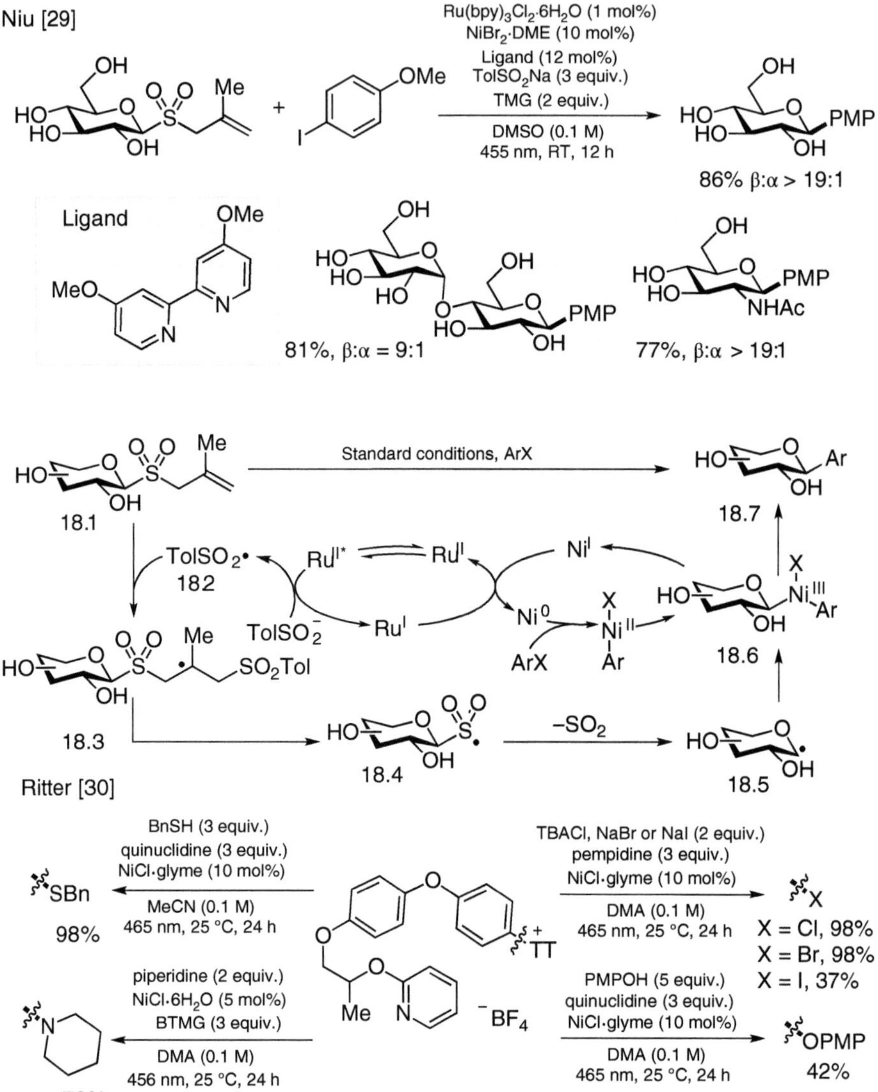

Scheme 8.18 Unprotected aryl C-glycosides and carbon–heteroatom bond coupling with electron-rich aryls.

Very recently, Niu, Wang, and coworkers reported a simple acidic condition to couple the glycosyl sulfinates with pyridine motifs through Minisci-type process successfully (Scheme 8.20) [32]. The glycosyl radicals **20.4** were generated under acidic conditions, which could be then trapped by protonated pyridines **20.5**. Re-aromatization of **20.6** with the assistance of *t*BuOOH, giving *tert*-butoxyl radical, initiates the next round of the process; meanwhile, the coupled product **20.7** could be formed smoothly.

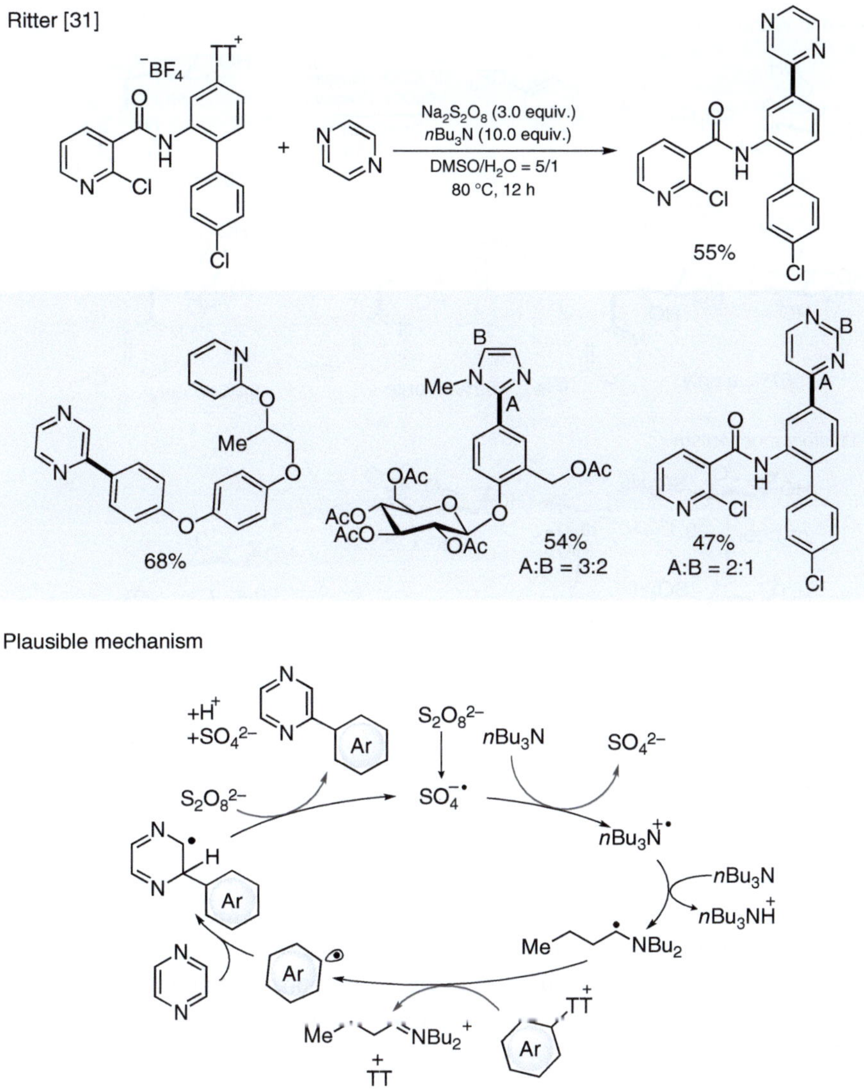

Scheme 8.19 Late-stage heteroarylation of hetero(aryl)sulfonium salts activated by α-amino alkyl radicals.

8.7 Summary and Conclusions

Recent advancements in late-stage functionalization via C—S bond activation have underscored the versatility and power of various catalytic strategies. Transition-metal catalysis, electrosynthesis, photoredox catalysis, metallaphotoredox catalysis, and traditional radical methods have all emerged as potent tools for selectively activating carbon–sulfur bonds in complex molecules and drug derivatives.

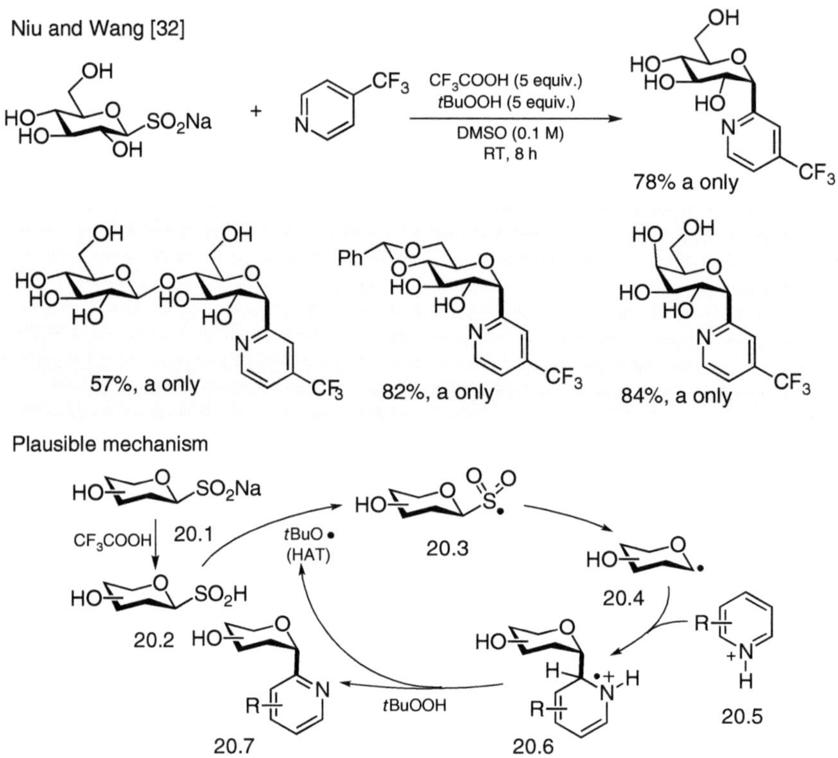

Scheme 8.20 Installation of glycosyl units onto pyridine rings directly by the Minisci reaction.

The ability to manipulate C—S bonds at a late stage in the synthesis of bioactive molecules and pharmaceuticals offers significant advantages over traditional methods, allowing for the direct installation of diverse functional groups with minimal disruption to the overall molecular structure. This approach not only streamlines the synthetic routes but also enhances the atom- and step economy, making it an attractive strategy for the modern synthetic chemists in both academia and industry.

References

1 Castellino, N.J., Montgomery, A.P., Danon, J.J., and Kassiou, M. (2023). Late-stage functionalization for improving drug-like molecular properties. *Chemical Reviews* 123: 8127–8153.

2 Guillemard, L., Kaplaneris, N., Ackermann, L., and Johansson, M.J. (2021). Late-stage C—H functionalization offers new opportunities in drug discovery. *Nature Reviews Chemistry* 5: 522–545.

3 Bellotti, P., Huang, H.-M., Faber, T., and Glorius, F. (2023). Photocatalytic late-stage C—H functionalization. *Chemical Reviews* 123: 4237–4352.

4 Gong, L., Sun, H.-B., Deng, L.-F. et al. (2019). Ni-catalyzed Suzuki–Miyaura cross-coupling of α-oxo-vinylsulfones to prepare C-aryl glycals and acyclic vinyl ethers. *Journal of the American Chemical Society* 141: 7680–7686.

5 Berger, F., Plutschack, M.B., Riegger, J. et al. (2019). Site-selective and versatile aromatic C—H functionalization by thianthrenation. *Nature* 567: 223–228.

6 Lansbergen, B., Granatino, P., and Ritter, T. (2021). Site-selective C—H alkylation of complex arenes by a two-step aryl thianthrenation-reductive alkylation sequence. *Journal of the American Chemical Society* 143: 7909–7914.

7 Zhao, D., Petzold, R., Yan, J. et al. (2021). Tritiation of aryl thianthrenium salts with a molecular palladium catalyst. *Nature* 600: 444–449.

8 Wu, J., Wang, Z., Chen, X.-Y. et al. (2020). Para-selective borylation of monosubstituted benzenes using a transient mediator. *Science China. Chemistry* 63: 336–340.

9 Chen, X., Huang, Y., Zhou, J., and Wang, P. (2020). Pd-catalyzed site-selective borylation of simple arenes *via* thianthrenation. *Chinese Journal of Chemistry* 38: 1269–1272.

10 Tang, S., Zhao, X., Yang, L. et al. (2022). Copper-catalyzed carboxylation of aryl thianthrenium salts with CO_2. *Angewandte Chemie, International Edition* 61: e202212975.

11 Chen, S., Guo, X., Hou, H. et al. (2023). Thioethers as dichotomous electrophiles for site-selective silylation via C—S bond cleavage. *Angewandte Chemie, International Edition* 62: e202303470.

12 Deng, L.-F., Wang, Y., Xu, S. et al. (2023). Palladium catalysis enables cross-coupling-like S_N2-glycosylation of phenols. *Science* 382: 928–935.

13 Karipal Padinjare Veedu, D., Connal, L.A., and Malins, L.R. (2023). Tunable electrochemical peptide modifications: unlocking new levels of orthogonality for side-chain functionalization. *Angewandte Chemie, International Edition* 62: e202215470.

14 Kong, X., Chen, Y., Liu, Q. et al. (2023). Selective fluorosulfonylation of thianthrenium salts enabled by electrochemistry. *Organic Letters* 25: 581–586.

15 Aukland, M.H., Šiaučiulis, M., West, A. et al. (2020). Metal free photoredox-catalysed formal C—H/C—H coupling of arenes enabled by interrupted Pummerer activation. *Nature Catalysis* 3: 163–169.

16 Pulis, A.P. and Procter, D.J. (2016). C—H coupling reactions directed by sulfoxides: teaching an old functional group new tricks. *Angewandte Chemie, International Edition* 55: 9842–9860.

17 Chen, C., Wang, Z.-J., Lu, H. et al. (2021). Generation of non-stabilized alkyl radicals from thianthrenium salts for C—B and C—C bond formation. *Nature Communications* 12: 4526.

18 Cabrera-Afonso, M.J., Granados, A., and Molander, G.A. (2022). Sustainable thioetherification via electron donor–acceptor photoactivation using thianthrenium salts. *Angewandte Chemie, International Edition* 61: e202202706.

19 Zhang, C., Zuo, H., Lee, G.Y. et al. (2022). Halogen-bond-assisted radical activation of glycosyl donors enables mild and stereoconvergent 1,2-*cis*-glycosylation. *Nature Chemistry* 14: 686–694.

20 Dang, Q.-D., Deng, Y.-H., Sun, T.-Y. et al. (2024). Catalytic glycosylation for minimally protected donors and acceptors. *Nature* 632: 313–319.

21 Dewanji, A., van Dalsen, L., Rossi-Ashton, J.A. et al. (2023). A general arene C—H functionalization strategy via electron donor–acceptor complex photoactivation. *Nature Chemistry* 15: 43–52.

22 Chan, A.Y., Perry, I.B., Bissonnette, N.B. et al. (2022). Metallaphotoredox: the merger of photoredox and transition metal catalysis. *Chemical Reviews* 122: 1485–1542.

23 Sang, R., Korkis, S.E., Su, W. et al. (2019). Site-selective C—H oxygenation via aryl sulfonium salts. *Angewandte Chemie, International Edition* 58: 16161–16166.

24 Ye, F., Berger, F., Jia, H. et al. (2019). Aryl sulfonium salts for site-selective late-stage trifluoromethylation. *Angewandte Chemie, International Edition* 58: 14615–14619.

25 Li, J., Chen, J., Sang, R. et al. (2020). Photoredox catalysis with aryl sulfonium salts enables site-selective late-stage fluorination. *Nature Chemistry* 12: 56–62.

26 Chen, C., Wang, M., Lu, H. et al. (2021). Enabling the use of alkyl thianthrenium salts in cross-coupling reactions by copper catalysis. *Angewandte Chemie, International Edition* 60: 21756–21760.

27 Chen, X., Kuang, X., Wu, Y. et al. (2023). Thianthrene radical cation as a transient SET mediator: photoinduced thiocyanation and selenocyanation of arylthianthrenium salts. *Chinese Journal of Chemistry* 41: 1979–1986.

28 Cai, Y., Chatterjee, S., and Ritter, T. (2023). Photoinduced copper-catalyzed late-stage azidoarylation of alkenes via arylthianthrenium salts. *Journal of the American Chemical Society* 145: 13542–13548.

29 Zhang, C., Xu, S.-Y., Zuo, H. et al. (2023). Direct synthesis of unprotected aryl C-glycosides by photoredox Ni-catalysed cross-coupling. *Nature Synthesis* 2: 251–260.

30 Ni, S., Halder, R., Ahmadli, D. et al. (2024). C–heteroatom coupling with electron-rich aryls enabled by nickel catalysis and light. *Nature Catalysis* 7: 733–741.

31 Alvarez, E.M., Karl, T., Berger, F. et al. (2021). Late-stage heteroarylation of hetero(aryl)sulfonium salts activated by A-amino alkyl radicals. *Angewandte Chemie, International Edition* 60: 13609–13613.

32 Xu, S., Zhang, W., Li, C. et al. (2023). Generation and use of glycosyl radicals under acidic conditions: glycosyl sulfinates as precursors. *Angewandte Chemie, International Edition* 62: e202218303.

Part III

Molecular/Stereochemical Editing

9

Molecular Editing-Based Strategies

Javid Rzayev, Illia Ruzhylo, Philippe Jubault, and Tatiana Besset

Organic Chemistry Department, Univ Rouen Normandie, Univ Caen Normandie, ENSICAEN, CNRS, Institut CARMeN (UMR 6064), INSA Rouen Normandie, 1 rue Tesniere 76821 Mont-Saint-Aignan, Rouen, F-76000, France

9.1 Introduction

The field of organic chemistry has been completely reshaped over the past decades. A significant revolution emerged with major breakthroughs in late-stage functionalization and molecular diversification within the realm of organic synthesis. Alongside the groundbreaking advances in peripheral C—H bond functionalization, recent years have seen a renewed interest in molecular editing [1–11]. The molecular editing strategy encompasses an array of site-selective and chemoselective transformations from skeletal rearrangements (contraction, expansion), atom swapping, to atom insertion/deletion starting with readily available feedstocks or occurring at the late stage of a multi-step synthesis. In this chapter, the main advances made in the molecular editing strategy are highlighted. It will provide an overview of this active research field by covering transformations based on (1) atom insertion, (2) atom deletion, (3) atom swapping, (4) functional group transposition, and (5) stereochemical editing (Figure 9.1). Though not intended to be exhaustive, the chapter aims to illustrate the synthetic potential of this approach by a judicious selection of relevant examples. Note that this approach has also been successfully applied to the synthesis of radiolabeled derivatives but is beyond the scope of the chapter [12–14].

9.2 Atom Insertion

9.2.1 Carbon Atom Insertion

One of the first examples of molecular editing by carbon atom insertion was discovered over 140 years ago, in 1881, by Ciamician and Dennstedt [15]. They reported the synthesis of 3-halopyridines by carbon atom insertion into the pyrrole core using the *in situ*-generated dichlorocarbene from chloroform under

Late-Stage Functionalization and Diversification in Organic Synthesis: Methods and Applications, First Edition. Edited by Tatiana Besset.

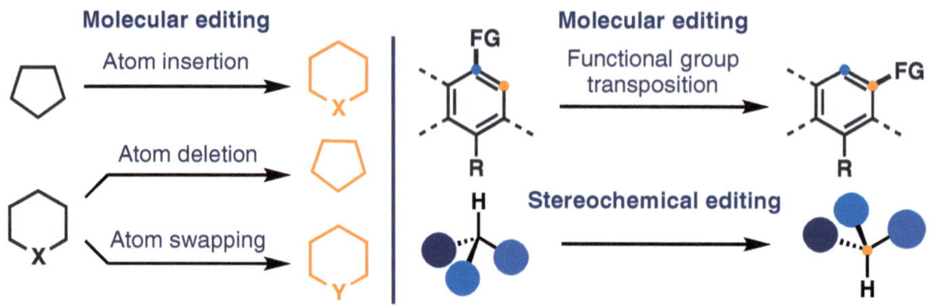

Figure 9.1 Overview of molecular editing-based strategies depicted in the chapter. FG = functional group.

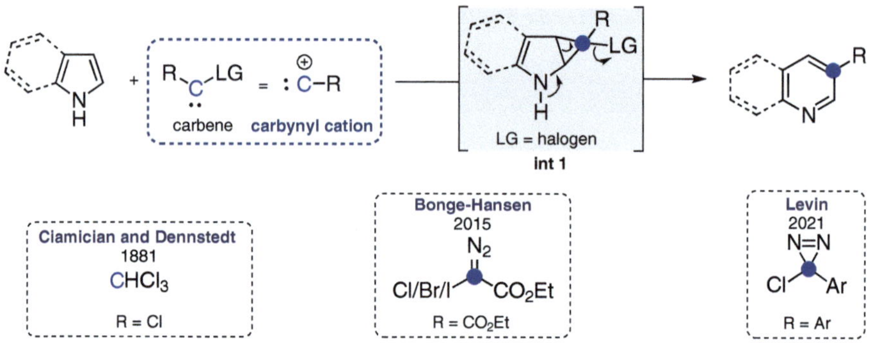

Scheme 9.1 Carbon atom insertion via Ciamician–Dennstedt rearrangement.

basic conditions (Scheme 9.1). The key to success for the desired transformation was the use of a reagent capable of forming a carbenic center bearing an appropriate leaving group (i.e., a carbynyl cation equivalent). In the case of chloroform, the unstable intermediate halogenocyclopropane **int 1** was formed, which would then rearrange into pyridine after elimination of HCl. This transformation is currently known as Ciamician–Dennstedt rearrangement. More than a century later, Bonge-Hansen's group revisited this approach and expanded it toward the formation of quinoline-3-carboxylates from indoles [16]. As a carbynyl cation equivalent, ethyl halodiazoacetates were employed, which in the presence of a dirhodium(II) catalyst would release N_2, resulting in the formation of reactive carbene species. In 2021, the use of α-chlorodiazirines as a carbynyl cation equivalent was reported by Levin's group for the generation of 3-aryl pyridines or quinolines from pyrroles or indoles, respectively, via a transition-metal free process [17]. Later, this approach was extended to carbon atom insertion into N—N bond of pyrazoles and indazoles, resulting in the formation of pyrimidines and quinazolines, respectively [18].

Recently, Glorius and coworkers designed the first photoredox ring expansion reaction using hypervalent iodonium diazo compounds as a carbynyl radical synthetic equivalent [19]. This approach allowed access to a large library of 2-substituted naphthalenes from indenes with moderate to high yields. It is worth mentioning that due to the presence of radical intermediates, the reaction mechanism differs from the classical Ciamician–Dennstedt rearrangement (Scheme 9.2).

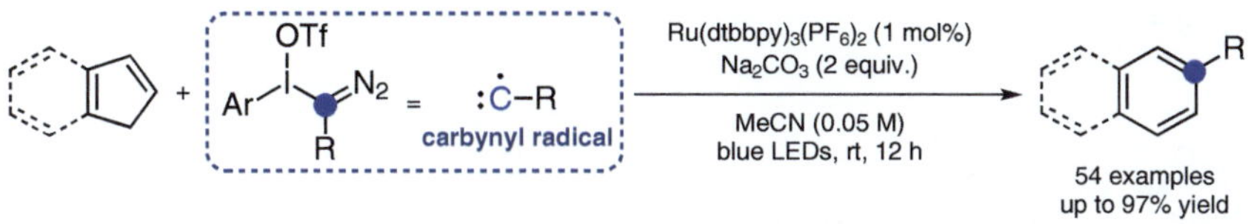

Scheme 9.2 Carbon atom insertion using carbynyl radical synthetic equivalent.

Another classical transformation is the Buchner ring expansion [20]. The reaction was first discovered in 1885 by Buchner and Curtius, allowing access to cycloheptatrienes from benzene derivatives under thermal or photochemical condition. The first step was the formation of a reactive carbene species from ethyl diazoacetate, capable of performing a cyclopropanation of an aromatic ring, leading to the **int 2**. Then, a 7-membered ring was formed via cyclopropane ring opening (Scheme 9.3).

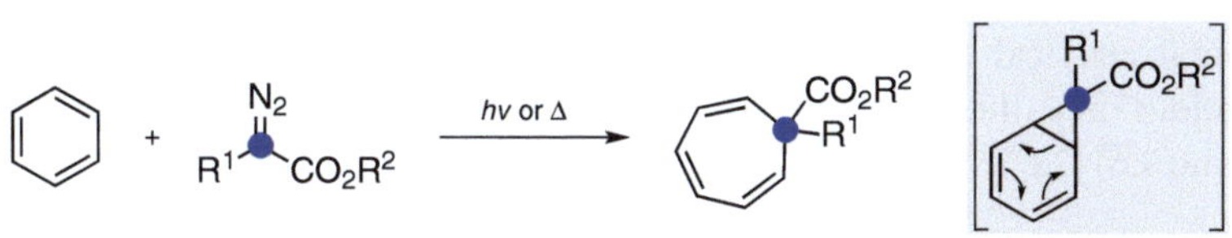

int 2

Scheme 9.3 Buchner ring expansion.

This approach has gained considerable attention over the years, and numerous variations of this strategy have been developed, differing primarily in the methods used to generate the reactive carbene species, which are necessary for the formation of intermediate **int 2**. Typically, the use of transition metal-based catalysts, such as rhodium [21], copper [22], silver [23], or even iron [24] complexes, was required to increase both reactivity and selectivity. Notably, this approach was not restricted to C—C bond insertion and was successfully applied to C—N [25], N—N [26], N—O [27], and N—S [28] bond insertion. To avoid the direct use of diazo compounds, methods based on their *in situ* generation from hydrazine derivatives have been developed, making the reaction safer (Scheme 9.4) [29].

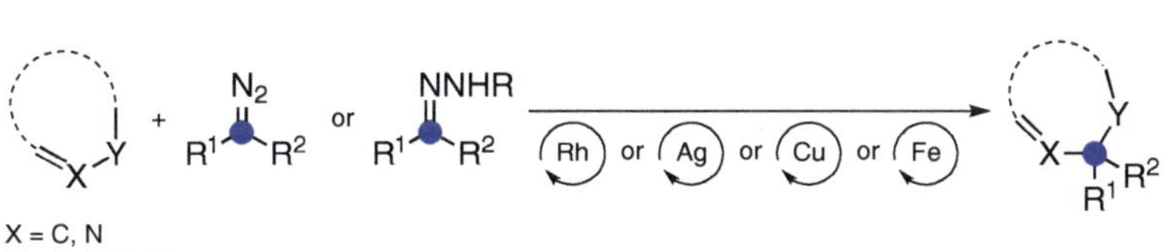

Scheme 9.4 Carbene insertion into X–Y bond.

In 2018, the groups of R.-S. Liu and Y. Liu independently demonstrated the potential of using ynamide as a carbene precursor for gold(I)-catalyzed carbon atom insertion into N—O bond of 1,2-benzisoxazoles [30, 31]. Interestingly, Y. Liu also showed that this transformation was also possible with propargyl esters (Scheme 9.5).

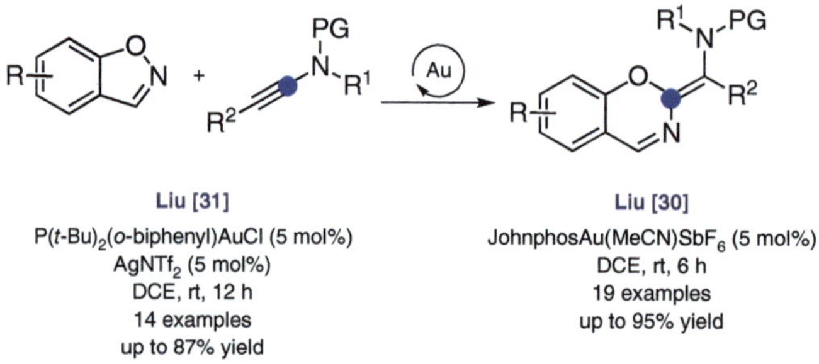

Scheme 9.5 Gold-catalyzed carbon atom insertion using ynamides as a carbene precursor.

9.2.2 Nitrogen Atom Insertion

There are several classical examples of N atom insertion into ketones to form substituted amides like the venerable Beckmann rearrangement and Schmidt reaction (Scheme 9.6) [32, 33].

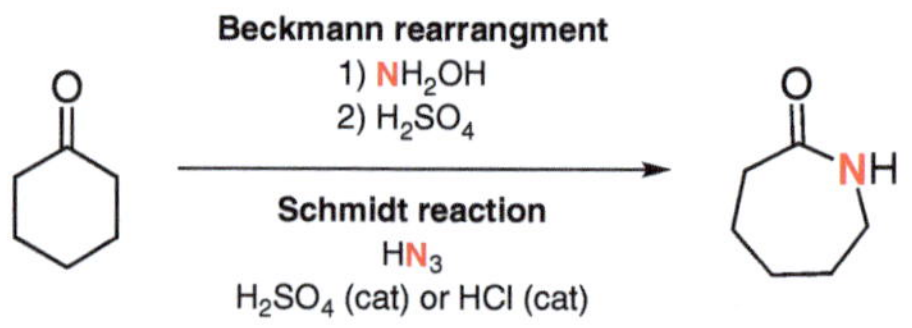

Scheme 9.6 Venerable Beckmann rearrangement and Schmidt reaction.

The photochemical preparation of 1,2-diazepines from 1-iminopyridinium ylides was first discovered by Streith and coworkers in 1968 [34]. Under UV irradiation (Scheme 9.7), pyridinium ylides undergo 1,3-dipolar ring closure to form a diaziridine **int 3**, followed by isomerization to afford 1,2-diazepines [35].

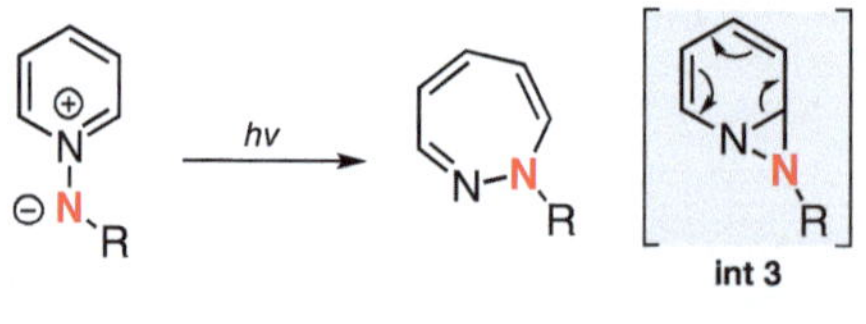

Scheme 9.7 Formation of 1,2-diazepines starting from pyridinium ylides.

Recently, the group of Ghiazza and Moreau revisited this approach and reported a way to access to 1,2-diazepines from pyridine derivatives via a one-pot procedure [36]. First, the 1-amino pyridinium salt (**int 4**), which was unreactive upon UV irradiation, was formed *in situ* from the corresponding pyridine (Scheme 9.8). Then, upon acylation, the reactive 1-iminopyridinium ylide **int 5** was produced, which under irradiation formed the desired 1,2-diazepine.

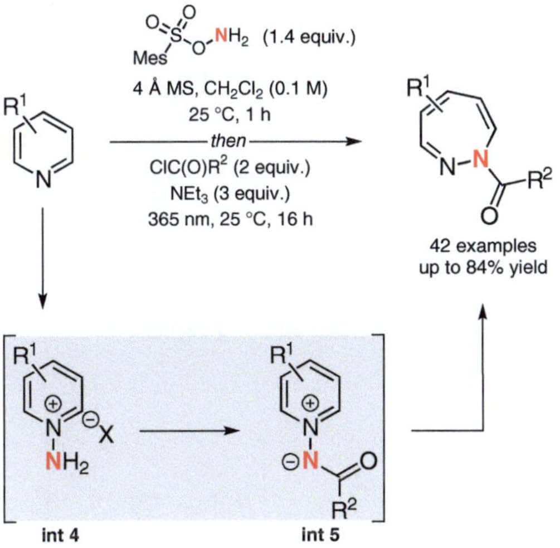

Scheme 9.8 Formation of 1,2-diazepines starting from pyridines.

The pioneering work on using nitrenes as N atom source in ring expansion reactions was reported by Kumar and coworkers in 1984 [37]. First, the reactive phthalimidonitrene species was *in situ* formed after the aziridination of the double bond of cyclopentadienes in the presence of $Pb(OAc)_4$. Then, upon treatment with sodium hydride, the desired pyridine product was obtained. Several years later, the substrate scope was expanded toward *N*-phenylsulfonylindole and unprotected indoles, resulting in the formation of quinazolines with up to 70% yields (Scheme 9.9) [38].

In 2022, the group of Morandi reported the one-step synthesis of quinazolines from silyl-protected indoles using the *in situ*-generated iodonitrene as an electrophilic nitrene equivalent (Scheme 9.10) [39]. First, aziridine (**int 6**) was formed via a (2+1) cycloaddition reaction, followed by iodobenzene elimination, leading to the desired ring-expansion product, namely, the quinazoline. One year later, it was discovered that in the presence of a strong base such as LiHMDS, this transformation could be applied to unprotected indoles and even pyrroles, leading to the formation of quinazolines and pyrimidines, respectively [40]. In the same year, the N atom insertion into the double bond of indenes was reported affording isoquinolines (up to 66% yield) [41]. This transformation proceeded through the formation of an aziridine intermediate **int 7**, which had an acidic C—H proton at the benzylic position. Notably, it was also demonstrated that cyclopentadienes could be converted to pyridines under similar conditions.

Recently, the use of iodonitrene species was reported in an oxidative ring expansion of cyclopentenones and 1-indanones, leading to the formation of 2-pyridones

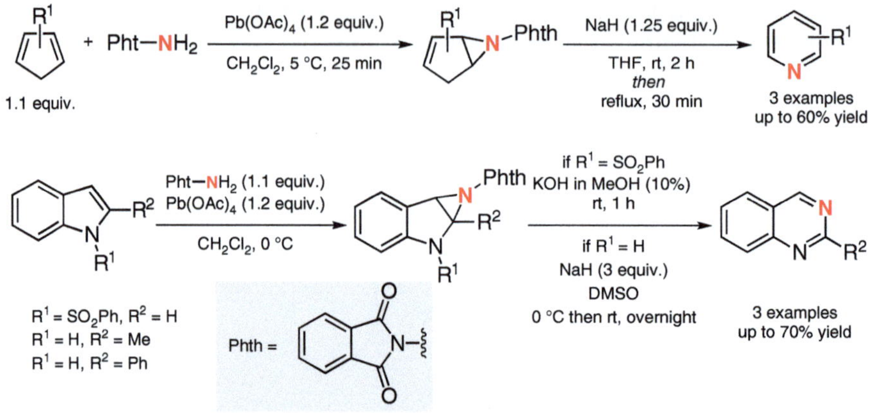

Scheme 9.9 Using phthalimidonitrene as a nitrogen atom source in ring expansion reactions.

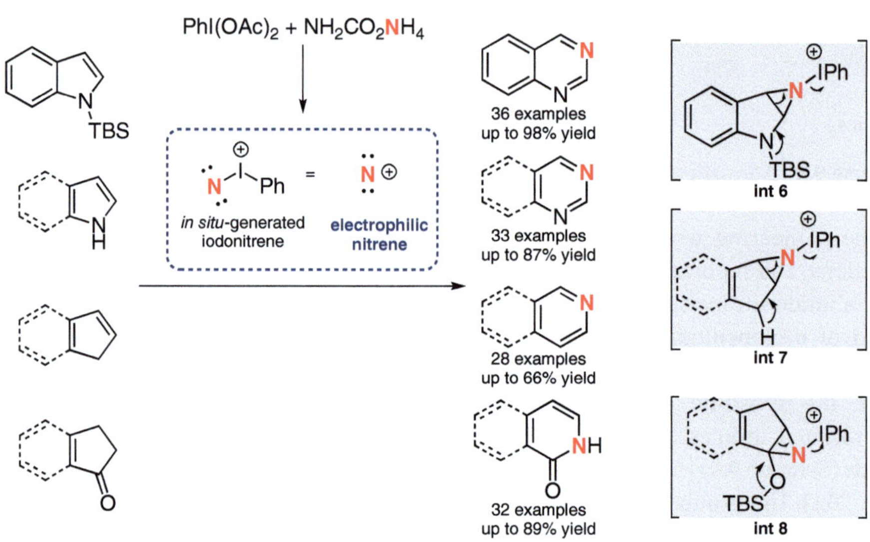

Scheme 9.10 Using iodonitrene as a nitrogen atom source in ring expansion reactions.

or isoquinolinones, respectively [42]. This one-pot, two-step transformation started with the formation of a silyl enol ether, which underwent an aziridation reaction with iodonitrene species, leading to the intermediate **int 8**. The latter then underwent a ring opening along with the release of iodobenzene and the cleavage of silyl group, which acted as an electrofuge. Finally, upon tautomerization, the desired 2-pyridones or isoquinolinone was obtained with up to 89% of yield. Interestingly, in the case of 1-indanone, which led to isoquinolinone, the regioselectivity of N atom insertion was opposite to that observed during the traditional Beckmann rearrangement.

Sharma and coworkers designed the ring expansion reaction using a precursor of sulfenylnitrene (Scheme 9.11) [43]. Using this strategy, nitrogen insertion into pyrroles, indoles, and imidazoles was possible, affording the corresponding pyrimidines, quinazolines, and 1,3,5-triazenes, respectively, with up to 95% of yield. The reaction mechanism is similar to the previously discussed examples with nitrenes. First, the (2+1) cycloaddition reaction afforded the aziridine **int 9**, followed by ring opening, elimination of thiolate, and aromatization.

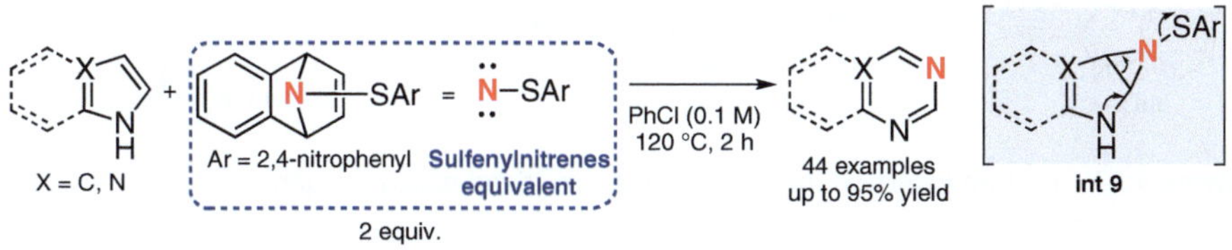

Scheme 9.11 Using sulfenylnitrene as a nitrogen atom source in ring expansion reactions.

A photochemical approach of nitrogen atom insertion into the imidazole to access 1,3,5-triazole was reported by Hayashi and coworkers [44]. First, the hydroperoxide of imidazole (**int 10**) was generated under irradiation using methylene blue as a photosensitizer. Then, upon heating in the presence of ammonium acetate, the desired 1,3,5-triazole was formed (Scheme 9.12).

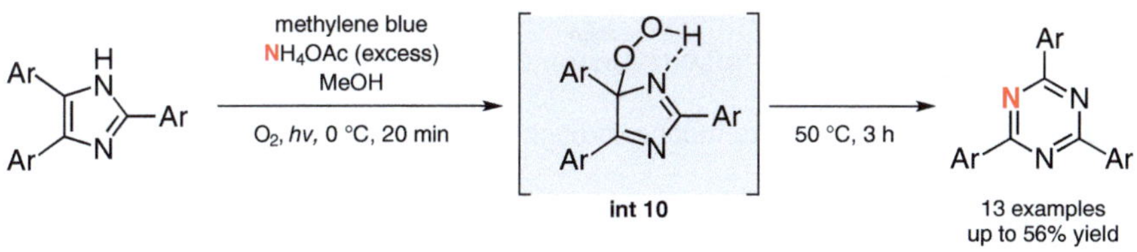

Scheme 9.12 N atom insertion into imidazoles via sensitized photo-oxygenation.

Subsequently, the same group demonstrated that this approach could also be applied to produce quinazolines from indoles using slightly modified reaction conditions [45]. However, the reaction mechanism is slightly different from the previous case. After the initial formation of hydroperoxide intermediate (**int 11a**), a ring opening step led to the formation of the dicarbonyl intermediate (**int 11b**). The latter was then transformed to quinazoline via ring closure with ammonia, which was first described by Bischler (Scheme 9.13) [46].

Another approach for N atom insertion proceeding via a dicarbonyl derivative as an intermediate was discovered by Fields in 1964, with the conversion of indene into isoquinoline upon ozonolysis in aqueous ammonia [47]. Inspired by this work, Miller and Frincke generalized this approach, and a one-pot three-step synthesis of isoquinolines from indenes was developed (Scheme 9.14) [48]. First, upon ozonolysis followed by reductive work-up with dimethyl sulfide and sodium bicarbonate

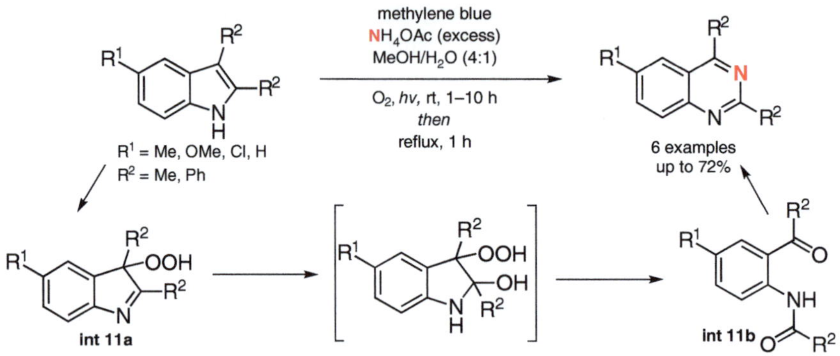

Scheme 9.13 N atom insertion into indoles via sensitized photo-oxygenation.

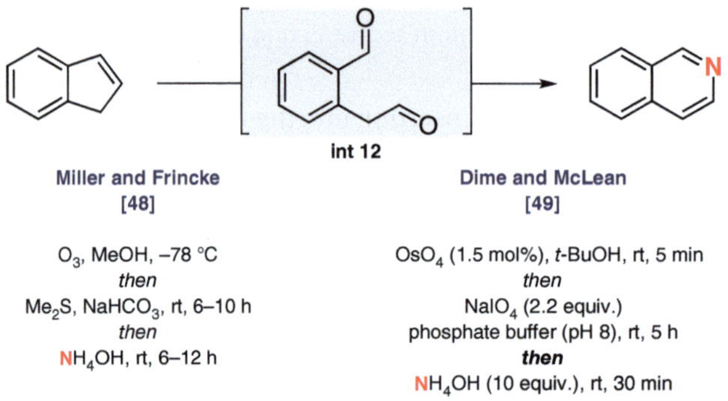

Scheme 9.14 N atom insertion into indene through a dicarbonyl intermediate.

dicarbonyl, the indene was converted into the intermediate **int 12**. The latter reacted with ammonia to provide the corresponding isoquinoline. One year later, the same transformation was depicted by Dime and McLean, using osmium tetraoxide with sodium metaperiodate to reach the dicarbonyl intermediate **int 12** [49].

Several approaches for N atom insertion via radical intermediates were developed. Wei and coworkers reported the cobalt-catalyzed radical N atom insertion reaction into cyclopentenes, cyclobutenes, or indenes, resulting in the formation of pyridine, pyrrole, or isoquinoline derivative, respectively (Scheme 9.15) [50]. Later, in 2024, the group of Zhu showed that manganese catalysts can also promote N atom insertion under similar conditions [51].

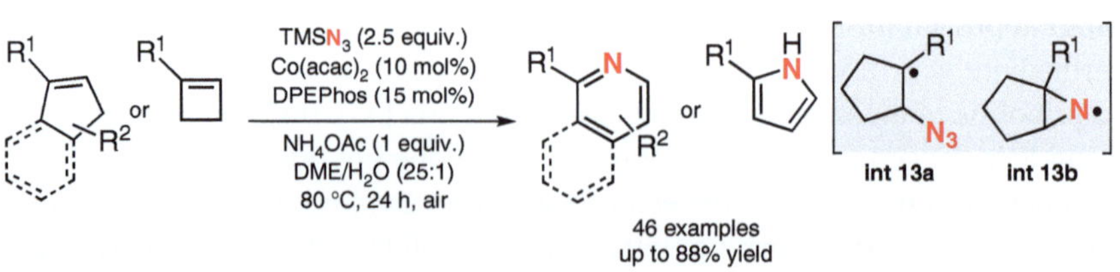

Scheme 9.15 Cobalt-catalyzed radical N atom insertion in C—C bond.

Later, Wei and coworkers reported a copper-catalyzed oxidative N atom insertion into π-extended phenol derivatives to afford the corresponding benzazepines via the formation of the intermediate **int14** (Scheme 9.16) [52].

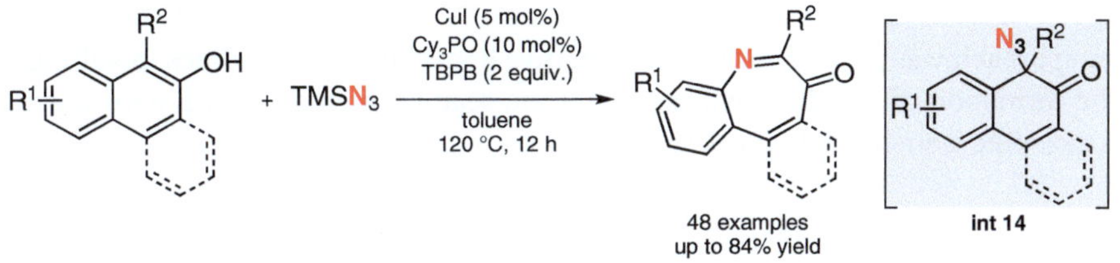

Scheme 9.16 Copper-catalyzed radical N atom insertion in C—C bond.

In 2022, Cheng and coworkers developed an electrochemical process to convert indene into isoquinoline using ammonia gas by means of the formation of aziridine intermediate **int 15** (Scheme 9.17a) [53]. Two years later, the group of Ackermann depicted a similar approach starting from silyl-protected indoles to get the corresponding quinazolines using ammonium carbonate as a nitrogen atom source (Scheme 9.17b) [54].

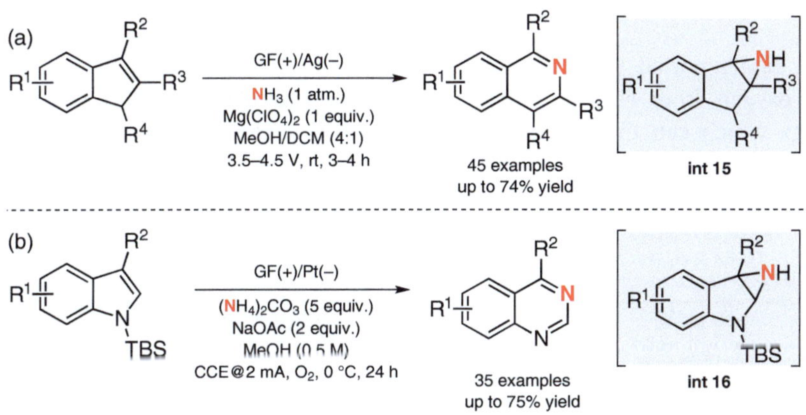

Scheme 9.17 Electrochemical process for the N atom insertion into indenes (a) and protected indoles (b).

9.2.3 Oxygen Atom Insertion

The Baeyer–Villiger oxidation, first reported in 1899, is the very first example of oxygen atom insertion [55]. While initially, Baeyer and Villiger used peroxymonosulfuric acid, nowadays, *meta*-chloroperbenzoic acid (*m*CPBA) or trifluoroperacetic acid is commonly used (Scheme 9.18).

In 2020, the group of Sarlah reported a two-step approach to insert an oxygen atom onto an aromatic ring to afford the desired oxepines with a yield of up to 69% over two steps via the formation of the polycyclic intermediate **int 17** [56]. Later, the

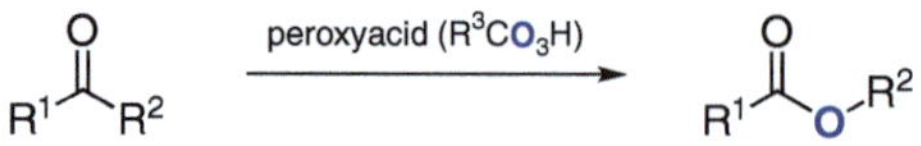

Scheme 9.18 Baeyer–Villiger oxidation.

same approach was applied for the dearomative epoxidation of pyridines, resulting in the formation of pyridine oxides, which are readily interconverting with their 1,4-oxazapine isomers (Scheme 9.19) [57].

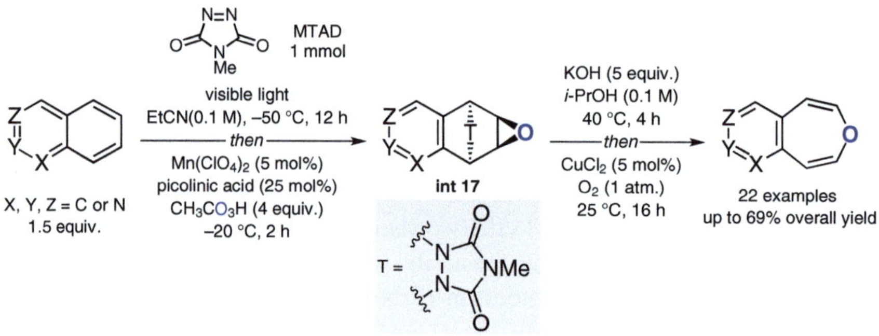

Scheme 9.19 Preparation of oxepines.

Qin and Jiao reported the functionalization of Ar—C(sp^2) bond of styrenes using the oxygen from air [58]. The final product was obtained with a sulfonyl group, which can then be easily removed using magnesium in methanol (Scheme 9.20).

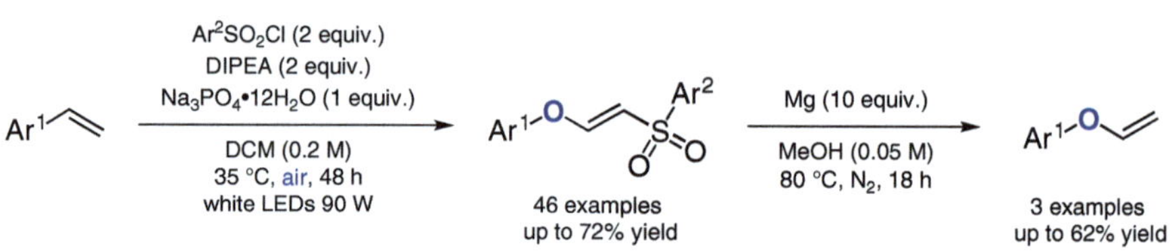

Scheme 9.20 Oxygen atom insertion into Ar–C(sp^2) bond of styrenes.

9.2.4 Other Heteroatom Insertion

In 2012, the group of Radius discovered the ring expansion of *N*-heterocyclic carbenes (NHC) using hydrosilanes, wherein the silicon atom was inserted into the carbene carbon–nitrogen bond [59]. They successfully inserted phenyl silane, diphenyl silane, and triphenyl silane motifs into the NHC core (Scheme 9.21).

In 2016, Yorimitsu and coworkers designed a nickel-catalyzed boron atom insertion into C(sp^2)—O bond of benzofuranes (Scheme 9.22a) [60]. Few years later, the group of Dong reported the boron atom insertion into more challenging

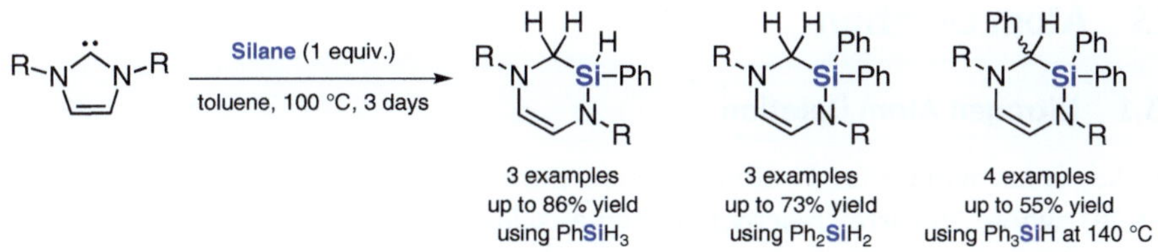

Scheme 9.21 Silicon atom insertion into *N*-heterocyclic carbenes.

C(sp^3)—O bonds(Scheme 9.22b), providing access to 5-, 6-, and 7-membered ring boron-containing cycles with up to 82% of yield [61].

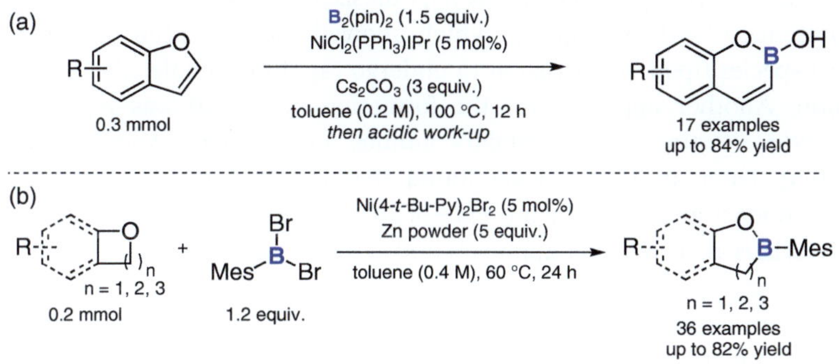

Scheme 9.22 Nickel-catalyzed boron atom insertion into C—O bond of benzofurans (a) and oxetanes (b).

In 2017, Yorimitsu and coworkers reported the insertion of various heteroatoms into C(sp^2)—O bond of benzofurans [62]. First, a manganese-catalyzed ring opening of benzofuranes with an organolithium reagents took place with a concomitant arylation or alkylation at the C2-position. Then, the resulting dianionic species (**int 19**) were trapped with various multivalent-heteroatom electrophiles. Using this methodology, Si, B, P, Ge, and Ti atoms were successfully inserted with up to 81% of yield (Scheme 9.23).

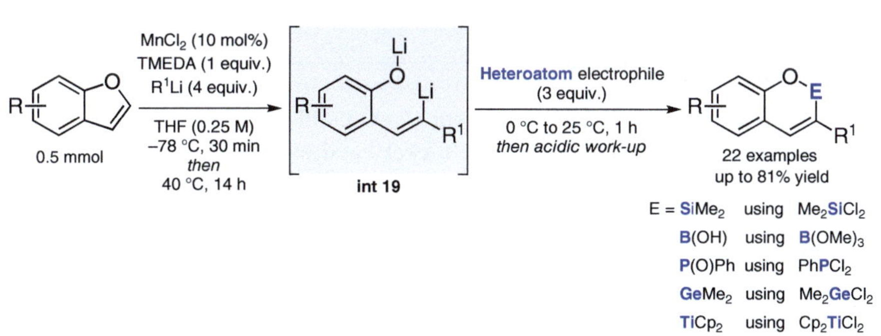

Scheme 9.23 Manganese-catalyzed different heteroatoms insertion into C—O bond.

9.3 Atom Deletion

9.3.1 Nitrogen Atom Deletion

The key intermediate for N atom deletion transformations is 1,1-diazene species (**int 20**), whose "bizarre" reactivity intrigues scientists since the beginning of 20th century. Historically, it was commonly formed via the oxidation of 1,1-disubstituted hydrazines. However, in the majority of cases, due to high reactivity, the intermediate (**int 20**) formed a dimeric tetrazene product [63]. But, first discovered by Busch and Weiss [64], and then studied in detail by Hinman and Hamm [65], the oxidation of dibenzyl hydrazine with mercury oxide surprisingly provided bibenzyl – N atom deletion product. It was hypothesized that such behavior was due to the presence of a benzyl group, which allowed the formation of stable enough radical species upon N_2-extrusion to undergo rapid, intramolecular C—C bond formation. Another approach for the formation of **int 20** was reported by Lemal in 1965, starting from secondary amines and utilizing Angeli's salt ($Na_2[N_2O_3]$) [66]. Later, in 1998, Glover found that **int 20** could be obtained starting from secondary amines in the presence of an anomeric amide [67]. In 2021, Levin and coworkers revisited this approach and designed a single-step N atom deletion protocol using anomeric amides, allowing the conversion of benzylic secondary amines into dibenzyl derivatives [68]. In the same year, the group of Lu revised their initial work on thermal rearrangement of sulfamoyl azide [69] and designed a one-pot two-step N atom deletion protocol from N-heterocycles with 3- to 20-membered rings [70]. First, sulfamoyl azide intermediate was formed, which upon thermal rearrangement provided the key 1,1-diazene intermediate (**int 20**). In the same year, the group of Antonchick reported an additional approach to achieve 1,1-diazene **int 20** from various secondary amines using an *in situ*-generated iodonitrene species [71]. One more way to achieve **int 20** was designed by Lu and coworkers [72]. They discovered that the reaction of *O*-diphenylphosphinylhydroxylamine with secondary amines provided a triazonium intermediate which in the presence of base-released ammonium generate **int 20** (Scheme 9.24).

9.3.2 Oxygen Atom Deletion

Reductive deoxygenation of ethers to form C—C bond was developed by Shi in 2017 [73]. The transformation features nickel catalysis, and it was limited to symmetric benzylic ethers. In the case of non-symmetrical ethers, the desired product was obtained as a major product, however, along with two other symmetric by-products (Scheme 9.25a). In 2023, Tobisu and coworkers reported a dual-catalyzed conversion of allyl ester into the corresponding ketones [74]. Reaction was limited to aryl and heteroaryl allyl esters and gave a mixture of allyl ketone and conjugated ketone as a product (Scheme 9.25b).

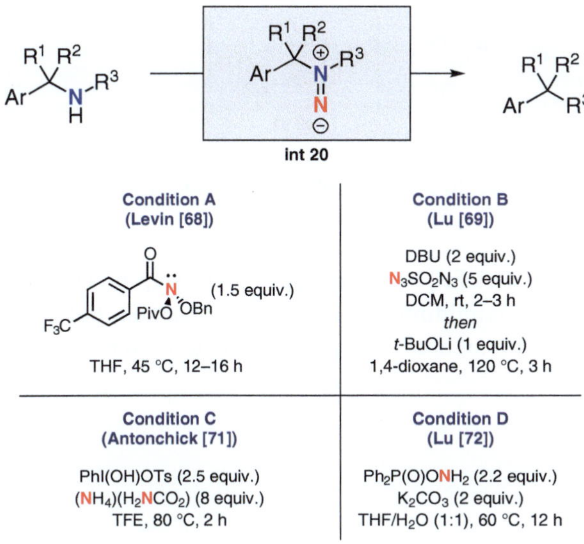

Scheme 9.24 Four ways for nitrogen atom deletion from secondary amines via 1,1-diazene intermediate.

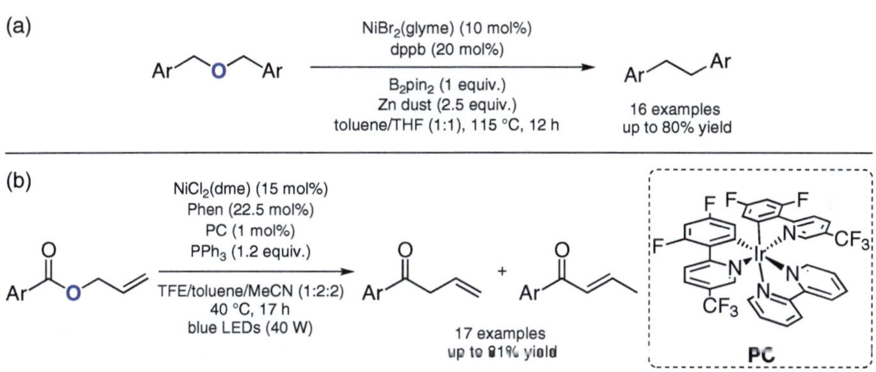

Scheme 9.25 Nickel-catalyzed reductive deoxygenation of ethers (a) and allyl esters (b).

9.3.3 Carbon Atom Deletion

Transformation of pyrimidines to pyrazole in the presence of a huge excess of hydrazine was first discovered by van der Plas and Jongejan in 1968 [75]. The reaction required harsh reaction conditions (190 °C), and the scope was extremely limited. However, when a pyrimidinium salt with iodide or methyl sulfate counterion was used, the required reaction conditions were milder. In 2022, Sarpong and coworkers revisited this approach and discovered that an *in situ*-formed pyrimidinium triflate with a small excess of hydrazine afforded the corresponding

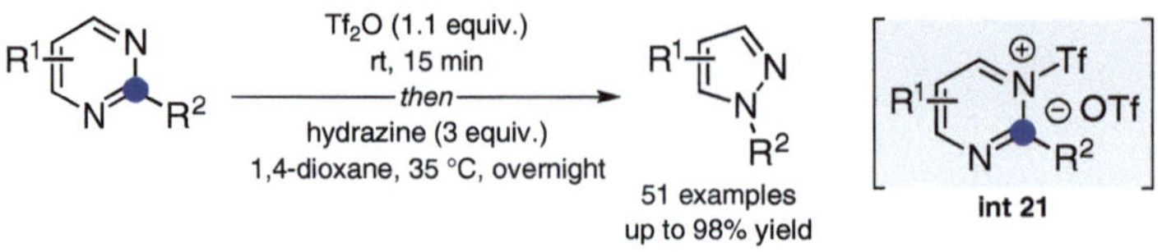

Scheme 9.26 Transformation of pyrimidines to pyrazoles via the formation of a pyrimidinium salt.

pyrazoles at 35 °C with up to 98% of yield [76]. This one-pot two-step procedure started with the formation of the pyrimidinium salt **int 21** (isolatable but not isolated), which would then react with an excess of hydrazine to afford the desired pyrazoles (Scheme 9.26).

9.4 Atom Swapping

9.4.1 Nitrogen-to-Carbon Swapping

There are two main approaches for pyridine activation: organometallic approach and electrophilic approach (Scheme 9.27). The former was explored by Mindiola et al., who performed a N-to-C atom swapping for simple pyridine derivatives, utilizing a titanium alkylidyne reagent via the formation of the intermediate **int 22**. However, this method required solvent-level quantities of substrate, which rendered it unsuitable for late-stage functionalization of complex molecules [77].

Further studies tried to harness the nucleophilic nature of the pyridinyl substrate, which, upon activation with electrophilic reagents via N-arylation (**int 23**), N-oxidation (**int 24**), or N-sulfonylation (**int 25**), was more prone to react with a nucleophile, leading to ring opening and enabling subsequent transformations.

Morofuji and Kano reported the transformation of pyridines to *para*-substituted benzenes via a three-step sequence. After initial pyridine activation through N-arylation, a nucleophilic attack by a secondary amine led to the formation of a streptocyanine intermediate **int 26**, which further underwent a (5+1) annulation with acetophenone-derived enolates to yield diaryl ketones [78]. Alternatively, reacting the streptocyanine intermediate **int 26** with a sulfonium ylide derived from trimethylsulfonium salts not only allowed the N-to-C atom swap but also introduced an amino substituent at the *ortho* position of the pyridinyl nitrogen [79]. Recently, Feuillastre, Audisio et al. reported an alternative two-step one-pot approach to diaryl ketones from pyridines via a Zincke imine intermediate (**int 27**) [14].

The pioneering work on N-to-C single atom swapping via N-oxidation was reported for simple benzo[*h*]quinoline using dimsyl anion (deprotonated dimethyl sulfoxide) by Takeuchi and Hamada in 1977 [80]. This approach was recently independently revisited by Sorensen's and Song's scientific groups and expanded to pyridines and quinolines. The transformation followed an Addition of Nucleophile, Ring Opening and Ring Closure (ANRORC) sequence mechanism with a

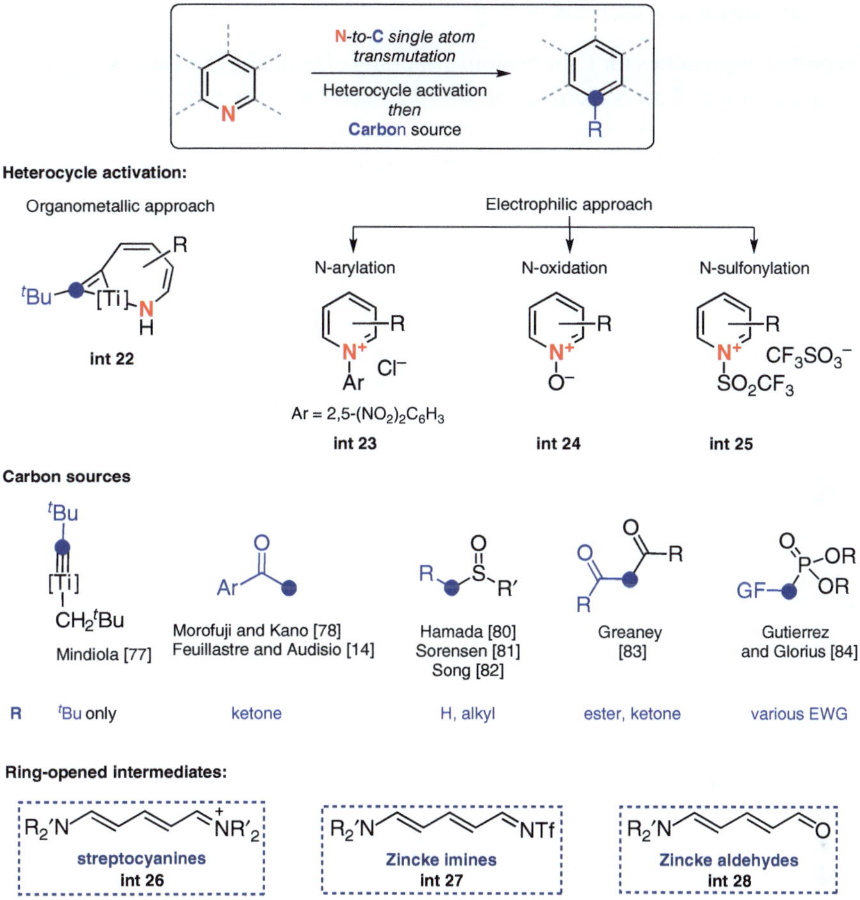

Scheme 9.27 N-to-C single atom transmutation of pyridine-related heterocycles.

nucleophilic attack of the anion generated after the deprotonation of methyl phenyl or dialkyl sulfoxide. This transformation exhibited good functional group tolerance and replaced the N-oxide moiety with either unsubstituted, alkyl-substituted, or isotopically labeled carbon atoms [81, 82].

Greaney et al. utilized pyridine activation via N-sulfonylation with triflate anhydride to transform pyridines into benzoic acid esters. This process was executed via a one-pot ANRORC process, utilizing malonate-derived nucleophiles [83].

Recently, Gutierrez and Glorius et al. developed a method allowing for the installation of various functional groups in the benzene scaffold, including ester, ketone, amide, nitrile, and phosphate ester via an N-to-C single atom swapping of pyridines. This approach proceeded via a sequence of pyridine ring opening, imine hydrolysis to Zincke aldehydes (**int 28**), Wittig-type olefination, electrocyclization, and finally aromatization to achieve the desired transmutation [84].

9.4.2 Carbon-to-Nitrogen Swapping

The reported approaches for C-to-N atom swapping of benzene derivatives required the pre-installation of an N atom in the starting material in the form of azido group (Scheme 9.28). The pioneering work was reported by Sundberg in 1972. However, the reaction suffered from harsh reaction conditions, along with low yields and selectivity [85].

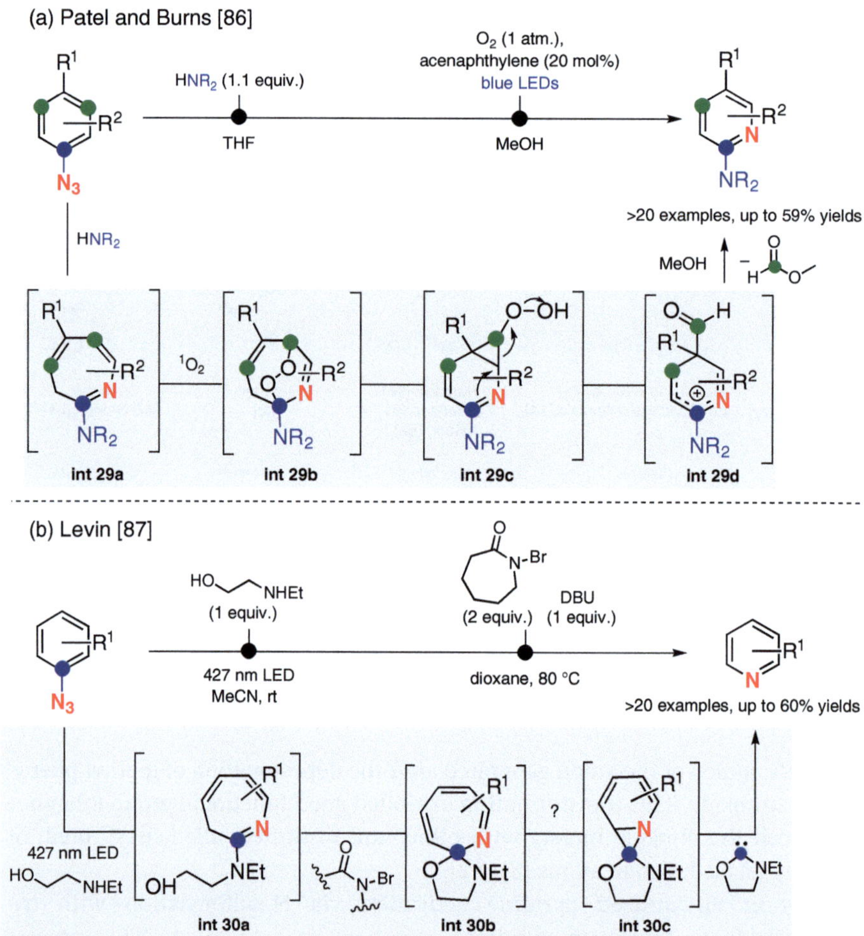

Scheme 9.28 Single N-to-C atom swapping of aryl azides to pyridines through deletion of *meta*-carbon (a) or *ipso*-carbon (b).

In 2022, Patel and Burns revisited this approach and reported a methodology for transforming arylazides into 2-aminopyridines via a two-step, one-pot reaction sequence with the deletion of a carbon atom at the *meta* position to the azido group (Scheme 9.28a). Initially, in the presence of a secondary amine, N atom insertion occurred, forming a 3*H*-azepine (**int 29a**). Under blue light irradiation and with the aid of a photosensitizer, this intermediate underwent a [4+2] cycloaddition with singlet oxygen, forming an endoperoxide (**int 29b**), which upon ring opening and electrocyclization yielded a cyclopropyl-fused dihydropyridine **int 29c**. Subsequent cyclopropane ring opening and loss of hydroxide generated a Wheland intermediate **int 29d**, which, upon methanol-mediated deformylation, produced the final 2-aminopyridine and methyl formate by-product [86].

In 2023, Levin reported a complementary methodology to perform transformation of arylazides into pyridines via a two-step, one-pot reaction sequence with the deletion of the carbon atom directly linked to the azido group (Scheme 9.28b). Initially, aryl azides were photochemically converted to 3*H*-azepines (**int 30a**), which then underwent an oxidatively triggered *C*2-selective chelotropic carbon extrusion through a spirocyclic azanorcaradiene intermediate **int 30c** affording the pyridine products [87].

In 2023, Levin reported the transformation of quinolines into the corresponding quinazolines by replacing the *C*3 carbon with a nitrogen atom (Scheme 9.29) [88]. First, the corresponding quinoline N-oxides were photochemically converted to 3,1-benzoxazepines (**int 31**). The latter, upon ozonolysis, underwent an oxidative cleavage to give the reactive intermediate **int 32**, containing electrophilic acyl and imidic anhydride functionalities, activated for nucleophilic attack by an ammonia nitrogen source, ultimately furnishing the desired quinazolines. During this process, the carboxylate subunit containing the *C*3 carbon was displaced as a leaving group, while concurrently deleting the R^3 substituent.

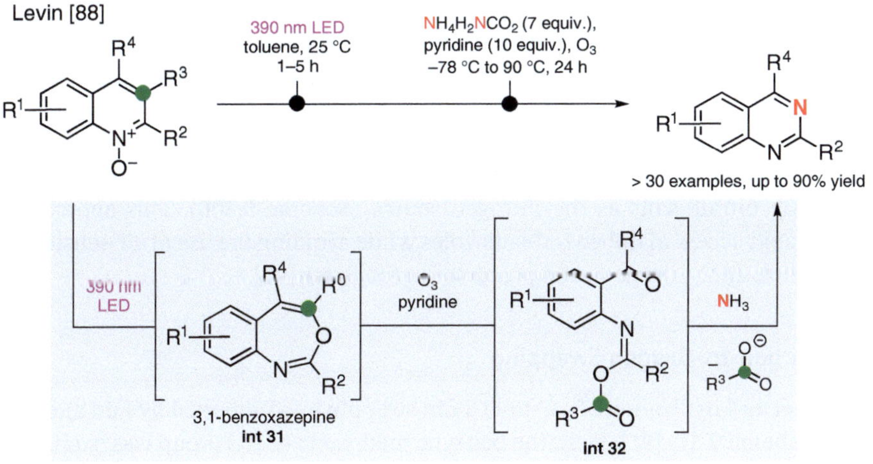

Scheme 9.29 Single N-to-C atom swapping of quinoline derivatives.

In 2024, Morandi et al. reported a C-to-N atom swapping in *N*-alkyl indoles, leading to the formation of benzimidazoles via the *in situ* generation of iodonitrenes (Scheme 9.30a). This transformation complements previous studies, where, under similar reaction conditions, *N*-H and *N*-silyl indoles undergo N atom insertion to produce quinazolines. The reaction exploited the inherent reactivity of the indole scaffold, starting with oxidative cleavage, followed by an oxidative amidation, a

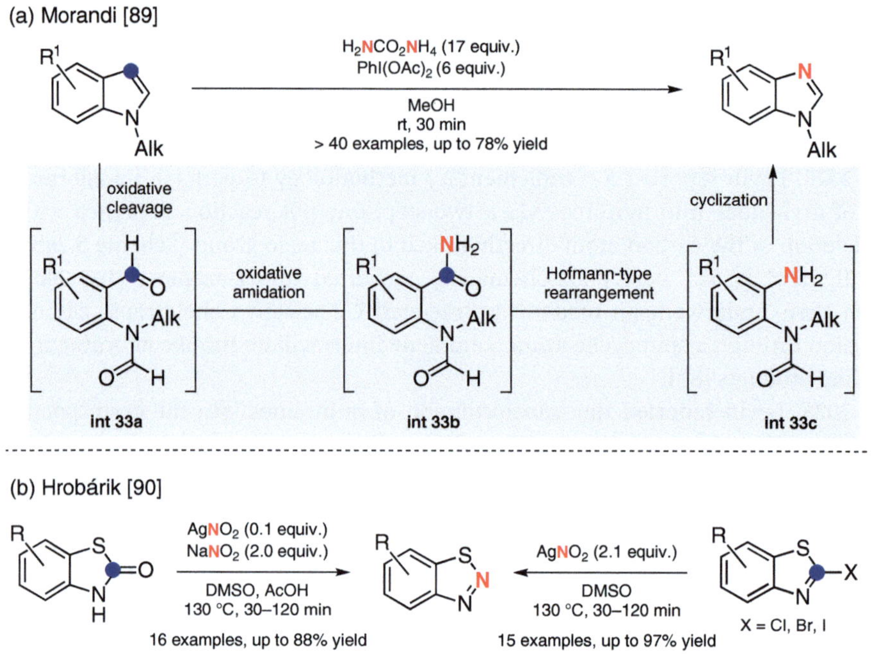

Scheme 9.30 Single N-to-C atom swapping in indoles (a) and benzothiazole derivatives (b).

Hofmann-type rearrangement, and cyclization. The reaction tolerated a wide range of functional groups and was applicable for the late-stage functionalization of drug-like molecules [89].

Hrobárik reported the Ag(I)-catalyzed skeletal editing of benzothiazolinones and 2-halogen-substituted benzothiazoles to the corresponding isobenzothiadiazoles using nitrate salts as the nitrogen source (Scheme 9.30b). This approach provided rapid access to isobenzothiadiazoles while avoiding the use of air-sensitive 2-aminobenzothiols, the common precursor to the latter [90].

9.4.3 Carbon-to-Oxygen Swapping

The concept and realization of a C-to-O atom swapping was reported by Luu and Li in 2021 (Scheme 9.31) [91]. First, the benzylic methylene (CH_2) group was oxidized to the corresponding aromatic ketone (**int 34**), which, upon Baeyer–Villiger oxidation, introduced the oxygen atom that replaced the carbon one. Finally, the newly developed nickel-catalyzed ring contraction of the intermediate lactone (**int 35**) via decarbonylation furnished the cyclic diarylethers.

9.4.4 Heteroatom-to-Heteroatom Swap

There are several remarkable examples of heteroatom swapping within an aromatic ring under mild reaction conditions (Scheme 9.32a). In 2024, Park

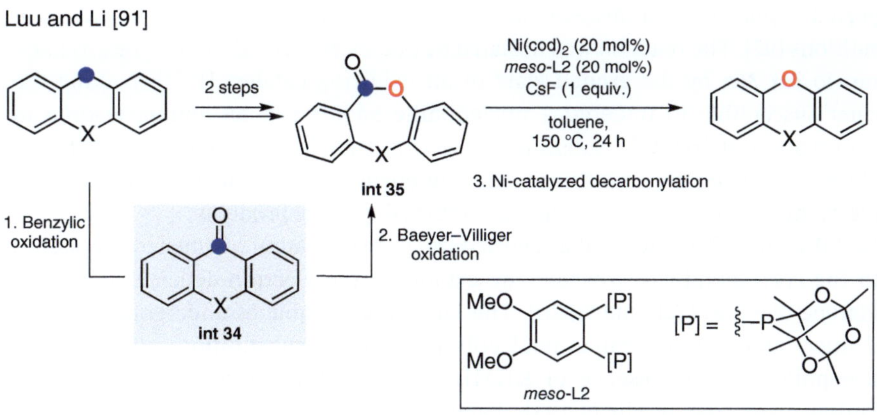

Scheme 9.31 Single C-to-O atom swapping.

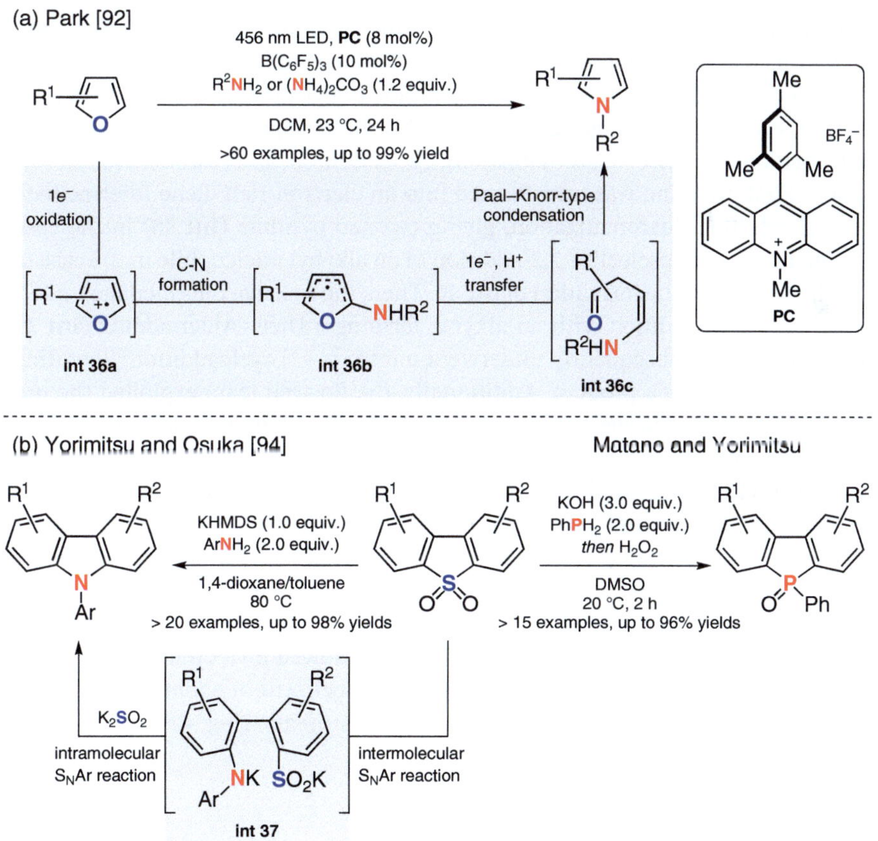

Scheme 9.32 Atom swapping between heteroatoms in furans (a) and dibenzothiophenes (b).

reported a single-step transformation of furans to pyrroles under photochemical conditions [92]. The reaction was initiated by one-electron oxidation of the aromatic furan to **int 36a** by the excited state of an oxidizing catalyst [PC*], making the former susceptible to undergo a nucleophilic addition of an amine, leading to the formation of **int 36b**. Subsequent electron transfer from the reduced [PC^-] and proton transfer generated a singlet, ring-opened intermediate (**int 36c**), which underwent a Paal–Knorr-type reaction to furnish pyrrole products.

In 2015, Yorimitsu and Osuka reported the transformation of dibenzothiophenes into the corresponding carbazoles by exploiting the electron-deficient nature of thiaarene dioxides (Scheme 9.32b). The latter was capable of undergoing sequential inter- and intramolecular nucleophilic aromatic substitution with nitrogen nucleophiles in the presence of KHMDS base via the intermediacy of **int 37**, affording carbazole products [93]. Later, this transformation was extended to phenylphosphine nucleophiles, leading to the formation of the corresponding dibenzophosphole oxides after oxidative work-up [94].

9.4.5 A Swap of Two Atoms by Two Atoms

There are only a few examples where two atoms from the starting aromatic compound are displaced by another two atoms. In 2024, the Studer [95] and Boswell [96] teams independently reported the transformation of pyridines into benzene derivatives via *C=N* to *C=C* swapping (Scheme 9.33). Both approaches share the same transformation sequence: dearomatization, cycloaddition, and rearomative retrocyclization. First, pyridine was transformed into an electron-rich diene intermediate via a cycloaddition dearomatization, giving oxazino pyridine (**int 38**) in the case of Studer, or via a regioselective 1,2-addition of an alkynyl nucleophile in the case of Boswell, resulting in the formation of **int 40**. Then, the reactive intermediate underwent a [4+2] cycloaddition with an alkyne, forming a Diels–Alder adduct (**int 39** and **int 41**), which subsequently underwent a retro [4+2] cycloaddition, liberating the substituted benzene product. Additionally, the Boswell team exploited the use of an *in situ*-generated aryne, which led to the formation of naphthalene derivatives after a cycloaddition-retro-cycloaddition cascade.

In 2024, Bouffard reported a method for transforming benzene derivatives into more functionalized forms by swapping *C=C* bonds (Scheme 9.34) [97]. The process started with an unactivated benzene derivative, which underwent a formal [3+2] cycloaddition with the *in situ*-generated 1,3-diazaazoniaallene cations. This reaction produced thermally stable dearomatized compounds (**int 42**). Next, a [4+2] cycloaddition with alkynes or arynes took place, followed by a cycloreversion cascade. This sequence resulted in the formation of new benzene or naphthalene derivatives, where a 2-carbon fragment from the original aromatic ring was replaced by a fragment from the dienophile.

9.5 Functional Group Transposition

In 2024, Lumb and Edelmann reported an elegant approach for the transformation of *para*-substituted phenols into more synthetically challenging *meta*-substituted

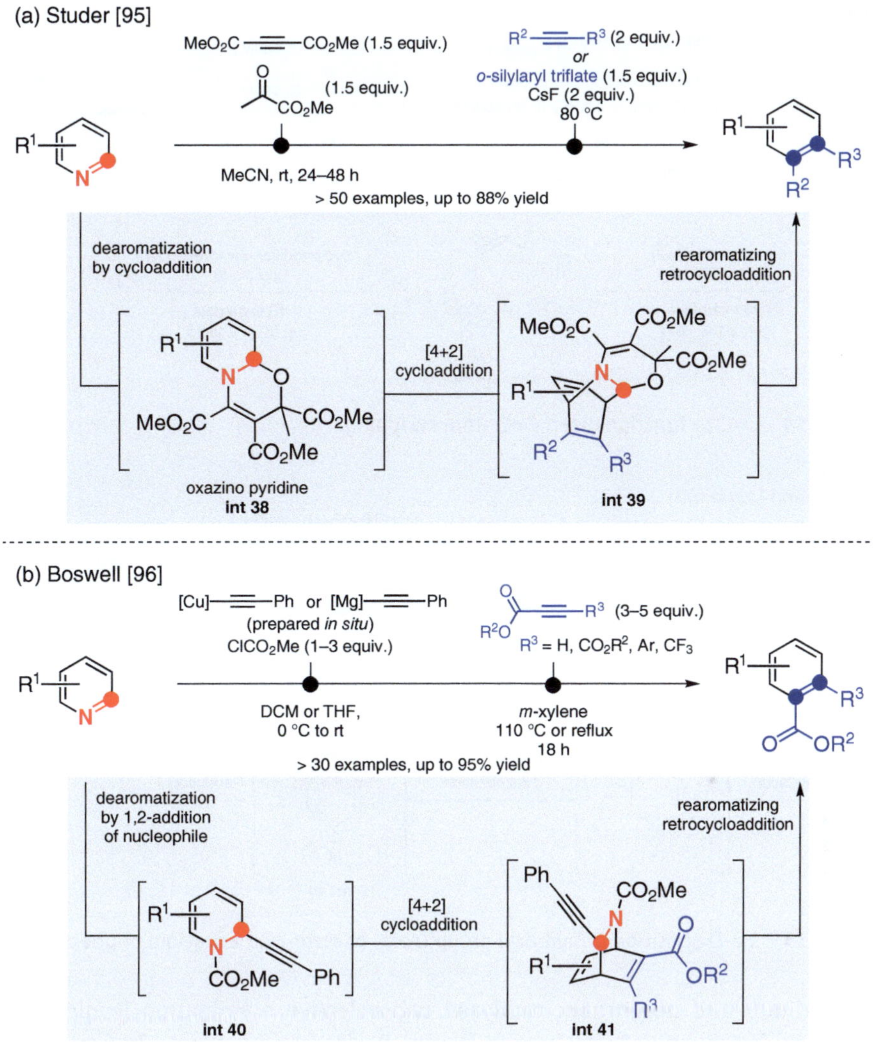

Scheme 9.33 *C*=*N* to *C*=*C* atom swapping through oxazino pyridines (a) or *C*2-alkylated dehydropyridines (b).

isomers using a one-pot three-phase transformation sequence (Scheme 9.35) [98]. First, the phenolic substrate was oxidized with 2-iodoxybenzoic acid (IBX) to the corresponding *ortho*-benzoquinone (**int 43**). Then, the latter was regioselectively diazotized at the C1-position with 2,4,6-triisopropylbenzenesulfonyl hydrazide (TBSH), leading to intermediate **int 44**, followed by the reduction to the isomeric phenolic product using a combination of hypophosphorous acid (H_3PO_2) and catalytic amounts of copper(I) salts.

Several methodologies for the 1,2-disposition of carbonyl group were developed. In 2021, Dong reported one- or two-pot protocol that transposed a ketone to the vicinal carbon (Scheme 9.36a). This approach first converted the starting ketone to the corresponding alkenyl triflate (**int 45**), which could then undergo

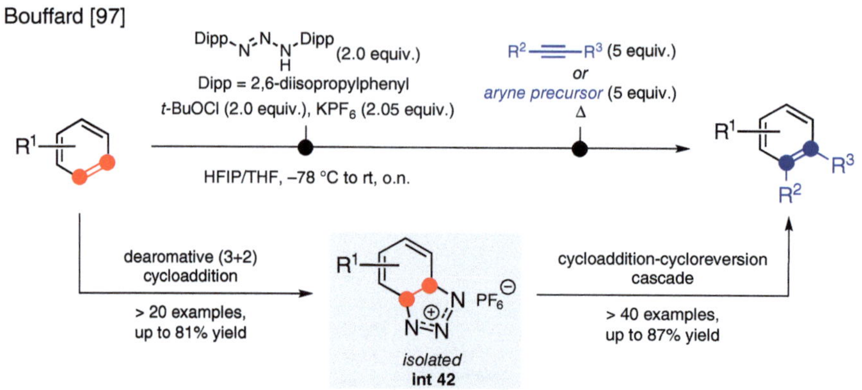

Scheme 9.34 *C*=*C* to functionalized *C*=*C* atom swapping.

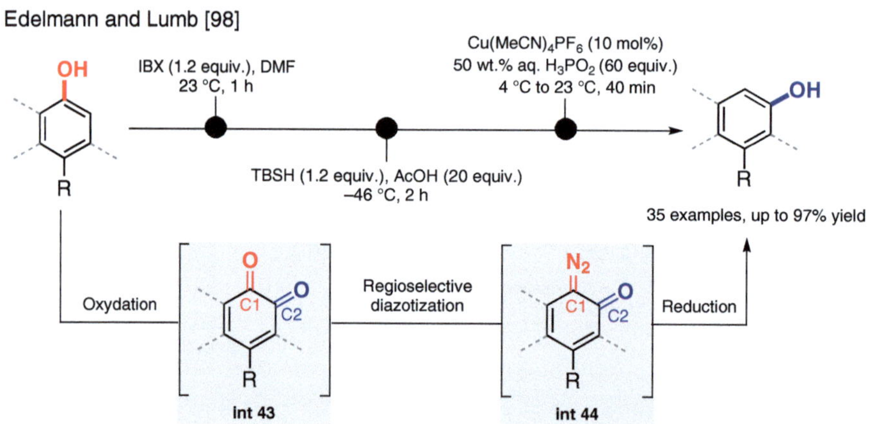

Scheme 9.35 1,2-Disposition of hydroxyl group (*para*- to *meta*-isomerization) of phenols.

the palladium- and norbornene-catalyzed regioselective α-amination leading to **int 46**. The latter, upon hydrolysis, released the isomeric ketone. This method showed broad functional group tolerance and was successfully applied for late-stage functionalization of natural products derivatives [99].

In 2023, the 1,2-disposition of carbonyl group of cyclic ketones was reported by Bhawal and Morandi using a combination of inexpensive pyrrolidine and elemental sulfur catalysts (Scheme 9.36b). The process was thermodynamically driven and favored the formation of ketone with the reduced number of 1,3-diaxial interactions. This process was applied for the isomerization of various steroids and allowed a rapid access to novel steroids with “unnatural” oxidation patterns [100].

Dixon reported an interesting example of the 1,2-disposition of a carbonyl group, enabling the transformation of tertiary amides into α-amino ketones (Scheme 9.36c). First, the starting amide was reduced to an enamine (**int 48**) through an iridium-catalyzed process that involved a silylated hemiaminal intermediate **int 47**. Then, selective oxidation with *m*CPBA produced the α-amino ketone [101].

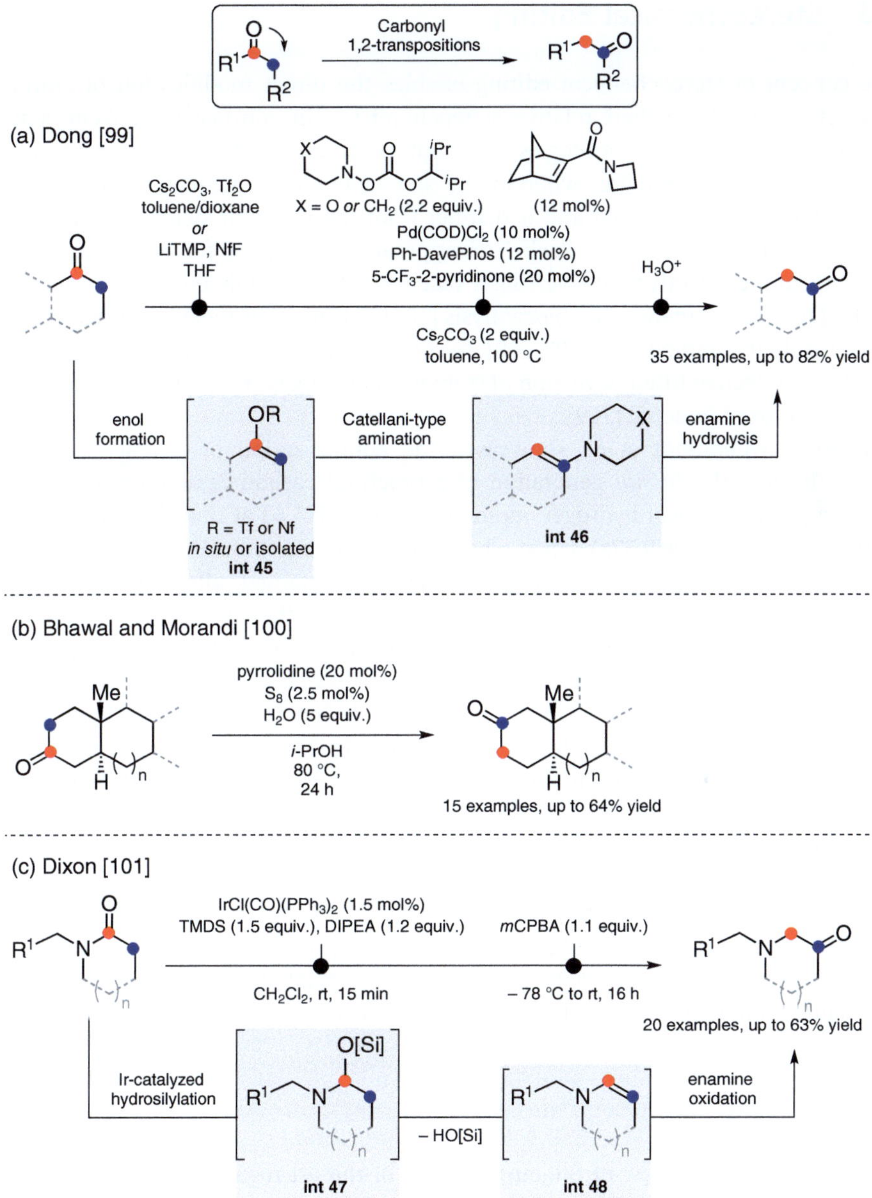

Scheme 9.36 1,2-Disposition of carbonyl group through palladium catalysis Catellani-type process (a), reversible isomerization in the presence of sulfur and amine catalysts (b) and iridium-catalyzed deoxygenation of amides (c).

9.6 Stereochemical Editing

The concept of stereochemical editing enables the direct modification of chiral products to fine-tune their relative stereochemistry. In contrast to conventional stereoselective synthetic strategies, this approach establishes a fundamentally distinct synthetic paradigm, wherein the stereochemical outcome of the newly C—C bond can be modified and is distinct from the inherent stereoselectivity of the C—C bond-forming event itself. This approach was applied for deracemization [102], meaning selectively converting a racemate into a single enantiomer, using photo- [103, 104], organo-, and biocatalysis [105] or transition-metal catalysis [106], as well as their combinations [107, 108]. In contrast, epimerization processes, which enable the selective interconversion of stereoisomers, are comparatively scarce and will be discussed in detail [109].

In 2022, Wendlandt et al. showcased the epimerization of a single carbon center through the *in situ* generation of a prochiral carbon-centered radical **int 49** via a photochemical hydrogen atom abstraction (HAA) at unactivated tertiary stereocenters (Scheme 9.37a). Upon a hydrogen atom donation (HAD) process, the thermodynamic product is generally preferentially obtained, allowing an overall stereochemical editing [110]. This transformation was then expanded to various diols (not depicted here), resulting in the transformation of *cis*-diols to corresponding *trans*-diols [111, 112]. Interestingly, in the presence of methyl boronic acid, a reverse transformation from *trans*-diols to *cis*-diols was achieved through the formation of a transient cyclic boronic ester [113]. In 2023, Lautens et al. reported a stereoselective epimerization of secondary aliphatic iodides, namely, the epimerization of the C—I bond from the *exo* to the *endo* face of [2.2.1] bicyclic aliphatic derivatives, via a combination of palladium and blue light catalysis (Scheme 9.37b). After the oxidative addition of palladium catalyst into the C—I bond (**int 50**), a photochemically promoted homolytic cleavage of the C—Pd bond, resulted in a Pd(I)/carbon-centered radical pair (**int 51**). Finally, halide abstraction occurred, leading to the desired epimerized product after a formal stereoinvertive reductive elimination and regeneration of a Pd(0) species [114].

9.7 Conclusion

In summary, an overview of the current state of the art regarding synthetic tools for late-stage functionalization with molecular editing strategy has been provided. This approach is an efficient tool for offering diversification in organic synthesis by leveraging the existing pool of molecules and allowing the straightforward synthesis of challenging and even previously impossible molecules via alternative synthetic paths. By fitting with the circular economy idea, any advances in this emerging field will be significant and of high interest for researchers and professionals in academia and industry working in the areas of organic synthesis, natural products chemistry, medicinal chemistry, drug discovery, and molecular imaging, among others.

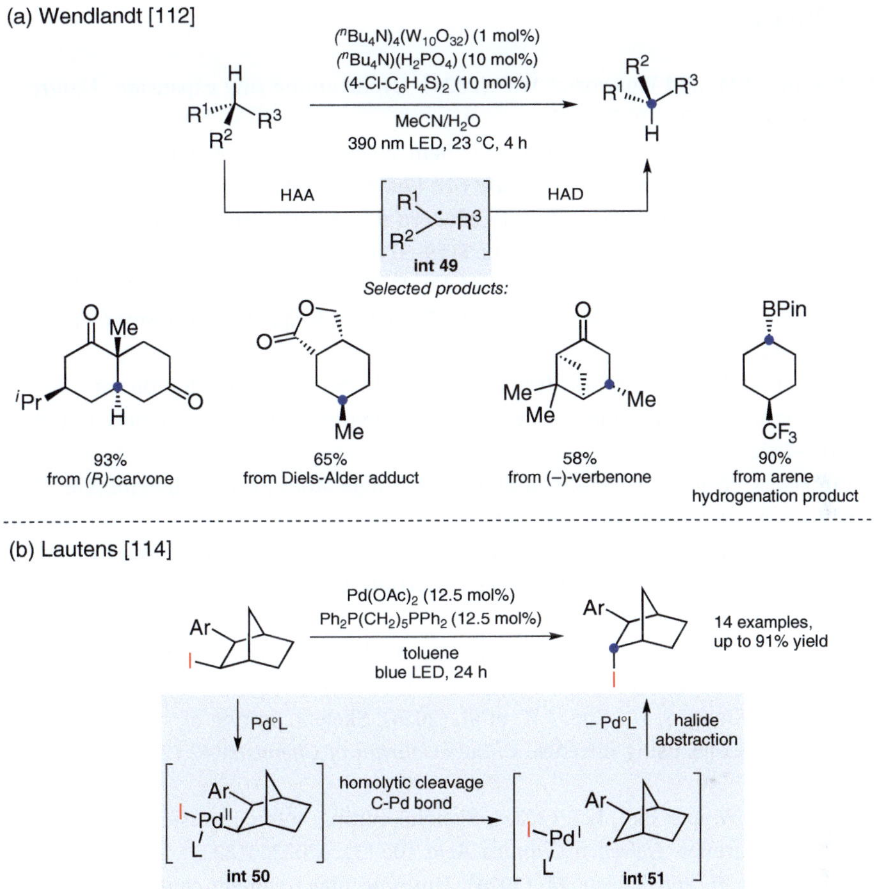

Scheme 9.37 Stereochemical editing of alkanes (a) and aliphatic iodides (b).

Acknowledgments

This work was partially supported by Normandie Université (NU), the Région Normandie, the Centre National de la Recherche Scientifique (CNRS), Université de Rouen Normandie (URN), INSA Rouen Normandie, Université Caen Normandie, ENSICAEN, Labex SynOrg (ANR-11-LABX-0029), the graduate school for research XL-Chem (ANR-18-EURE-0020 XL CHEM) and Innovation Chimie Carnot (I2C). Javid Rzayev thanks the French National Research Agency (ANR-21-CE07-0035) for a doctoral fellowship. The French National Research Agency (ANR-21-CE07-0035 and ANR-22-CE92-0083) and the project NextFluoChem under the program «Normandie Recherche – Labels d'excellence» (n°23E02521) are gratefully acknowledged for generous financial support.

References

1 Wootton, J.M. and Unsworth, W.P. (2024). Cyclic amine ring expansion. *Nature Chemistry* 16 (12): 1932–1933.

2 Peplow, M. (2023). 'Almost magical': chemists can now move single atoms in and out of a molecule's core. *Nature* 618 (7963): 21–24.

3 Moreau, X. and Ghiazza, C. (2024). Modern dearomative enlargement of heteroaromatic rings. *Synthesis* 56 (20): 3119–3130.

4 Ma, C., Lindsley, C.W., Chang, J., and Yu, B. (2024). Rational molecular editing: a new paradigm in drug discovery. *Journal of Medicinal Chemistry* 67 (14): 11459–11466.

5 Yuan, T. and Shi, L. (2024). Recent advances in carbon atom addition for ring-expanding single-atom skeletal editing. *Organic Chemistry Frontiers* 11 (24): 7318–7332.

6 Jurczyk, J., Woo, J., Kim, S.F. et al. (2022). Single-atom logic for heterocycle editing. *Nature Synthesis* 1 (5): 352–364.

7 Gupta, A., Bhatti, P., Laha, J.K., and Manna, S. (2024). Skeletal editing by hypervalent iodine mediated nitrogen insertion. *Chemistry – A European Journal* 30 (56): e202401993.

8 Liu, Z., Sivaguru, P., Ning, Y. et al. (2023). Skeletal editing of (hetero)arenes using carbenes. *Chemistry – A European Journal* 29 (42): e202301227.

9 Xu, Y.-A., Xiang, S.-H., Che, J.-T. et al. (2024). Skeletal editing of cyclic molecules using nitrenes. *Chinese Journal of Chemistry* 42 (21): 2656–2667.

10 Joynson, B.W. and Ball, L.T. (2023). Skeletal editing: interconversion of arenes and heteroarenes. *Helvetica Chimica Acta* 106 (3): e202200182.

11 Shimazumi, R. and Tobisu, M. (2024). Unimolecular fragment coupling: a new bond-forming methodology via the deletion of atom(s). *JACS Au* 4 (5): 1676–1695.

12 Zhong, H., Egger, D.T., Gasser, V.C.M. et al. (2023). Skeletal metalation of lactams through a carbonyl-to-nickel-exchange logic. *Nature Communications* 14 (1): 5273.

13 Tolchin, Z.A. and Smith, J.M. (2024). 15NRORC: an azine labeling protocol. *Journal of the American Chemical Society* 146 (5): 2939–2943.

14 Feng, M., Norlöff, M., Guichard, B. et al. (2024). Pyridine-based strategies towards nitrogen isotope exchange and multiple isotope incorporation. *Nature Communications* 15 (1): 6063.

15 Ciamician, G.L. and Dennstedt, M. (1881). Ueber die Einwirkung des Chloroforms auf die Kaliumverbindung Pyrrols. *Berichte der Deutschen Chemischen Gesellschaft* 14 (1): 1153–1163.

16 Mortén, M., Hennum, M., and Bonge-Hansen, T. (2015). Synthesis of quinoline-3-carboxylates by a Rh(II)-catalyzed cyclopropanation-ring expansion reaction of indoles with halodiazoacetates. *Beilstein Journal of Organic Chemistry* 11 (1): 1944–1949.

17 Dherange, B.D., Kelly, P.Q., Liles, J.P. et al. (2021). Carbon atom insertion into pyrroles and indoles promoted by chlorodiazirines. *Journal of the American Chemical Society* 143 (30): 11337–11344.

18 Hyland, E.E., Kelly, P.Q., McKillop, A.M. et al. (2022). Unified access to pyrimidines and quinazolines enabled by N—N cleaving carbon atom insertion. *Journal of the American Chemical Society* 144 (42): 19258–19264.

19 Wu, F.-P., Chintawar, C.C., Lalisse, R. et al. (2024). Ring expansion of indene by photoredox-enabled functionalized carbon-atom insertion. *Nature Catalysis* 7 (3): 242–251.

20 Buchner, E. and Curtius, T. (1885). Ueber die Einwirkung von Diazoessigäther auf aromatische Kohlenwasserstoffe. *Berichte der Deutschen Chemischen Gesellschaft* 18 (2): 2377–2379.

21 Fleming, G.S. and Beeler, A.B. (2017). Regioselective and enantioselective intermolecular buchner ring expansions in flow. *Organic Letters* 19 (19): 5268–5271.

22 Qi, M., Suleman, M., Fan, J. et al. (2022). Cu(I)-catalyzed synthesis of spiro[isoquinoline-4,2′-[1,3]oxazin]-3-ones via ring expansion reactions of isoxazoles with 4-diazoisoquinolin-3-ones. *Tetrahedron* 128: 133092.

23 Lovely, C.J., Browning, R.G., Badarinarayana, V., and Dias, H.V.R. (2005). A silver-catalyzed Büchner reaction. *Tetrahedron Letters* 46 (14): 2453–2455.

24 Mbuvi, H.M. and Keith Woo, L. (2009). Addition of carbenes derived from aryldiazoacetates to arenes using chloro(tetraphenylporphyrinato)iron as catalyst. *Journal of Porphyrins and Phthalocyanines* 13 (01): 136–152.

25 Yadav, J.S., Reddy, B.V.S., Gupta, M.K. et al. (2004). First example of ring expansion of activated quinolines and isoquinolines: novel benzoazepines. *Chemical Communications* 18: 2124–2125.

26 Koronatov, A.N., Rostovskii, N.V., Khlebnikov, A.F., and Novikov, M.S. (2018). Rh(II)-catalyzed ring expansion of pyrazoles with diazocarbonyl compounds as a method for the preparation of 1,2-dihydropyrimidines. *Journal of Organometallic Chemistry* 83 (16): 9210–9219.

27 Manning, J.R. and Davies, H.M.L. (2008). Efficient route to 2*H*-1,3-oxazines through ring expansion of isoxazoles by rhodium carbenoids. *Tetrahedron* 64 (29): 6901–6908.

28 Li, L., Ning, Y., Chen, H. et al. (2024). Dearomative insertion of fluoroalkyl carbenes into azoles leading to fluoroalkyl heterocycles with a quaternary center. *Angewandte Chemie, International Edition* 63 (1): e202313807.

29 Liu, S., Yang, Y., Song, Q. et al. (2024). Tunable molecular editing of indoles with fluoroalkyl carbenes. *Nature Chemistry* 16 (6): 988–997.

30 Xu, W., Zhao, J., Li, X., and Liu, Y. (2018). Selective [5+1] and [5+2] cycloaddition of ynamides or propargyl esters with benzo[*d*]isoxazoles via gold catalysis. *Journal of Organometallic Chemistry* 83 (24): 15470–15485.

31 Jadhav, P.D., Lu, X., and Liu, R.-S. (2018). Gold-catalyzed [5+2]- and [5+1]-annulations between ynamides and 1,2-benzisoxazoles with ligand-controlled chemoselectivity. *ACS Catalysis* 8 (10): 9697–9701.

32 Beckmann, E. (1886). Zur Kenntniss der Isonitrosoverbindungen. *Berichte der Deutschen Chemischen Gesellschaft* 19 (1): 988–993.

33 Schmidt, K.F. (1924). Über den Imin-Rest. *Berichte der Deutschen Chemischen Gesellschaft* 57 (4): 704–706.

34 Streith, J. and Cassal, J.-M. (1968). Photochemical synthesis of ethyl 1*H*-diazepine-1-carboxylate. *Angewandte Chemie, International Edition* 7 (2): 129–129.

35 Streith, J., Luttringer, J.P., and Nastasi, M. (1971). Photochemical synthesis of 1,2-diazepines. V. Synthesis and rearrangements of 1,2-diazepines. *Journal of Organometallic Chemistry* 36 (20): 2962–2967.

36 Boudry, E., Bourdreux, F., Marrot, J. et al. (2024). Dearomatization of pyridines: photochemical skeletal enlargement for the synthesis of 1,2-diazepines. *Journal of the American Chemical Society* 146 (4): 2845–2854.

37 Narasimhan, K. and Kumar, P.R. (1984). Addition of phthalimido nitrene to substituted cyclopentadienes. *Heterocycles* 22 (6): 1369–1375.

38 Kumar, P.R. (1987). Addition of phthalimidonitrene to substituted indoles. *Heterocycles* 26 (5): 1257–1262.

39 Reisenbauer, J.C., Green, O., Franchino, A. et al. (2022). Late-stage diversification of indole skeletons through nitrogen atom insertion. *Science* 377 (6610): 1104–1109.

40 Reisenbauer, J.C., Paschke, A.-S.K., Krizic, J. et al. (2023). Direct access to quinazolines and pyrimidines from unprotected indoles and pyrroles through nitrogen atom insertion. *Organic Letters* 25 (47): 8419–8423.

41 Finkelstein, P., Reisenbauer, J.C., Botlik, B.B. et al. (2023). Nitrogen atom insertion into indenes to access isoquinolines. *Chemical Science* 14 (11): 2954–2959.

42 Botlik, B.B., Weber, M., Ruepp, F. et al. (2024). Streamlining the synthesis of pyridones through oxidative amination of cyclopentenones. *Angewandte Chemie, International Edition* 63 (38): e202408230.

43 Ghosh, B., Kafle, P., Mukherjee, R. et al. (2025). Sulfenylnitrene-mediated nitrogen-atom insertion for late-stage skeletal editing of N-heterocycles. *Science* 387 (6729): 102–107.

44 Maeda, K. and Hayashi, T. (1971). The formation of triaryl-*s*-triazine in the chemiluminescence reaction of triarylimidazole and in the photo-oxygenation of triarylimidazole in the presence of ammonia. *Bulletin of the Chemical Society of Japan* 44 (2): 533–536.

45 Maeda, K., Mishima, T., and Hayashi, T. (1974). The formation of substituted quinazolines from substituted indoles by the sensitized photo-oxygenation in the presence of ammonium acetate. *Bulletin of the Chemical Society of Japan* 47 (2): 334–338.

46 Bischler, A. and Barad, D. (1892). Zur Kenntniss der Phenmiazinderivate. *Berichte der Deutschen Chemischen Gesellschaft* 25 (2): 3080–3097.

47 Fremery, M.I. and Fields, E.K. (1964). Amozonolysis of cycloolefins. *Journal of Organometallic Chemistry* 29 (8): 2240–2243.

48 Miller, R.B. and Frincke, J.M. (1980). Synthesis of isoquinolines from indenes. *Journal of Organometallic Chemistry* 45 (26): 5312–5315.

49 Dime, D.S. and McLean, S. (1981). Synthesis of isoquinolines from indenes. *Journal of Organometallic Chemistry* 46 (24): 4999–5000.

50 Wang, J., Lu, H., He, Y. et al. (2022). Cobalt-catalyzed nitrogen atom insertion in arylcycloalkenes. *Journal of the American Chemical Society* 144 (49): 22433–22439.

51 Wang, Z., Xu, H., Han, X. et al. (2024). Manganese-catalyzed cycloalkene ring expansion synthesis of azaheterocycles. *Organic Letters* 26 (40): 8559–8564.

52 He, Y., Wang, J., Zhu, T. et al. (2024). Nitrogen atom insertion into arenols to access benzazepines. *Chemical Science* 15 (7): 2612–2617.

53 Liu, S. and Cheng, X. (2022). Insertion of ammonia into alkenes to build aromatic N-heterocycles. *Nature Communications* 13 (1): 425.

54 Zhang, B.-S., Homölle, S.L., Bauch, T. et al. (2024). Electrochemical skeletal indole editing via nitrogen atom insertion by sustainable oxygen reduction reaction. *Angewandte Chemie, International Edition* 63 (41): e202407384.

55 Baeyer, A. and Villiger, V. (1899). Einwirkung des Caro'schen Reagens auf Ketone. *Berichte der Deutschen Chemischen Gesellschaft* 32 (3): 3625–3633.

56 Siddiqi, Z., Wertjes, W.C., and Sarlah, D. (2020). Chemical equivalent of arene monooxygenases: dearomative synthesis of arene oxides and oxepines. *Journal of the American Chemical Society* 142 (22): 10125–10131.

57 Siddiqi, Z., Bingham, T.W., Shimakawa, T. et al. (2024). Oxidative dearomatization of pyridines. *Journal of the American Chemical Society* 146 (4): 2358–2363.

58 Qin, Q., Zhang, L., Wei, J. et al. (2024). Direct oxygen insertion into C—C bond of styrenes with air. *Nature Communications* 15 (1): 9015.

59 Schmidt, D., Berthel, J.H.J., Pietsch, S., and Radius, U. (2012). C—N bond cleavage and ring expansion of *N*-heterocyclic carbenes using hydrosilanes. *Angewandte Chemie, International Edition* 51 (35): 8881–8885.

60 Saito, H., Otsuka, S., Nogi, K., and Yorimitsu, H. (2016). Nickel-catalyzed boron insertion into the C2—O bond of benzofurans. *Journal of the American Chemical Society* 138 (47): 15315–15318.

61 Lyu, H., Kevlishvili, I., Yu, X. et al. (2021). Boron insertion into alkyl ether bonds via zinc/nickel tandem catalysis. *Science* 372 (6538). 175–182.

62 Tsuchiya, S., Saito, H., Nogi, K., and Yorimitsu, H. (2017). Manganese-catalyzed ring opening of benzofurans and its application to insertion of heteroatoms into the C2—O bond. *Organic Letters* 19 (20): 5557–5560.

63 Urry, W.H., Kruse, H.W., and McBride, W.R. (1957). Novel organic reactions of the intermediate from the two-electron oxidation of 1,1-dialkylhydrazines in acid. *Journal of the American Chemical Society* 79 (24): 6568–6569.

64 Busch, M. and Weiss, B. (1900). Ueber as-dibenzylhydrazine. *Berichte der Deutschen Chemischen Gesellschaft* 33 (2): 2701–2711.

65 Hinman, R.L. and Hamm, K.L. (1959). The oxidation of 1,1-dibenzylhydrazines. *Journal of the American Chemical Society* 81 (13): 3294–3297.

66 Lemal, D.M. and Rave, T.W. (1965). Diazenes from Angeli's salt. *Journal of the American Chemical Society* 87 (2): 393–394.

67 Glover, S.A. (1998). Anomeric amides – structure, properties and reactivity. *Tetrahedron* 54 (26): 7229–7271.

68 Kennedy, S.H., Dherange, B.D., Berger, K.J., and Levin, M.D. (2021). Skeletal editing through direct nitrogen deletion of secondary amines. *Nature* 593 (7858): 223–227.

69 Zou, X., Zou, J., Yang, L. et al. (2017). Thermal rearrangement of sulfamoyl azides: reactivity and mechanistic study. *Journal of Organometallic Chemistry* 82 (9): 4677–4688.

70 Qin, H., Cai, W., Wang, S. et al. (2021). N-atom deletion in nitrogen heterocycles. *Angewandte Chemie, International Edition* 60 (38): 20678–20683.

71 Hui, C., Brieger, L., Strohmann, C., and Antonchick, A.P. (2021). Stereoselective synthesis of cyclobutanes by contraction of pyrrolidines. *Journal of the American Chemical Society* 143 (45): 18864–18870.

72 Guo, T., Li, J., Cui, Z. et al. (2024). $C(sp^3)$—$C(sp^3)$ bond formation through nitrogen deletion of secondary amines using *O*-diphenylphosphinylhydroxylamine. *Nature Synthesis* 3 (7): 913–921.

73 Cao, Z.-C. and Shi, Z.-J. (2017). Deoxygenation of ethers to form carbon–carbon bonds via nickel catalysis. *Journal of the American Chemical Society* 139 (19): 6546–6549.

74 Shimazumi, R., Tanimoto, R., and Tobisu, M. (2023). Nickel/photoredox dual-catalyzed conversion of allyl esters to ketones via the formal deletion of oxygen. *Organic Letters* 25 (34): 6440–6445.

75 van der Plas, H.C. and Jongejan, H. (1968). Ring transformations in reactions of heterocyclic compounds with nucleophiles (III): conversion of pyrimidine and some of its methyl derivatives by hydrazine and by methylhydrazine sulfate into pyrazoles and methylpyrazoles. *Recueil des Travaux Chimiques des Pays-Bas* 87 (9): 1065–1072.

76 Bartholomew, G.L., Carpaneto, F., and Sarpong, R. (2022). Skeletal editing of pyrimidines to pyrazoles by formal carbon deletion. *Journal of the American Chemical Society* 144 (48): 22309–22315.

77 Fout, A.R., Bailey, B.C., Tomaszewski, J., and Mindiola, D.J. (2007). Cyclic denitrogenation of *N*-heterocycles applying a homogeneous titanium reagent. *Journal of the American Chemical Society* 129 (42): 12640–12641.

78 Morofuji, T., Kinoshita, H., and Kano, N. (2019). Connecting a carbonyl and a π-conjugated group through a *p*-phenylene linker by (5+1) benzene ring formation. *Chemical Communications* 55 (59): 8575–8578.

79 Morofuji, T., Inagawa, K., and Kano, N. (2021). Sequential ring-opening and ring-closing reactions for converting para-substituted pyridines into meta-substituted anilines. *Organic Letters* 23 (15): 6126–6130.

80 Hamada, Y. and Takeuchi, I. (1977). Syntheses of nitrogen-containing heterocyclic compounds. 26. Reaction of benzo[f or h]quinolines and their N-oxides with methylsulfinyl carbanion. *Journal of Organic Chemistry* 42 (26): 4209–4213.

81 Falcone, N.A., He, S., Hoskin, J.F. et al. (2024). N-oxide-to-carbon transmutations of azaarene N-oxides. *Organic Letters* 26 (20): 4280–4285.

82 Wang, T., Li, C., Mi, J. et al. (2024). Skeletal editing of pyridine and quinoline N-oxides through nitrogen to carbon single atom swap. *CCS Chemistry* 7 (2): 1–11.

83 Conboy, A. and Greaney, M.F. (2024). Synthesis of benzenes from pyridines via N to C switch. *Chem* 10 (6): 1940–1949.

84 Wu, F.-P., Lenz, M., Suresh, A. et al. (2024). Nitrogen-to-functionalized carbon atom transmutation of pyridine. *Chemical Science* 15 (37): 15205–15211.

85 Sundberg, R.J., Suter, S.R., and Brenner, M. (1972). Photolysis of 0-substituted aryl azides in diethylamine. Formation and autoxidation of 2-diethylamino-1*H*-azepine intermediates. *Journal of the American Chemical Society* 94 (2): 513–520.

86 Patel, S.C. and Burns, N.Z. (2022). Conversion of aryl azides to aminopyridines. *Journal of the American Chemical Society* 144 (39): 17797–17802.

87 Pearson, T.J., Shimazumi, R., Driscoll, J.L. et al. (2023). Aromatic nitrogen scanning by ipso-selective nitrene internalization. *Science* 381 (6665): 1474–1479.

88 Woo, J., Stein, C., Christian, A.H., and Levin, M.D. (2023). Carbon-to-nitrogen single-atom transmutation of azaarenes. *Nature* 623 (7985): 77–82.

89 Paschke, A.-S., Brägger, Y., Botlik, B. et al. (2024). Carbon-to-nitrogen atom swap enables direct access to benzimidazoles from drug-like indoles. *Nature Chemistry* 17: 1750–1756.

90 Nociarová, J., Purkait, A., Gyepes, R., and Hrobárik, P. (2024). Silver-catalyzed skeletal editing of benzothiazol-2(3*H*)-ones and 2-halogen-substituted benzothiazoles as a rapid single-step approach to benzo[1,2,3]thiadiazoles. *Organic Letters* 26 (3): 619–624.

91 Luu, Q.H. and Li, J. (2022). A C-to-O atom-swapping reaction sequence enabled by Ni-catalyzed decarbonylation of lactones. *Chemical Science* 13 (4): 1095–1100.

92 Kim, D., You, J., Lee, D.H. et al. (2024). Photocatalytic furan-to-pyrrole conversion. *Science* 386 (6717): 99–105.

93 Bhanuchandra, M., Murakami, K., Vasu, D. et al. (2015). Transition-metal-free synthesis of carbazoles and indoles by an SNAr based "aromatic metamorphosis" of thiaarenes. *Angewandte Chemie, International Edition* 54 (35): 10234–10238.

94 Onoda, M., Koyanagi, Y., Saito, H. et al. (2017). Synthesis of dibenzophosphole oxides from dibenzothiophene dioxides and phenylphosphine by two successive SNAr reactions. *Asian Journal of Organic Chemistry* 6 (3): 257–261.

95 Cheng, Q., Bhattacharya, D., Haring, M. et al. (2024). Skeletal editing of pyridines through atom-pair swap from CN to CC. *Nature Chemistry* 16 (5): 741–748.

96 Boswell, B.R., Zhao, Z., Gonciarz, R.L., and Pandya, K.M. (2024). Regioselective pyridine to benzene edit inspired by water-displacement. *Journal of the American Chemical Society* 146 (29): 19660–19666.

97 Pradhan, S., Mohammadi, F., and Bouffard, J. (2023). Skeletal transformation of unactivated arenes enabled by a low-temperature dearomative (3+2) cycloaddition. *Journal of the American Chemical Society* 145 (22): 12214–12223.

98 Edelmann, S. and Lumb, J.-P. (2024). A para- to meta-isomerization of phenols. *Nature Chemistry* 16 (7): 1193–1199.

99 Wu, Z., Xu, X., Wang, J., and Dong, G. (2021). Carbonyl 1,2-transposition through triflate-mediated a-amination. *Science* 374 (6568): 734–740.

100 Brägger, Y., Green, O., Bhawal, B.N., and Morandi, B. (2023). Late-stage molecular editing enabled by ketone chain-walking isomerization. *Journal of the American Chemical Society* 145 (36): 19496–19502.

101 Shennan, B.D.A., Sánchez-Alonso, S., Rossini, G., and Dixon, D.J. (2023). 1,2-Redox transpositions of tertiary amides. *Journal of the American Chemical Society* 145 (40): 21745–21751.

102 Huang, M., Pan, T., Jiang, X., and Luo, S. (2023). Catalytic deracemization reactions. *Journal of the American Chemical Society* 145 (20): 10917–10929.

103 Shi, Q. and Ye, J. (2020). Deracemization enabled by visible-light photocatalysis. *Angewandte Chemie, International Edition* 59 (13): 4998–5001.

104 Ričko, S., Bitsch, R.S., Kaasik, M. et al. (2023). Enantioconvergent 6π electrocyclization enabled by photoredox racemization. *Journal of the American Chemical Society* 145 (38): 20913–20926.

105 Aranda, C., Oksdath-Mansilla, G., Bisogno, F.R., and de Gonzalo, G. (2020). Deracemisation processes employing organocatalysis and enzyme catalysis. *Advanced Synthesis and Catalysis* 362 (6): 1233–1257.

106 Zhang, J., Wang, K., and Zhu, C. (2024). Deracemization of atropisomeric biaryls enabled by copper catalysis. *JACS Au* 4 (2): 502–511.

107 Kutta, R.J., Großkopf, J., van Staalduinen, N. et al. (2023). Multifaceted view on the mechanism of a photochemical deracemization reaction. *Journal of the American Chemical Society* 145 (4): 2354–2363.

108 Hölzl-Hobmeier, A., Bauer, A., Silva, A.V. et al. (2018). Catalytic deracemization of chiral allenes by sensitized excitation with visible light. *Nature* 564 (7735): 240–243.

109 Tan, G. and Glorius, F. (2023). Stereochemical editing. *Angewandte Chemie, International Edition* 62 (11): e202217840.

110 Zhang, Y.-A., Palani, V., Seim, A.E. et al. (2022). Stereochemical editing logic powered by the epimerization of unactivated tertiary stereocenters. *Science* 378 (6618): 383–390.

111 Lahdenperä, A.S.K., Dhankhar, J., Davies, D.J. et al. (2024). A chiral hydrogen atom abstraction catalyst for the enantioselective epimerization of meso-diols. *Science* 386 (6717): 42–49.

112 Zhang, Y.-A., Gu, X., and Wendlandt, A.E. (2022). A change from kinetic to thermodynamic control enables trans-selective stereochemical editing of vicinal diols. *Journal of the American Chemical Society* 144 (1): 599–605.

113 Oswood, C.J. and MacMillan, D.W.C. (2022). Selective isomerization via transient thermodynamic control: dynamic epimerization of trans to cis diols. *Journal of the American Chemical Society* 144 (1): 93–98.

114 Arora, R., Oechsner, R.M., Jans, C. et al. (2023). Flipping the switch on palladium-catalyzed carboiodination: accessing kinetic and thermodynamic products. *ACS Catalysis* 13 (10): 6562–6567.

Index

Late-Stage Functionalization and Diversification in Organic Synthesis: Methods and Applications, First Edition. Edited by Tatiana Besset.

f